# Lecture Notes in Electrical Engineering

Volume 213

For further volumes:
http://www.springer.com/series/7818

Punitha P. Swamy · Devanur S. Guru
Editors

# Multimedia Processing, Communication and Computing Applications

Proceedings of the First International Conference, ICMCCA, 13–15 December 2012

Springer

*Editors*
Punitha P. Swamy
Master of Computer Applications
PES Institute of Technology
Bangalore
Karnataka
India

Devanur S. Guru
Department of Studies in Computer Science
University of Mysore
Mysore
Karnataka
India

ISSN 1876-1100 ISSN 1876-1119 (electronic)
ISBN 978-81-322-1142-6 ISBN 978-81-322-1143-3 (eBook)
DOI 10.1007/978-81-322-1143-3
Springer New Delhi Heidelberg New York Dordrecht London

Library of Congress Control Number: 2013939041



Printed on acid-free paper

Springer is part of Springer Science+Business Media (www.springer.com)

# Foreword

I write this foreword with a deep satisfaction to the proceedings of the First "International Conference on Multimedia Processing, Communication and Computing Applications-ICMCCA 2012" held in Bengaluru during December 13–15, 2012, which has the central theme "Multimedia Processing and its Application". Our research experiences in related areas for the last decade have inspired us to conduct the ICMCCA 2012.

I feel, I am honored having served as the program chair for the First "International Conference on Multimedia Processing, Communication and Computing Applications-ICMCCA 2012". This conference was planned to provide a platform for researchers both from academia and industries where they can discuss and exchange their research thoughts to have better future research plans, particularly in the field of Multimedia Data Processing.

Soon after we notified a call for original research papers, there has been a tremendous response from the researchers. There were 106 papers submitted, out of which, we could accommodate only 32 papers based on the reports of the reviewers. Each paper was blindly reviewed by at least two experts from the related areas. The overall acceptance rate is about 30 %. The conference is aimed at: Image Processing, Signal Processing, Pattern Recognition, Document Processing, Biomedical Processing, Computer Vision, Biometrics, Data Mining and Knowledge Discovery, Information Retrieval and Information Coding, all in all Multimedia Data Processing. For all these areas, we got a number of papers reflecting their right combinations. I hope that the readers will appreciate and enjoy the papers published in this proceeding.

We could make this conference a successful one, though it was launched with a relatively short notice. It was because of the good response from the research community and the good effort put in by the reviewers to support us with timely reviews. The authors of all the papers submitted deserve my acknowledgment. The Proceedings are published and indexed by Springer-LNEE, which is known for bringing out this type of Proceedings. Special thanks to them. The proceedings will be available globally from 2–3 months after this conference.

I strongly believe that you will find the proceedings to be very helpful for your future research. Let us all join together with our total support for the progress of Research and Development in the area of Multimedia Data Processing.

Dr. D. S. Guru
Professor, DoS in CS
Program Chair-ICMCCA 2012
University of Mysore

# About the Conference

This year's theme for the planned biannual conference, First "International Conference on Multimedia Processing, Communication and Computing Applications-ICMCCA 2012" is chosen to be 'Multimedia Processing and its Applications'.

Multimedia processing has been an active research area contributing in many frontiers of today's science and technology. Multimedia processing covers a very broad area of science and technology and is widely used in many disciplines such as Medical Diagnosis, Digital Forensic, Object Recognition, Image and Video Analysis, Robotics, Military, Automotive Industries, Surveillance and Security, Quality Inspection etc.

The advent of digital multimedia acquisition systems combined with decreasing storage and processing cost, allows more and more people to invent and explore new ideas to analyze and benefit from collection of digital images and other visual content available. Organizing and managing humungous multimedia data is imperative for accessing, browsing, and retrieving useful data in "real time". People's interest to have their own domain specific, digital libraries; analytical and authenticating systems; and knowledge mining tools has burgeoned and hence requires a dedicated easy access to the data, to understand and interpret the data belonging to different domains. There is considerable overlapping work being carried out across the nation at many medical hospitals and institutions, defense labs, forensic labs, academic institutions, IT companies, security and surveillance domains etc. These requirements have indeed forced the design of specialized multimedia systems/image databases. Years of research contribute to such working and acceptable systems in an ever demanding user centric world.

Hence, the objective of the conference is to bring together all researchers working in the areas of Multimedia processing and address issues pertaining to multimedia information acquisition, analysis, understanding, storage, and retrieval. The scientific discussions planned during the conference will hopefully stimulate avenues for collaborative research and socially useful and easily usable products.

Dr. P. Punitha
Professor and Head
Department of MCA, PESIT, Organising Chair-ICMCCA 2012

# Organizing Committee

### Senior Organizing Panel

- Prof. Dr. M. R. Doreswamy, Chairman, PES Group, Bangalore, India
- Prof. D. Jawahar, CEO, PES Group, Bangalore, India
- Prof. Ajoy Kumar, COO, PESIT, Bangalore, India
- Prof. Dr. R. Chandrasekhar, MCA Department, PESIT, Bangalore, India
- Prof. Dr. V. K. Agrawal, CORI, PESIT, Bangalore, India
- Prof. Dr. Jamadagni, CEDT, IISc, Bangalore, India
- Prof. Dr. H. P. Kincha, EE Department, IISC, Bangalore, India
- Prof. Dr. C. V. Srikrishna, PESIT, Bangalore, India

### Finance Chair

- Prof. Dr. K. N. B. Murthy, Principal and Director, PESIT, Bangalore, India

### Local Organizing Committee (MCA Department, PESIT, Bangalore)

- Dr. D. Uma
- Dr. Neelam Bawane
- Mrs. S. Veena
- Mr. B. S. Chengappa
- Mr. V. Srikanth
- Mr. P. Sreenivas
- Mrs. A. Lekha
- Mrs. H. M. Premalatha
- Mr. Tamal Dey
- Mrs. L. Meena
- Mr. C. J. Siddegowda
- Mrs. Raghavi Bhujang
- Ms. R. Geetha
- Mrs. Dhanya Bibin
- Mrs. Suma S.

- Mrs. Jayanthila Devi
- Mr. Santosh S. Katti
- Mr. P. S. Kannan
- Mrs. Sree Rekha

# Programme Committee

- Dr. Achuthsankar S. Nair, Kerala University, India
- Mr. Anuj Goyal, Carnegie Mellon University, USA
- Dr. Arti Arya, PESSE, India
- Dr. N. Avinash, PESIT, India
- Dr. Bapu B. Kiranagi, HCL Technologies, Detroit, USA
- Dr. Basavaraj Anami, KLE, India
- Dr. Cathal Gurrin, Dublin City University, Ireland
- Dr. David Vallet, Universidad Autenoma de Madrid, Spain
- Dr. M. S. Dinesh, Siemens, India
- Dr. R. Dinesh, HCL Technologies, India
- Dr. Feng Hue, King's College, London
- Dr. Frank Hopfgartner, Dublin City University, Ireland
- Dr. L. Ganesan, Alagappa Chettiar College of Engineering, India
- Dr. George Thallinger, Joanneum Research, Graz, Austria
- Dr. Gowri Srinivasa, PESSE, India
- Dr. B. S. Harish, SJCE, India
- Dr. G. Hemantha Kumar, University of Mysore, India
- Dr. Iaonnis Arapakis, Yahoo Labs, Barcelona
- Dr. Ivan Cantador, Autonomous University Madrid, Spain
- Dr. Klaus Schoeffmann, Klagenfurt University, Austria
- Dr. Kumar Rajamani, GE Global Research, India
- Dr. Lalitha Rangarajan, University of Mysore, India
- Dr. V. N. Manjunath Aradhya, SJCE, India
- Dr. Martin Halvey, Glasgow Caledonian University, Scotland
- Dr. Mohan Kankanahalli, NUS, Singapore
- Dr. N. Y. Murali, MIT, India
- Dr. H. S. Nagendraswamy, University of Mysore, India
- Dr. B. K. Nagesh, Ministry of Higher Education, Oman
- Dr. Nawali Naushath, Ministry of Higher Education, Oman
- Dr. Naveen Onkarappa, Barcelona, Spain
- Dr. Pablo Bermejo, University of Castilla-La Mancha, Spain
- Dr. H. N. Prakash, Shimoga, India
- Dr. H. S. Prashanth, PESIT, Bangalore

- Dr. K. Ramar, Einstein College of Engineering, India
- Dr. D. R. Ramesh Babu, DSCE, India
- Dr. Reede Ren, University of Glasgow, UK
- Dr. Robert Villa, University of Sheffield, UK
- Dr. B. H. Shekar, Mangalore University, India
- Dr. Sahana Gowda, BIT, India
- Dr. P. Shivakumar, NSU, Singapore
- Dr. Srikanta Murthy, PESSE, India
- Dr. M. G. Suraj, Chikmagalur, India
- Dr. Thierry Urruty, Poitiers, France
- Dr. V. Vaidehi, Anna University, India
- Dr. T. Vasudeva, MIT, India
- Dr. Venkat N. Gudivada, Marshall University, VA, USA
- Dr. Vijay Kumar, MSRIT, Bangalore

# Technical Program Committee

**General Chair**

- Dr. P. Punitha, MCA Department, PESIT, India

**Publicity Chair**

- Dr. Ram P. Rustagi, IS Department, PESIT, India

**Publication Chair**

- Dr. D. S. Guru, Department of Studies in CS, University of Mysore, India

**Programme Co-chairs**

- Dr. Arun A. Ross, WVU, USA
- Dr. Bapu B. Kiranagi, HCL, Detroit, USA
- Dr. L. Ganesan, Alagappa Chettiar College of Engineering, India
- Dr. Joemon M. Jose, UG, Scotland
- Mr. Subir Saha, Co-founder, Yotto Labs, India

**Advisory PC Co-chairs**

- Dr. Anil K. Jain, MSU, USA
- Dr. S. K. Chang, University of Pittsburgh, USA
- Dr. Manabu Ichino, Tokyo Denki University, Japan
- Dr. P. Nagabushan, University of Mysore, India
- Dr. Sankar K. Pal, ISI, Kolkata

# Reviewers' List

- Dr. Achuthsankar S. Nair, University of Kerala, India
- Dr. Anil K. Jain, MSU, USA
- Dr. Arun A. Ross, West Virginia University, USA
- Dr. N. Avinash, PES Institute of Technology, India
- Dr. Bapu B. Kiranagi, HCL, USA
- Dr. Basavaraj Anami, KLE Institutions, India
- Dr. Basavaraj Dhandra, Gulbarga University, India
- Dr. Chethana Hegde, RNSIT, India
- Dr. David Vallet, Universidad Autónoma de Madrid, Spain
- Dr. M. S. Dinesh, GE Research Lab, India
- Dr. Dinesh Ramegowda, HCL Technologies, India
- Dr. D. L. Elham, University of Mysore, India
- Dr. Frank Hopfgartner, Dublin City University, Ireland
- Dr. George Thallinger, Joanneum Research, Austria
- Dr. S. Girish, PES, India
- Dr. Gowri Srinivasa, PESSE, India
- Dr. D. S. Guru, University of Mysore, India
- Dr. B. S. Harish, SJCE, India
- Dr. Hemant Misra, Philips Research, India
- Dr. Jitesh Kumar, Mando Infotech, India
- Dr. Kaushik Roy, West Bengal University, India
- Dr. V. N. Manjunath Aradhya, SJCE, India
- Dr. Manjunath Ramachandra, Philips Research, India
- Dr. Manjunath Shantharamu, JSS Government College, India
- Dr. K. B. Nagasundara, MIT, India
- Dr. H. S. Nagendraswamy, University of Mysore, India
- Dr. B. K. Nagesh, Ministry of Higher Education, Oman
- Dr. Nitendra Nath, Takata, Japan
- Dr. Nawali Noushad, Ministry of Higher Education, Oman
- Dr. Pavan Vempaty, Takata, Japan
- Dr. H. N. Prakash, University of Mysore, India
- Dr. Punitha Swamy, PES Institute of Technology, India
- Dr. Rajamani Kumar, GE Research Lab, India

- Dr. Ram P. Rustagi, PES Institute of Technology, India
- Dr. K. Ramar, Einstein College of Engineering, India
- Dr. D. R. Ramesh Babu, Dayananda Sagar College of Engineering, India
- Dr. C. N. Ravi Kumar, SJCE, India
- Dr. Reede Ren, University of Glasgow, UK
- Dr. Sahana D. Gowda, BIT, India
- Dr. Sankalp Kallakuri, Mando Infotech, India
- Dr. Shanmukhappa Angadi, BEC, India
- Dr. Shivakumara Palaiahnakote, National University of Singapore, Singapore
- Dr. K. Srikanta Murthy, PESSE, India
- Dr. Sudarshan Patil Kulkarni, Mando Infotech, India
- Dr. M. G. Suraj, Chikmagalur, India
- Dr. Thierry Urruty, Poitiers, France

# Messages

I am pleased to convey my warm greetings to all the participants of the International Conference "ICMCCA 2012" organized by Department of MCA, PESIT.

I congratulate Department of MCA, PESIT for hosting its First International Conference on "Multimedia Processing, Communication and Computing Applications-ICMCCA 2012" during December 13–15, 2012.

I learn that many technocrats, educationalists, and academic personnel from National level and International level involved in technical education congregate and exchange their experiences in the respective domain. This conference brings together academic luminaries and I am delighted such a significant event is happening in the precincts of PES Institutions.

I wish the organizing committee of Department of MCA the very best in their endeavor. I eagerly look forward to meet all the delegates and speakers from various parts of the country and world during this conference.

Thank you and wish the conference a grand success.

Prof. Dr. M. R. Doreswamy
Founder Chairman
PES Group of Institutions, Bengaluru

I am happy that Department of MCA, PESIT is hosting the First International Conference "ICMCCA 2012" in December 2012. I take this opportunity to extend my greetings and best wishes to all participants.

There are many evidences suggesting that images and videos are required for a variety of reasons, like, illustration of text articles (academia), conveying information or emotions that are difficult to describe in words (face/emotion analysis), displaying detailed data for analysis (medical images), and formal recording of design data for later use (architectural plans) etc.

I am sure the deliberations of the conference comprising of special panel discussions, technical paper presentations, and keynote addresses will bring out lot of value to effect desired changes in learning experience.

Prof. D. Jawahar
CEO
PES Group of Institutions, Bengaluru

I am delighted to note that Master of Computer Applications (MCA) Department is arranging First International Conference "ICMCCA 2012" at PES Institute of Technology, Bengaluru.

It is learnt that more than 100 papers have been received from all over the world. Approximately, 30 % of papers have been accepted for presentation.

All the staff members of MCA Department have been working hard for the months together for the success of this conference. I sincerely convey my heartfelt congratulations to all members, both teaching as well as non-teaching who have put efforts for this event.

I hope that all local as well as foreign delegates will have useful deliberations during the conference. I also wish their stay at PES Institute of Technology is joyful and useful.

Dr. K. N. B. Murthy
Principal and Director
PESIT, Bengaluru

I am happy to know that Department of MCA, PES Institute of Technology, Bangalore is organizing the First "International Conference on Multimedia Processing, Communication and Computing Applications ICMCCA 2012", during December 13–15, 2012.

It is a matter of great pride that PES Institute of Technology, Bengaluru, which is one of the top technical institutions in the state and which has the vision to "Create Professionally Superior and Ethically Strong Global Manpower" is holding this prestigious International Conference.

With the main theme being "Multimedia Processing and its Applications", this International Conference will provide the platform for the presentation of research papers and innovative ideas leading to drastic revolutions in the use of multimedia in the disciplines such as Medical Diagnosis, Digital Forensic, Object Recognition, Image and Video Analysis, Robotics, Military, Automotive Industries, Surveillance and Security, and Quality Inspection etc.

I am happy to note that the department is bringing out a Souvenir to commemorate this mega event. I wish this Souvenir to be a rich source of information and knowledge which will be highly useful for further research activities in this filed.

I wish the organizers of this International Conference all the best in this endeavor and I wish the participants all success.

26.10.12

Dr. S. A. Kori
Registrar
Visvesvaraya Technological University, Belgaum

# Invited Talks

**Building Practical Vision Systems: Case Studies in Surveillance and Inspection**


**Abstract**

It is becoming increasingly clear that it is humanly impossible to browse, navigate, and search a deluge of data from cameras and other sensors in a variety of applications including retail surveillance, railroad inspection, and Unmanned Aerial Vehicle (UAV) video surveillance. The practical systems that we built, although in pursuit of different business objectives, share a common goal, which is to intelligently and efficiently analyze and extract the most important actionable information from an overwhelming amount of data, while being able to effectively ignore a large portions of uneventful and/or noisy data. I will summarize a variety of computer vision, machine learning, and system optimization techniques that we used to successfully address different technical and business challenges, and to deliver differentiating performance to meet our customers' expectations.

Contributors, Lisa Brown, Jon Connell, Ankur Datta, Quanfu Fan, Rogerio Feris, Norman Haas, Jun Li, Ying Li, Sachiko Miyazawa, Juan Moreno, and Hoang Trinh.

**Brief Biodata**

Sharath Pankanti is a Research Staff Member and Manager with Exploratory Computer Vision Group at IBM T. J. Watson Research Center. He received a B.S. degree in Electrical and Electronics Engineering from College of Engineering, Pune in 1984, M.Tech. in Computer Science from Hyderabad Central University in 1988, and Ph.D. degree in Computer Science from the Michigan State University in 1995. He has published over 100 peer-reviewed publications and has contributed to over 50 inventions related to biometrics, privacy, object detection, and recognition. Dr. Pankanti is an IEEE Fellow. His experience spans a number of safety, productivity, and security focused projects involving biometrics-, multi-sensor surveillance, and driver assistance technologies.

Dr. Sharath Pankanti
Manager
Exploratory Computer Vision
IBM T. J. Watson Center
Yorktown Heights
NY, 10598
USA

## Statistical Parametric Speech Synthesis

**Abstract**

In this talk, I will give a quick overview on the progress of text-to-speech synthesis starting from formant synthesis (parametric synthesis) to the current trends of statistical parametric synthesis. I will first emphasize the role of text processing in building a text-to-speech system. I will then describe the mathematical formulations of statistical parametric synthesis techniques and review different implementations in the current state of the-art text-to-speech systems. I will also discuss the research issues involved in building text-to-speech systems in the context of Indian languages.

**Brief Biodata**

Kishore S. Prahallad is an Assistant Professor at the International Institute of Information Technology (IIIT) Hyderabad. He is associated with IIIT Hyderabad since March 2001, and started the speech activities in Language Technologies Research Center (LTRC) at IIIT-H. He served as a Research Scientist from 2001 to 2008, Senior Research Scientist from 2008 to 2010, and as the Coordinator of MSIT Division from 2004 to 2009 at IIIT-H.

He did B.E. in Computer Science and Engineering (CSE) from Deccan College of Engineering and Technology, Osmania University in 1998, M.S. (by Research) in 2001 from Department of CSE, Indian Institute of Technology (IIT) Madras, and Ph.D. in 2010 from Language Technologies Institute (LTI), School of Computer Science, Carnegie Mellon University (CMU), USA.

His research interests are in speech and language processing, specifically in text-to-speech synthesis, speech recognition, dialog modeling, and voice conversion. His current projects include building virtual assistant(s) on mobile devices specifically in the context of India and Indian languages, collection of speech and text corpora in ~1,000 Indian languages using crowd sourcing, automatic annotation of speech data at phonetic and prosodic level, audio search and summarization, and second language acquisition using audio books.

Dr. Kishore Prahallad
Assistant Professor
IIIT-Hyderabad, Gachibowli, Hyderabad-32
India

**Personal Life Archives: A Research Challenge**

**Abstract**
Our daily life started becoming digital over a decade ago. Now much of what we do is digitally recorded and accessible and this trend is accelerating, with the potential to bring astonishing benefits. In this talk, I will introduce the concept of the Personal Life Archive, a surrogate memory, that can operate in synergy with our own organic memories. These Personal Life Archives will allow those who choose, to store a lifetime of experiences in a private store; they are set to revolutionize our personal lives, healthcare, learning, and productivity. Since the inspiring essay 'As We May Think' by Vannevar Bush in the 1940s, the concept of a Personal Life Archive (the memex) has been imagined, yet only now, with the convergence of technical progress in three fields; data sensing, data storage, and information retrieval/data mining, are we in a position to finally realize Bush's vision. A surrogate memory can be achieved through the use of cheap and ubiquitous wearable, software and environmental sensors as well as a new generation of personal search engine that operates on the individual scale, as opposed to the WWW scale. In this talk, I will summarize progress to date on the development of Personal Life Archives and outline the technical challenges that need to be addressed by the Information Retrieval and HCI communities before we can truly enter the era of the Personal Life Archive.

**Brief Biodata**

Cathal Gurrin is the SFI Stokes Lecturer at the School of Computing at Dublin City University, Ireland and a Visiting Researcher at the University of Tromso, Norway. Cathal is the Director of the Human Media Archives Research Group at Dublin City University and a Collaborating Investigator in the CLARITY Centre for Sensor Web Technologies. His research interests focus on Information Access to Personal Life Archives, Multimodal Human–Computer Interaction and Information Retrieval in general. He has been an active lifelogger since mid-2006 and has amassed a large archive of over ten million sensecam images and is actively developing search and organization technologies for Personal Life Archives. He is the author of more than 100 academic publications and is regarded as a leading researcher in the field of Personal Life Archives.

For more details, please visit http://www.iiit.ac.in/~kishore

Dr. Cathal Gurrin
SFI Investigator and Lecturer
School of Computing/CLARITY CSET
Dublin City University, Dublin, Ireland

**Video and Sensor Observations for Health Care**

**Abstract**

Analyzing human activity is the key to understanding and searching surveillance videos. I will discuss current results from a set of studies to automatically analyze video of human activities to improve geriatric health care. Beyond just recognizing human activities observed on video and sensor information, we mine a long-term archive of observations, and link the observational analysis to medical records. This research explores the statistical patterns between a patient's daily activity and his/her clinical diagnosis. I will discuss some of the technical details of the work, as well as issues related to this type of health research. The main goal of this work is to develop an intelligent visual surveillance system based on efficient and robust activity analysis. The current results represent a feasibility demonstration of exploiting long-term human activity patterns though video analysis.

**Brief Biodata**

Dr. Alex G. Hauptmann is a Principal Systems Scientist in the School of Computer Science at Carnegie Mellon University, with a joint appointment in the Language Technologies Institute. His research interests are in multimedia analysis and indexing, speech recognition interfaces to multimedia system and language in general. He is currently spending most. of his time on the Informedia project. The following are specific projects building under the Informedia umbrella Caremedia, Assistance for use of home medical devices, Analysis of Large Multimedia Data Sets, and Tools and Ontologies for multimedia analysis.

Over the years, his research interests have led him to pursue and combine several different areas of research: man–machine communication, natural language processing and multimedia understanding: In the area of man–machine communication, he is interested in the tradeoffs between different modalities, including gestures and speech, and in the intuitiveness of interaction protocols. In natural language processing, his desire is to break through the bottlenecks that are currently preventing larger scale natural language applications.

The latter theme was also the focus of his thesis, which investigated the use of machine learning on large text samples to acquire the knowledge needed for semantic natural language understanding. In the field of multimedia understanding, he is intrigued by the potential of combining natural language technology with clever interfaces in video retrieval applications.

Dr. Alex G. Hauptmann
School of Computer Science
Carnegie Mellon University
5000 Forbes Ave
Pittsburgh, PA, 15213-3890
USA

## Shannon, Fourier and Compressing Multimedia

**Abstract**
This talk will review the backdrop of compressing multimedia by looking at the notion of entropy proposed by Shannon, which determines the limit of lossless compression. The talk will further discuss Fourier transforms as one of the basic tools to separate out basis functions in a signal, thus enabling a lossy compression. A case study of use of fractals in compressing Chaos Game Representation images in Bioinformatics will be presented.

**Brief Biodata**
Dr. Achuthsankar S. Nair heads the Department of Computational Biology and Bioinformatics and also the State Inter University Centre of Excellence in Bioinformatics, University of Kerala. He had his B.Tech. (Electrical Engineering) from College of Engineering, Thiruvananthapuram, M.Tech. (Electrical Engineering) from IIT Bombay, M.Phil. (Computer Speech and Language Processing) from the Department of Engineering, University of Cambridge, UK, and Ph.D. from University of Kerala.

Since 1987, he has taught in various Engineering Colleges, Universities and Institutes both within India and abroad. During 2001–2004, he served as Director of Centre for Development of Imaging Technology (C-DIT), Government of Kerala. In 2006, he was a Visiting Professor in University of Korea, Seoul.

He has a dozen of popular science books on IT in Malayalam, including one on the internet in 1996 and one on free software in 2002. His latest books are *Data Structures Using C* (Printice Hall), *Scilab* (S. Chand, New Delhi), and *Bioinformatics* (DC Books). He has also authored a science fiction novel for children.

His current research interests include use of digital signal processing (DSP) in biosequence analysis. 14 Candidates have taken Ph.D. under his supervision and he is currently guiding another dozen researchers. He has a modest number of research publications in International and Indian Journals. One of his contributions on electromechanical model for the Transistor is cited in the classic text book: *Hughes's Electrical Technology* (7th Edition) published by Orient Longman, UK.

He is recipient of Young Scientist Award of Government of Kerala (1991), Cambridge Barclay Scholarship (1991), ISTE National Award for Young Engineering Teachers (1994). He was a member of the Executive Committee of the first State Higher Education Council of Government of Kerala.

Dr. Achuthsankar S. Nair
Professor and Head
State Inter University Centre of Excellence in Bioinformatics
University of Kerala
Trivandrum, Kerala

# About Department of MCA, PESIT

It is my privilege to introduce the host department of this event, Department of MCA, PESIT, Bengaluru.

The Department of MCA was established in 1997. The department has been offering Master of Computer Applications program under Visvesvaraya Technological University. Since inception, the department has successfully made significant improvements, in empowering graduates having no computing knowledge, with the expert skills needed to work in the global IT sector. The students of the department have always scored top ranks every year including 1st Rank. The department started with an initial student intake of 30 students and from last 5 years the intake is 120. The total number of students currently enrolled in this course is 360. The employee strength of the department is 36 members with 24 teaching faculty and 12 non-teaching staff members.

**Department Vision is**

- To excel in imparting quality education, create ethically strong, creative, analytical, technically superior, knowledgeable, innovative and inquisitive minds.

**Department Mission is**

- To become dynamic and vigorous knowledge hub with an exposure to state of art computer applications and to empower students in becoming skilled and ethical entrepreneurs while endorsing free and open source software learning and usage. Also, to promote and adapt professional development in a perpetual demanding environment and nurture **M**egaminds for **C**ompetent **A**ccomplishments.

## Faculty

The department has 5 faculty members with Ph.D. degree and 8 members currently pursuing their Ph.D. studies. The faculty members specialize in different areas of computer science ranging across image processing and computer vision, computer networks and wireless communication, data mining, software engineering, and Graph theory. The department has 3 professors, who have extensive experience in guiding projects and research and are guiding around 10 Ph.D.s toward their doctoral degree. The faculty regularly publish their research findings in book chapters, journal articles, and in conference proceedings. The HOD is identified as a key resource person (KRP) and has a rich industry and academic experience in India and abroad to further enrich the strengths in teaching and research of the department. The department also has a research centre recognized by VTU and well-facilitated computer labs. **The number of publications has increased from ONE in 2007–2008 to NINETEEN in 2011–2012.**

Faculty are given financial support for paper presentations in various conferences and allowed to go abroad to present their papers. Two faculty of the department have utilized this facility. Publication incentives are given to the faculty who has published papers in international journals which have impact factor of more than one and is listed in **Scopus**. One faculty of the department has received this incentive. Deputation to CORI Lab is allowed for faculty with the thrust and enthusiasm toward research. One faculty from the department was deputed for one year. Students are also allowed to go to CORI Lab and work with various other projects involved and currently around 8–10, V semester MCA students are working on projects related to this. Book publishing incentives are given to the faculty who publish books with reputed publishers. Two faculty have received this incentive. In the case of funded project, 10 % of the total fund is given to the investigators and three faculty of the department have received this incentive. Faculty are also facilitated to work with industries for a certain stipulated period. Currently, one faculty is on deputation to an industry.

The department has initiated TEA (Technical but Eccentric yet Affable) Talks where the faculty of the department and the institution are encouraged to give an impromptu talk about their research finding and open research problems in their specialized domains.

## Infrastructure Facility

Technology is fast becoming ubiquitous and the educational environment is no exception. The department strives to make learning more creative, interactive, and information driven by using sophisticated delivery techniques. Eight different computer laboratories across the department house over a 120 computers for use by students and staff, well connected to internet and Wi-Fi infrastructure.

The department has **6 class rooms**, **1 seminar hall**, and **8 lab units**. The class rooms, seminar hall are fit with 6 projectors, respectively. The class rooms are also fit with an internet port that can be used by both students and faculty. The department has **16 faculty rooms** in which 8 are exclusive and the other 8 are shared. Each faculty is given a desktop from the college and each faculty room is equipped with either an inkjet or laser printer. The department also provides a public address system for the faculty while taking regular classes. Department of MCA has an **air-conditioned seminar hall** of **134.15 $m^2$** floor area Wi-Fi enabled that can accommodate 100 people.

## Computing Facility

The lab units have **120 systems** with one networked printer that is used by the students for their day to day lab work. The department has exclusively purchased IBM Rational Rose Suite for 30 standalone systems. The department has initiated a **FOSS-Free Open Source Software Lab**. This lab has 44 systems with a hardware configuration of Pentium Dual Core 3 GHz, 4 GB RAM, and HDD of 250 and 500 GB. The department has Ubuntu OS and a server.

The department has a server that is physically located at the **CISCO Lab** with a configuration of **1 × X3430, Intel 3420 Chipset**. Lecture notes, question banks and solutions are uploaded on to this server. The students also upload their class assignments on to their respective folders on this server.

The college has provided two facilities to track and maintain the infrastructure of the department, **Problem Change Management System** and **Asset Management System.** The college has provided an IPOMO facility that is used to keep track of the student's attendance.

## Library

The department has 764 books in its department library that have come from Central Library. Approximately 40 books have been donated to the department library from alumni, faculty, and students. The department has procured three journals.

## Support to Students

For the students, the department conducts SASP (Student Academic Support Program), GSDP (Gifted Student Support Program), and TSDP (Total Student Development Program). SASP is conducted in all the semesters for those students whose performance is poor (Test marks <25 and Attendance less than 75 %). GSDP is offered to students excelling in their studies and are supervised by professors to work on research and entrepreneur projects. Students are also encouraged to present their papers in conferences and financially supported. TSDP classes are conducted by subject experts from Industry/Institute. The department has collaboration with companies such as IBM and CISCO. IBM has offered courses on DB2, RAD. CISCO is currently offering courses on CCNA.

Preplacement training is being conducted for IV and V semester students by department and college. Internship process is handled at the department level to carry out VI semester dissertation. 40 % of internships get converted into job offers every year. The students have been given the opportunity to visit industries such as Infosys, CTS, CISCO, and Microsoft. Syllabus oriented Guest Lectures are conducted for the students in every semester as part of their curriculum. A Faculty Advisory Diary is maintained for each student to address student's grievances, to monitor the academic performance.

### Conferences/Seminars

For the benefit of faculty and students who wish to do the research, department has been organizing National Conferences/Seminars/Workshops every year. For this year, the department is organizing an "**International Conference on Multimedia Processing, Communication and Computing Applications**" (ICMCCA 2012) on 13–15 December 2012. Accepted research papers are being indexed by Springer-LNEE Journal.

### Extra Curricular Activities

The department has initiated an environmental awareness program called **Paryavaran**. It is conducted every year in the department by the students. There were invited lectures by eminent environmentalists such as Snake Shyam, Akshay Heblikar, and Ashok.

CDP's (Community development programmes) are being conducted by the department faculties every year for school students.

Dr. P. Punitha
Professor and Head
Department of MCA, PESIT

# Contents

# An Efficient Method for Indian Vehicle Number Plate Extraction

**M. N. Veena and T. Vasudev**

**Abstract** Vehicle number plate recognition system is the heart of an intelligent traffic system. Extracting the region of a number plate is the key component of the vehicle number plate recognition (VNPR) system. An efficient method is proposed in this paper to analyze using images which often contain vehicles and extract number plate, by finding vertical and horizontal edges from vehicle region. The proposed vehicle number plate detection (VNPD) method consists of 5 stages: (1) Assumption to consider probable region of number plate to concentrate on Region of Interest (ROI), (2) Extracted image from ROI to undergo noise removal and sharpening, (3) Finding vertical and horizontal edges from the image, (4) Applying Connected Component method to extract the number plate, (5) Finally, extracted number plate is subjected for slant correction by Radon transformations. Experimental results show the proposed method is very effective in coping with different conditions like poor illumination, varied distances from the vehicle and varied weather conditions.

**Keywords** Number plate · Sobel operator · Number plate extraction · Radon transformation

## 1 Introduction

The Intelligent Transportation Systems has impact on human life in terms of improving the safety and mobility. In the existing environment, it is necessary to incorporate new ideas and technologies. The vehicle number plate recognition

M. N. Veena (✉)
P. E. S. College of Engineering, Mandya, India
e-mail: veenadisha1@gmail.com

T. Vasudev
Maharaja Institute of Technology, Mysore, India
e-mail: banglivasu@yahoo.com

P. P. Swamy and D. S. Guru (eds.), *Multimedia Processing, Communication and Computing Applications*, Lecture Notes in Electrical Engineering 213,
DOI: 10.1007/978-81-322-1143-3_1, © Springer India 2013

(VNPR) system plays an important role in traffic surveillance systems, such as traffic law enforcement, real time monitoring, parking systems, road monitoring and security systems [1, 2]. Many researchers have suggested good number of techniques [3, 4] in the above application domain. The task of recognizing specific object in an image is one of the most difficult topics in computer vision and digital image processing. Recognizing the number plate of a vehicle from a natural image is one particular application case. The vehicle number plate detection is widely used for detecting speeding cars, security control in restricted areas, unattended parking zone, traffic law enforcement and electronic toll collection. Last few years have seen a continued increase in the need for the use of vehicle number plate recognition. Though good numbers of researchers have worked on this problem, still many issues are not addressed especially for generic solution. The number plate recognition system has plenty of challenging avenues for research because of its complexities like poor illumination, varied weather condition, dust, occlusion etc. This has motivated us to continue research on vehicle number plate recognition in a complex environment for an efficient solution.

Number plate standards vary from country to country. The existing vehicle number plate recognition (VNPR) system being applied are designed using domain knowledge of number plates specific to the country under consideration. This paper focuses on number plate extraction from Indian vehicle image using vertical and horizontal edges followed by slant correction process through Radon transformation. The rest of this paper is organized as follows. Next section composes a review of similar researches that have been implemented and tested for vehicle number plate detection. In Sect. 3, the proposed vehicle number plate extraction process and the slant detection and correction using Radon Transformation are discussed in detail. In Sect. 4, experimental results are reported. Finally, conclusion on the work and scope for future work is briefed in Sect. 5.

## 2 Review of Related Work

Literature survey indicates that quite a number of researchers have explored useful methods for locating number plate from the vehicle images. Some of the methods are based on normal features of number plates like color [5], shape [6], symmetry [7], texture of grayness [8], spatial frequency [9] and variance of intensity values [10]. An approach is suggested based on enhanced detection of boundary lines [11] using gradient filter. Another similar attempt for detection of bounding lines based on two pair of parallel lines using Hough transform [12] to designate number plate is reported. Other approaches are reported based on the morphology of objects in an image [10, 13]. These approaches focus on some salient properties of vehicle plates such as their brightness, contrast, symmetry, angles etc., in an image and locate the position of number plate regions. Two approaches are based on statistical properties of text [14, 15]. In these approaches, text regions are discovered using statistical properties of text like the variance of gray level, number of edges, edge densities in the region etc. These approaches were commonly used in finding text in images and used to detect and extract number plate areas as they contain alphabets and numerals.

In addition, vehicle number plate detection using artificial intelligence (AI) and genetic algorithm [16, 17] are proposed. These systems use edge detection and edge statistics and then AI techniques are applied to detect the location of the number plate area. The works reported in literature have some kind of limitations such as plate size dependency, color dependency, efficient only in certain conditions or environment like indoor images etc.

In the proposed work an attempt is made to detect vehicle number plate from the vehicle images obtained at different distance and different illumination conditions, and the methodology for detection of number plate is discussed in following section.

## 3 Proposed Methodology

Number plate detection in this work is primarily based on theory of edges [10]. A number plate in a vehicle image is viewed as irregularities in the texture of image. Any change in the texture (local characteristics) indicates the probable presence of a number plate. The proposed vehicle number plate detection algorithm consists of following stages: (1) Convert RGB image to grey image, (2) Horizontal and Vertical edge detection using sobel edge operator, (3) Detecting number plate region using recursive algorithm for connected component labelling operation, (4) Number plate extraction, (5) Number plate slant correction using Radon transformation. General flow of operations carried out for detecting number plate is shown in Fig. 1.

The proposed work of number plate detection is presented in two modules. The first module describes the number plate extraction process and the second module describes the slant detection and correction using Radon transformation.

### *3.1 Number Plate Extraction*

Number plate extraction process starts with accepting a vehicle image in RGB as input. In order to improve processing speed, the original RGB is converted into gray-level image. Next step is to enhance the gray-level image to remove the noise with preserving the sharpness of the image by median filter. The vehicle number plates are either rectangle or square in shape. This apriori knowledge is exploited in this method through finding vertical and horizontal edges from the vehicle image using Sobel edge operator [10]. This method finds the regions with high pixel variance values, because characters in the number plate include many edges. Using edge detection some unimportant regions are removed in which the

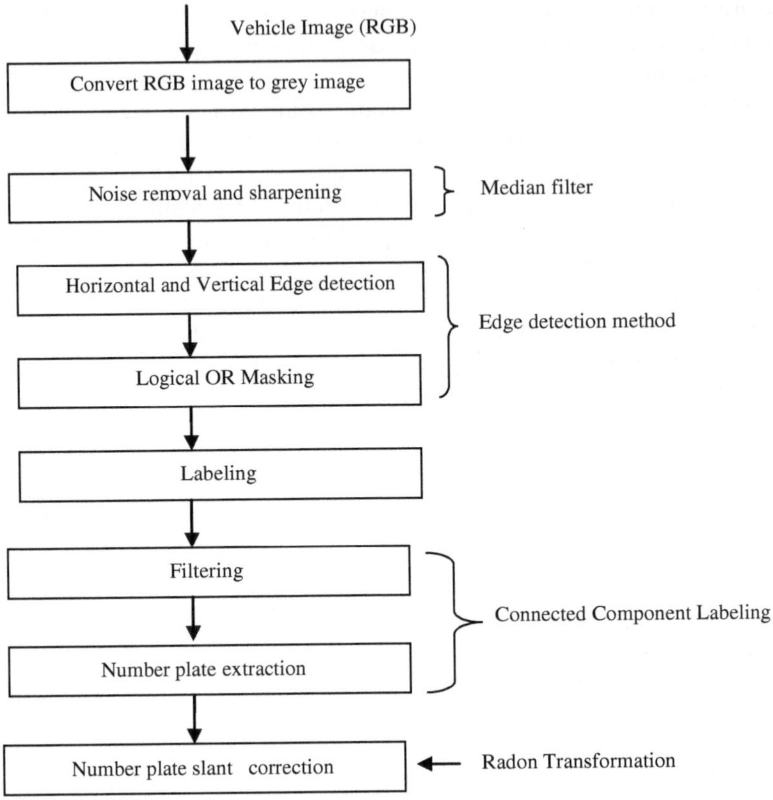


**Fig. 1** Proposed model for number plate detection

horizontal edges and vertical edges are obvious, the remaining areas are candidate regions for number plate. Sobel edge detection operation is carried out using 2 masks of size 3 × 3 as shown in Fig. 2. These masks are designed to detect horizontal and vertical edges separately. In addition, the masks are designed in such a way to convert the resulting image into a binary image, during the process of edge detection.

Next, a logical OR masking operation is performed to keep the background surrounding pixels preserved in the image. Further, each candidate region requires label for identification. This is achieved using recursive algorithm for connected component labeling [18], which works on one component at a time, to form candidate regions. In order to remove the regions other than number plate region

**Fig. 2** Sobel masks to detect: **a** Vertical edge, **b** Horizontal edge

**(a)**

| | | |
|---|---|---|
| -1 | 0 | -1 |
| -2 | 0 | -2 |
| -1 | 0 | -1 |

**(b)**

| | | |
|---|---|---|
| +1 | -2 | +1 |
| 0 | 0 | -0 |
| -1 | -2 | -1 |

from candidates, compute the geometrical property aspect ratio 'A' for available candidate regions in a given image. The aspect ratio also called as elongation or eccentricity is given by:

$$A = \frac{(c_{max} - c_{min}) + 1}{(r_{max} - r_{min}) + 1} \tag{1}$$

where c and r indicates column and row respectively, min and max are the minimum and maximum values of row and columns of the region under consideration. The objects whose measurement satisfies 2 < aspect ratio < 6 are considered as number plate region. The aspect ratio 'A' is used in filtering operation [19] to eliminate number plate like objects from image. Resultant object contain the number plate.

Figure 3a–g show the outcome of each step in number plate detection: (a) Initial RGB image (b) Gray image (c) Detecting horizontal edges (d) Detecting vertical edges (e) After image masking operation (f) Detection of candidate regions (g) Number plate detection.

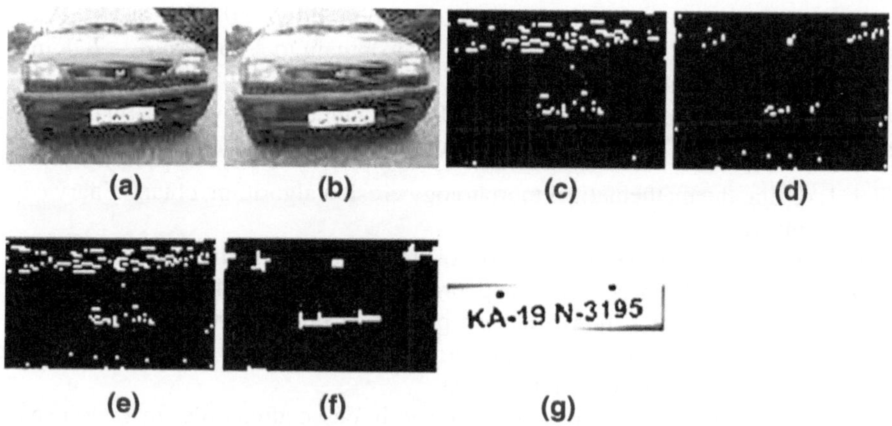


**Fig. 3** Steps for number plate detection

## 3.2 *Slant Correction Using Radon Transformation*

The extracted number plate may suffer from skew/slant during image capture process. The performance of character recognition decreases due to skewness. Hence, it is necessary to detect and correct the skewness in extracted number plate. The angle of skew in number plate is estimated using Radon transformation [20, 21]. The Radon Transformation in the 2-D plane is defined as:

$$R(\theta, \rho) = \iint_D f(x, y)\delta(\rho - x\cos\theta - y\sin\theta)dxdy \tag{2}$$

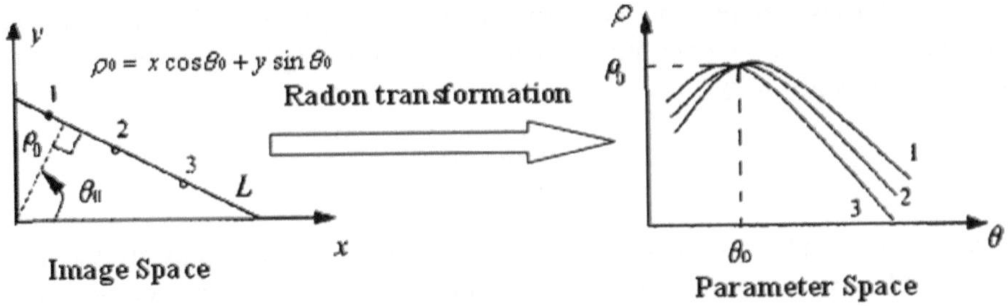


**Fig. 4** Sketch map of Radon transformation

where, D denotes the image plane. Eigen function is denoted by Dirac function $\delta(\rho - x\cos\theta - y\sin\theta)$. The pixels gray value at point $(x, y)$ is noted as $f(x, y)$. The distance from origin to straight line is noted as $\rho$ in $x - y$ domain, $\theta$ is defined as the angle between the vertical line and $x$ axis, which is from origin to straight line. The geometry of the Radon transform is showed in Fig. 4 [22]. It is noticed that the straight line is well determined when $\rho$ and $\theta$ are fixed values. In turn, each straight line in $x - y$ domain represents a point in $\rho - \theta$ domain. So Radon transformation maps the straight line in $x - y$ domain to a point in $\rho - \theta$ domain as indicated in Fig. 4.

The Radon transform is closely related to a common computer vision operation Hough transforms. The steps used in Radon transformation are as follows:

Step 1: Using the mathematical morphology erosion algorithm, obtain binary edge image.
Step 2: Compute the Radon transform of the image containing edges.
Step 3: Find the locations of strong projecting points in the Radon transform matrix. The locations of these projecting points correspond to the locations of straight lines in the origin. The longer the straight line is, the brighter the corresponding point.
Step 4: These projecting points are arranged in descending order in Radon space in view of the veracity.
Step 5: Select projecting points.
Step 6: Calculate the sum of each row and store in matrix R.
Step 7: Calculate
  for i = 2: length(R)
  E(i) = a* R(i) + (1 − a) * E(i − 1);
  Get the maximum of *E(i)*, $\alpha = 90° - i$ is horizontal rotation angle.

Step 8: Lossless rotation correction is performed to the slant plate in the horizontal direction.

The result of horizontal slant correction is shown in Fig. 5.

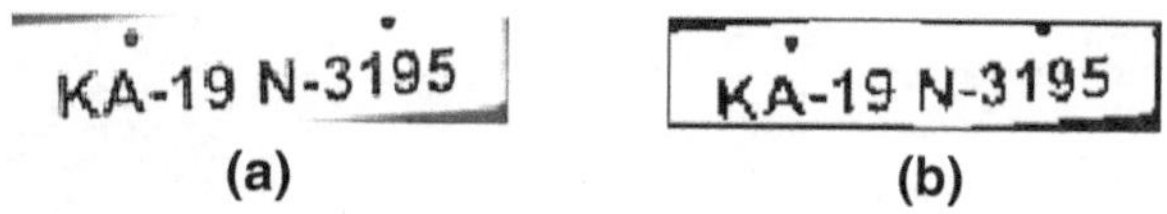


**Fig. 5** Slant correction of extracted number plate **a** Slant detection **b** Slant correction

## 4 Experimental Results

The experiments were conducted on Pentium Dual Core 2.10 GHz with 2 GB RAM on MATLAB R2009a environment. In the experiment about 120 images were used which are Indian styled number plates with the size as 490 × 367 pixels. These images were taken by mobile camera (Sony Ericsson K550i) and digital camera (Nikon) from different locations and under different weather conditions focusing the camera to the expected plate region. Satisfactory results have been obtained with the success rate of detection up to 85 %. In this method, failures are noticed when vehicle images were obtained in following cases: (a) from different viewpoints, (b) containing some special objects (stickers or stamps) (c) complex scenes (objects similar to number plates) (d) environment conditions (strong sun light or gloomy light) (e) oriented images.

The following figures show few successful and failure cases in the sequence: (a) Initial RGB image, (b) Gray image, (c) Detecting horizontal edges, (d) Detecting vertical edges, (e) After OR masking operation, (f) Detection of candidate regions (g) Number plate detection and (h) Number plate slant correction.

**Successful Cases:**
**Failure Cases:**

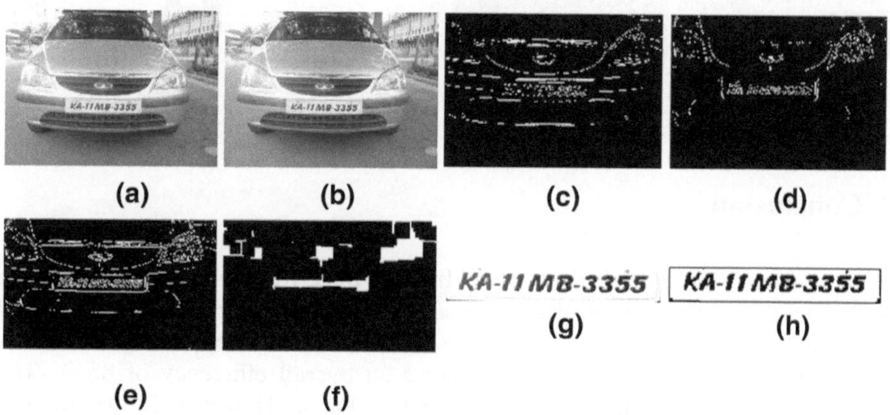

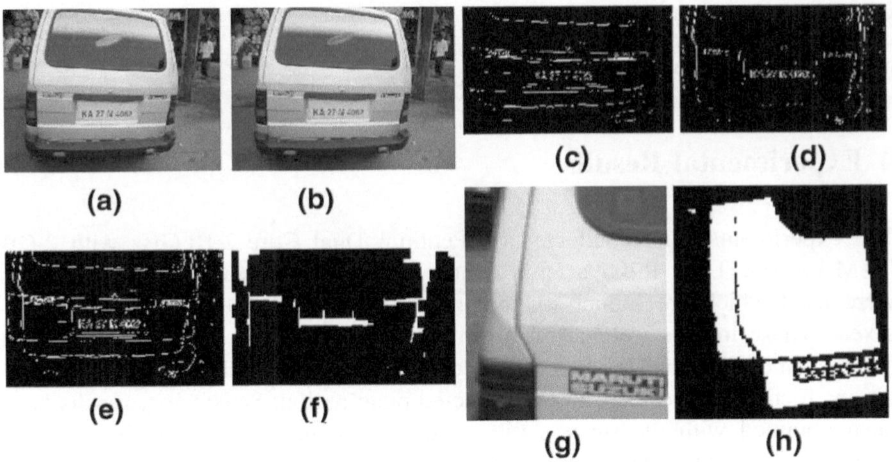


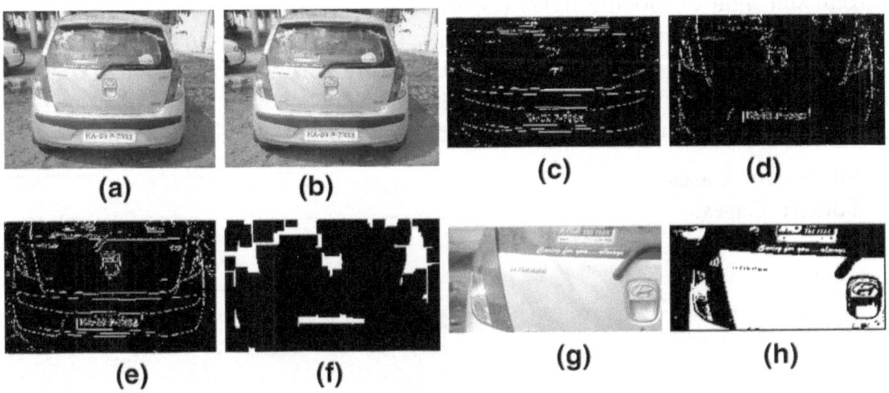


## 5 Conclusion

A method has been presented to extract number plate from a vehicle image using horizontal and vertical edges. This method essentially uses the theory of edges and connected components. The slant/skew in the extracted number plate detected uses Radon transformation. The detection shows an overall efficiency of 85 %. The failure cases are due to large orientation and occlusion. However, further work is in progress for better efficiency in detection and new method for slant detection is under investigation.

## References

1. Kato T, Ninomiya Y, Masaki I (2002) Proceeding vehicle recognition based on learning from sample images. IEEE Trans Intell Transp Syst 3(4):252–260
2. Maged Fahmy MM (1994) Automatic number-plate recognition neural network approach. In: Proceedings of vehicle navigation and information systems conference, Sept 1994, pp 99–101
3. Fucik O, Zemcik P, Tupec P, Crha L, Herout A (2004) The networked hoto-enforcement and traffic monitoring system of rollout steps includes: Unicam. In: Proceedings of the 1th IEEE international conference on the engineering of computer-based systems, pp 234–251
4. Connor S (2005) Britain will be first country to monitor every car journey, The independent http://www.matlab.net/0_index-shtml?x=496457, December 22
5. Wei-gang Z, Guo-jiang H, Xing J (2002) A study of locating vehicle license plate based on color feature and mathematical morphology, In: 6th international conference on signal processing, Beijing, vol 1, pp 748–751
6. Ahmed M, Sarfaz M, Zidouri A, AI-Khatib KG (2003) License plate recognition system. In: 10th IEEE international conference on electronics, circuits and systems, pp 898–901
7. Kim D-S, Chien S-I (2001) Automatic car license plate extraction using modified generalized symmetry transform and image wrapping. In: IEEE symposium on industrial electronics vol 3, pp 2022–2027
8. Anagnostopoulous CNE, Anagnostopoulous IEA, Kayafas VLAE (2006) A license plate recognition algorithm for intelligent transportation system applications. IEEE Trans Intell Transp Syst 7(3):377–392
9. Hsieh CY, Juan YS, Hung KM (2005) Multiple license plate detection for complex background. In: 19th IEEE international conference on advanced information networking and applications, vol 2, pp 389–392
10. Gonzalez RC, Woods RE (2002) Digital image processing, 2nd edn. Printice Hall, Englewood Cliffs
11. Duan TD, Duc DA, Due TLH (2004) Combining Hough transform and contour algorithm for detecting vehicles license plates. In: Proceedings of international symposium on intelligent multimedia, video and speech processing, pp 747–750
12. Remus B (2001) License plate recognition system. In: Proceedings of the 3rd international conference in information, communications and signal processing, pp 203–206
13. Bai HL, Liu CP (2004) A hybrid license plate extraction method based on edge statistics and morphology. In: Proceeding of the 17th international conference on pattern recognition
14. Clark P, Mirmehdi M (2000) Finding text regions using localised measures. In: Proceedings of the 11th British Machine Vision conference, pp 675–684
15. Clark P, Mirmehdi M (2000) Combining statistical measures to find image text regions. In: Proceedings of the 15th international conference on pattern recognition, pp 450–453
16. Bishop CM (1995) Neural networks for pattern recognition. Clarenden Press, Oxford
17. Parisi R, Di Claudio ED, Lucarelli G, Orlandi G (1998) Car plate recognition by neural networks and image processing. In: Proceedings of the 1998 IEEE international symposium on circuits and systems, pp 195–198
18. Shapiro LG, Stockman GC (2001) Computer vision, Prentice-Hall ISBN0-13-030796-3 New Jersy
19. Umbaugh SE (2005) Computer imaging: digital image analysis and processing. CRC Press, Florida, pp 45–50, 93–100, 168–177, 264
20. Sun D, Zhu C (2008) Skew angle detection of the vehicle license plate image and correct based on Radon. Transform J Microcomput Appl, China 29(2):114–115
21. Bai AY, Hu BH, Li CF (2010) A recognition system of china-style license plates based on mathematical morphology and neural network. Int J Math Models Methods Appl Sci 4(1): 66–73
22. Xu Jie X, Yin Y (2005) The Research on vehicle license plate location and segmentation in license plate paper, China, pp 35–37

23. Gao D-S, Jie Z (2000) Car license plates detection from complex scene. In: Proceedings of 5th IEEE international conference on signal processing, WCCC-ICSP, vol 2, pp 1409–1414
24. Di Stefano L, Bulgarelli A (1999) A simple and efficient connected components labeling algorithm. In: Proceedings of IEEE international conference on image analysis and processing, pp 322–327

# Facial Boundary Image Reconstruction Using Elliptical Approximation and Convex Hull

**A. Bindu, C. N. Ravi Kumar, S. Harsha and N. Bhaskar**

**Abstract** The Stretch of Biometrics' applications can only be limited by limiting ones' imagination! Biometrics is the Science and Technology of measuring and analyzing biological data for authentication purposes. In addition to verification the guiding force behind biometric verification has been convinience. Face enjoys a prime position in the realm of biometrics because it is very easily accessible when compared to the other biometrics. Efficient accomplishment of Face Recognition confronts innumerable hurdles in the form of variations in lighting conditions during image capture, Occlusions, damage in facial portions due to accidents etc. The application of Facial Image Inpainting also fails when the occlusions or the deformalities are present across the boundary of the object of interest(face), since the bounds for the application of the inpainting algorithm is not precisely defined. Hence recovery of the complete picture of a human face from partially occluded images is quite a challenge in Image Processing. The proposed FIREACH algorithm concentrates on the generation of a convex hull and a non linear elliptical approximation of the depleted and partially visible boundary of the human face, given different parameters to achieve an Efficient Boundary Recovery. The Boundary Recovery Algorithm is a pre-processing step which aids in setting up of a suitable platform for the proficient application of the Facial Image Inpainting.

A. Bindu (✉) · C. N. R. Kumar
Department of Computer Science and Engineering, SJCE, Mysore, India
e-mail: bindukiranmys@gmail.com

C. N. R. Kumar
e-mail: kumarcnr123@gmail.com

S. Harsha
Department of Computer Science and Engineering, ATMA, Bangalore, India
e-mail: harshahassan@gmail.com

N. Bhaskar
Department of Applied Mathematics, VVCE, Mysore, India
e-mail: bhasiyer@gmail.com

P. P. Swamy and D. S. Guru (eds.), *Multimedia Processing, Communication and Computing Applications*, Lecture Notes in Electrical Engineering 213,
DOI: 10.1007/978-81-322-1143-3_2, © Springer India 2013

**Keywords** Contour recovery · Convex hull · Nonlinear elliptical approximation · Region of interest · Identity retrieval

## 1 Introduction

Biometrics finds its application in Identity Authentication by means of analysis and measurement of the biological data. Face biometric enjoys a prime position for its ease of accessibility. Recognition of the face biometric is accomplished with the help of important features extracted from the subjects' face. But, efficient face recognition under some circumstances becomes very difficult viz., if facial region is severely disfigured during an accident [1], if the facial region is occluded with structural elements like beard, moustache, goggles [2–5] and other accessories, external occlusions like door, people etc. covering the prominent facial features at various degrees. Such hindrances are solved considerably by the application of Inpainting Algorithms [6–11] etc. Inpainting Algorithms aid in filling the missing image data in a visually credible manner such that the change is difficult to be percieved by a naïve unexpecting viewer [12]. But, there exist some extremities where it is totally impossible to get conducive results even after the application of Inpainting Algorithms. The proposed research work deals with such an extreme case, where the occlusion is present across the boundary of the region of interest. Under such a circumstance it becomes difficult to predict with precision the boundary of the region to be inpainted. Through the proposed work a modest effort has been initiated towards disoccluding the boundary of the region of interest thereby setting up a limiting condition for the application of the Inpainting Algorithm.

The proposed FIREACH Algorithm accomplishes its task of occluded face contour evolution with the help of 7 major steps, where the Step 1 involves detecting the face region by the application of the skin illumination compensation algorithm [13] which projects only the visible facial portions present in the image. Step 2 involves edge detection. Step 3 evolves the contour using Convex Hull for each of the visible face regions. Step 4 fits the ellipse around each of the visible face regions. Step 5 deals with the extension of the Convex Hull region into the depleted region until the boundary limits set by the ellipse fit. Step 6 involves finding the centroid followed by the Step 7 which evolves better contour of the facial regions, obtained by selectively choosing the contour points of the evolved face between the outputs obtained from Step 3 and Step 4.

## 2 Literature Review

The initiation in the work regarding manual face reconstruction (through surgery) dates back to a period unimaginable—"Reconstructive surgery techniques were being carried out in India around 800 BC by Sushruta, who is popularly known as the father of Surgery for his important contributions towards the field of plastic surgery and cataract surgery" [Wikipedia]. The research work defining the geographical boundaries for surgical reconstruction of marred face was proposed by Penn [1]. Nowadays, in the computer age the human kind is more reliant on getting the task automated by the machine. The research work initiated by Bertalmio [6] was a leading light towards opening a new avenue for image disocclusion [14] using the technique used by artists to retouch the damaged paintings. This new technique came to be widely known as Image Inpainting. In this technique the depleted region or the hole is filled by using the structure elements or isophotes (image details) present at the boundary of the hole (depleted image region). This inpainting technique was meddled by many scientists to achieve better results, but it was localised to the removal of small scratches in the image. Large depleted image regions could not be inpainted efficiently, for which [15] gave a solution in the form of Exemplar Based Inpainting. The Exemplar technique efficiently propogates both texture and structure information into the depleted region of the image. The evolution of the inpainting techniques was extrapolated to the 3D surfaces by Han and Zhu [16, Jin et al. [17]. The earliest work on line and edge extraction dates back to the early 60s. The process edge extraction for disocclusion using image segmentation and depth extraction was initiated by Nitzberg et al. [18] where the detection of T-junctions was accomplished and the connection was established amongst the related T-Junctions with the help of edge length and curvature minimizing energy function. The linking was accomplished by a new edge whose length and curvature are minimum. The drawback with this method was that it could be applied to highly segmented regions of interest with a few T Junctions available. Following the work proposed by Nitzberg et al. [18], Masnou and Morel [19] proposed a technique which involved evolving the deformities using level lines of larger objects in images relying on variational formulations and level sets. The method portrayed success without the necessity for T Junctions. The method by Masnou and Morel [19] had an overhead of finding the level lines of the region of interest and was efficient in removing minor scratches or deformities in the image. The contributory work in [6] came up with a technique which relied on manually selecting the region to be disoccluded using PDE which propogates the information in the direction of the isophotes and successfully removes minor scratches in the image. Bounded Variation based image model stated that the oscillation range of the image level lines are finitely constricted [20]. Following which the Total Variation tries to constrain and minimize the curve length of level lines over Bounded variation. But, it was later realized that a realistic inpainting output is achieved when the curvature is taken into account. Ballester et al. [21] stipulated an inpainting technique with a due consideration given to the curvature which was synonymous with Euler Elastica, in which prior models have a prime significance. The research contributions in [8] and [9] succeeding the previous inpainting contributions coined an inpainting approach with the nomenclature Curvature Driven Diffusion(CDD)—a computational scheme based on numerical PDE's, which is efficient of handling topologically complex Inpainting Domains. All the decisions realized in [8, 9, 11, 22–25] and [26] are primarily based on the finest guess or stated in a better way are based on Bayesian Inference (relies on the prior model of a specific class) of image objects. The major setback displayed by the prior models is their inefficiency in realizing the connectivity principle and also because of the fact that they create visible corners which is considerably eliminated by the usage of level sets. Level sets are efficient in segmenting the object of interest in the image, but the question arises what if the contour of the object visible in the image is not its actual one, but a pseudo contour of the object of interest remnant after occlusion? The problem was partially looked into in [27, 28] where the system using the level set approach requires the deformities to be regular! The question which lingers is ….. It is practically impossible to expect the deformities aprior in the case of realistic imagery with drastic illumination variations, erratic shapes of the objects occluding the object of interest etc.

With the backdrop of such complexities, disocclusion or invisible information recovery is to be accomplished. It is evident from the literature survey that the inpainting algorithms could be used. But what if the actual contour of the object of interest is unknown! Through the proposed FIREACH algorithm a humble effort has been put forth to tackle such a situation with face as the object of interest. Face is considered in our proposed work because of one major challenge it poses—occluded Face cannot be reconstructed from the information retrieved from the boundary of the hole. The Literature review related to Facial Inpainting are as follows: In the experimental results discussed by Hwang and Lee [29] and Mo et al. [30] rely on retrieving the missing portion of the face from a reference face image. The major drawback posed by these approaches is that the drastic variations in the illumination conditions in real life imagery and the various combinations of the reference facial images will not faithfully conform to the specific illumination and photographic conditions of the target image. The contribution in [31] proposed a facial inpainting technique which follows the principle of face recognition by classifying a query image into one or more categories in a database and extend the same principle for searching of faces with missing areas by using prior probabilistic distribution of facial structural information. The patch guided facial inpainting approach also has limitations. If the original face is partially covered by an occlusion the inpainted output does not yield a satisfactory outcome. The approach proposed in [31] assumes normal and uniform distribution of luminance across the face (which is not true in realistic imagery).

Literature Review of the related work to date percolates to the fact that the work done till date concentrates on occlusions of features like eyes and mouth, which are retrieved with reference to the database information. And the exuberant Survey of Literature makes it evident that contributions throwing light towards the evolution of the occluded face region contour has not been explored.

In a real life situation the occlusion will not always be oriented towards the eye or the mouth regions. For e.g.: if a person has met with an accident and his lower part of the face is completely lost! Under such circumstances it becomes necessary to evolve the contour of the face before the application of any reconstruction techniques. FIREACH is a humble step towards efficient accomplishment of Occluded Face Contour Evolution!

## 3 FIREACH Algorithm

### Step 1: Face Region Localization

The Face Region Localization in the captured color image is accomplished by using the Skin Illumination Compensation Model proposed in [13]. The Model uses the concept of region signatures to separate out the skin regions corresponding to the face.

### Step 2: Edge Detection and Sampling

Detected Visible Face Regions are Binarised to reduce the computational overhead. The Edge Maps of the detected Face regions are constructed by using the Canny Edge Detection Operator. The Edge points are periodically sampled to improvise the efficiency of the algorithm by reducing the computational time complexity

### Step 3: Convex Hull

The Convex Hull [32] for the Face edge map is successfully evolved out in this module by using the Aki Toussaint Heuristic. The efficiency of the Chans' Algorithm [33] is exploited affirmatively. Initial assumption is that the 'h' points on the convex hull is known apriori. Chan's Algorithm starts by shattering the input points into 'n/h' arbitray subsets, each of size 'h', and computing the convex hull of each subset using the Graham's scan [34]. The algorithm requires O((n/h)hlogh) = O(nlogh) time for finding out the arbitrary subsets.

Following the computation of 'n/h' subhulls, the general outline of Jarvis's march is followed, which involves wrapping a string around the 'n/h' subhulls. Initiating with the leftmost input point l, starting with p = l (p is the boundary pixel of the Convex Hull) and successively evolving the convex hull vertices in counter-clockwise order until the traversal returns back to the original leftmost point again [35].

The output of the step 3 results in success only if the depleted regions are very small (within 10 % of the visible face region). As the percentage of depletion exceeds beyond 10 % of the visible face region the output is not satisfactory!

### Step 4: Ellipse Fitting

Many unsuccessful attempts to make the fitting process computationally effective were succeeded efficiently by direct least squares based ellipse-specific method proposed by Fitzgibbon and Fischer [36], Fitzgibbon [37], Fitzgibbon et al. [38].

The Direct Least Square Fitting provides impressive results under noisy conditions with dependable computational efficiency. This method is stated to be a non-iterative ellipse-specific fitting.

Ellipse-a special case of a general conic, can be described by an implicit second order polynomial.

$$F\ (a,\ x) = a.\ x = ax^2 + bxy + cy^2 + dx + ey + f = 0 \tag{1}$$

with an Ellipse specific constraint

$$\left(b^2 - 4ac\right) < 0 \tag{2}$$

where a, b, c, d, e, f are coefficients of the ellipse which fits the detected face and (x, y) are the coordinates of points lying on the detected face. The polynomial F(x, y) is the algebraic distance of point (x, y) to the given conic.

By introducing vectors $a' = [a\ b\ c\ d\ e\ f]^{\mathbf{T}}$

$$x' = \left[x^2\ xy\ y^2\ x\ y\ 1\right] \tag{3}$$

it can be rewritten to the vector form

$$F_a(x) = x.\ a = 0 \tag{4}$$

The fitting of ellipse to the set of points $(x_i, y_i)$, $i = 1,\ldots,N.$, extracted from the boundary of the partially occluded face is accomplished by minimizing the sum of squared algebraic distances of the points to the ellipse fit which is represented by the coefficients a′:

$$\min_a \sum_{i=1}^{N} F(x_i, y_i)^2 = \min_a \sum_{i=1}^{N} (F_a(x_i))^2 = \min_a \sum_{i=1}^{N} (x_i \cdot a)^2 \tag{5}$$

To ensure ellipse-specificity of the solution, constraint in Eq. (2 )is taken into account.

**Step 5: Convex Hull Extension**

In the case of facial regions with occlusions: after the evolution of the ellipse fit for the facial region with occlusions the convex hull region is extended into the depleted region until the bounds of the contour created by the ellipse fit.

**Step 6: Centroid Estimation**

The Convex Hull for the detected face region is evolved in Step 3, and is succeeded by applying the Direct Least Squares Ellipse fit on the edge recognized detected face region in Step 4. The output of Step 3 and Step 4 are overlapped on their respective face regions and the Centroid for the ellipse is detected by using Eq. (6)

$$\text{Centroid(EllipseDirectFit)} = \text{mean(XY)} \tag{6}$$

where, X = Major Axis of the Ellipse and Y = Minor Axis of the Ellipse

**Step 7: Boundary Estimation**

The output of the Step 6 results in the evolved contour of the Face Region, along with the Convex Hull (B1) of the visible face region (in the occluded case) and the Ellipse fit of the region (B2) with the Centroid. It is evident that both the Convex Hull and the Ellipse Fit result in extraneous outliers which are to be eliminated. To accomplish this following steps are followed:

1. Move outwards from the centroid, until either B1 or B2 is confronted.
2. Once any one of the Boundaries is confronted while moving outwards from the centroid, set the corresponding value of the pixel (save the pixel coordinates in a vector) as the actual contour of the face region.
3. Continue this process starting from the Centroid in all possible orientations ranging from 0 to 360°.
4. Once the entire cycle from 0 to 360(degrees) is completed, the set of pixels whose corresponding coordinate values are saved in the vector depict the newly evolved Contour of the Disoccluded Face Region!

The Steps 1–7 are repeated iteratively on all the detected face regions in the image. Finally, the evolved contour is multiplied with the original image to get the exact face region.

## 4 Data Set

The prime goal of the proposed algorithm is to evolve the damaged/occluded contours of face biometric. The Face Database is suitably compiled to suit the application under consideration considering different degrees of occlusion across the face. The database consists of 2 sections, the first one comprises of images personally collected and the second section comprises of images collected from the World Wide Web. The first section includes 10 different variants per subject, with reference to the degree of occlusions and variations in the lighting conditions. Similar types of variants are collected for 100 different subjects with a uniform background, totaling the content of section 1–1,000 samples. For each of the 100 subjects, one of the sample collected is an ideal one without any occlusion and is considered as the ground truth for the respective subject. The second section in the dataset involves 50 random images collected from the World Wide Web (WWW). The face contours of the faces in the collected web images are considered as the ground truth and the occlusion is artificially induced at different degrees (10 different degrees of occlusions are induced for each of the facial components present in the web images).

The samples in the first section are captured in standard conditions with a uniform background and with the distance of the subject from the camera being maintained constant for all the samples. The ground truth samples are not taken into consideration for training or for testing.

## 5 Experiments and Results

The FIREACH Algorithm is designed to automatically select and assign the dataset samples randomly as training and testing categories in every iteration (for e.g.: 10 % training and 90 % testing, 60 % training and 40 % testing and the like) (Figs. 1, 2, 3, 4, 5, 6).

The Dataset totally includes section 1:1,000 samples captured individually by using a standard 12 Mega Pixel Nikon Digital Camera plus section 2:50 random image samples collected from the World Wide Web (WWW) for which the occlusions are manually induced resulting in 9 different degrees of occlusions for

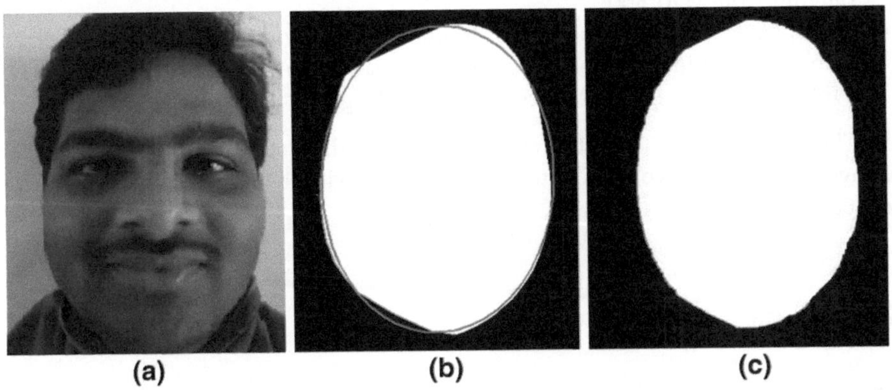


**Fig. 1** Case 1: FIREACH output for frontal face with no occlusion (used as the **ground truth** to evaluate the success of FIREACH for various degrees of occlusions present across the face samples). **a** Represents the original image, **b** represents the contours of the detected face region resulting after the application of step 3 and step 4 in the FIREACH Algorithm, **c** represents the evolved face contour output resulting from step 7 in FIREACH Algorithm

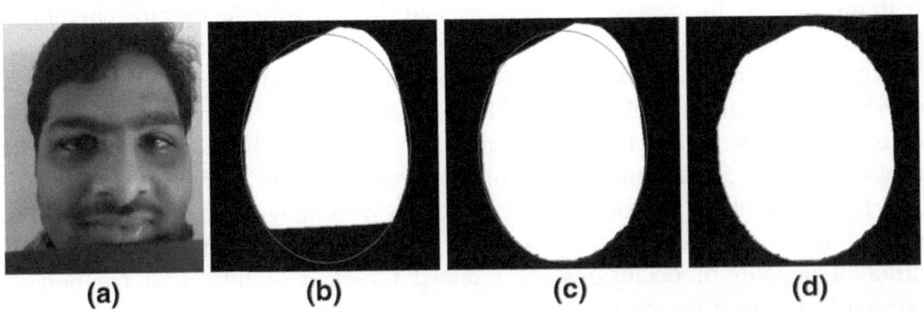


**Fig. 2** Case 2: FIREACH output when 10 % of the face occluded. **a** pertains to the sample collected with 10% occlusion in the sample; **b** the output after the application of Convex Hull and Ellipse Fitting with evident occlusions; **c** The output after extending the Convex hull into the depleted/occluded face regions until the bounds set by ellipse fit; **d** Signifies the output after the application of Step 7 in the FIREACH Algorithm

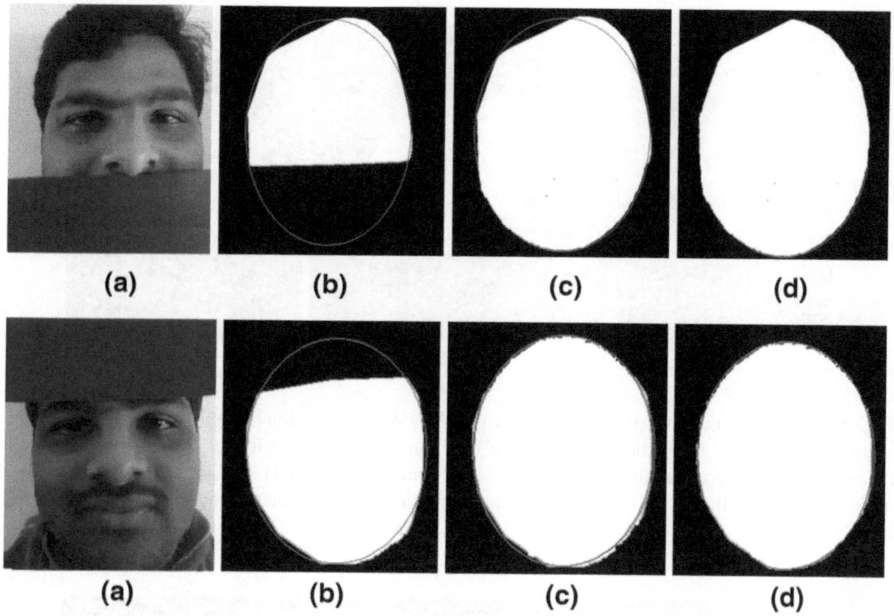


**Fig. 3** Case 3: FIREACH output when (30–40) % of the face occluded. **a** pertains to the sample collected with (30-40)% occlusion in the sample; **b** the output after the application of Convex Hull and Ellipse Fitting with evident occlusions; **c** The output after extending the Convex hull into the depleted/occluded face regions until the bounds set by ellipse fit; **d** Signifies the output after the application of Step 7 in the FIREACH Algorithm

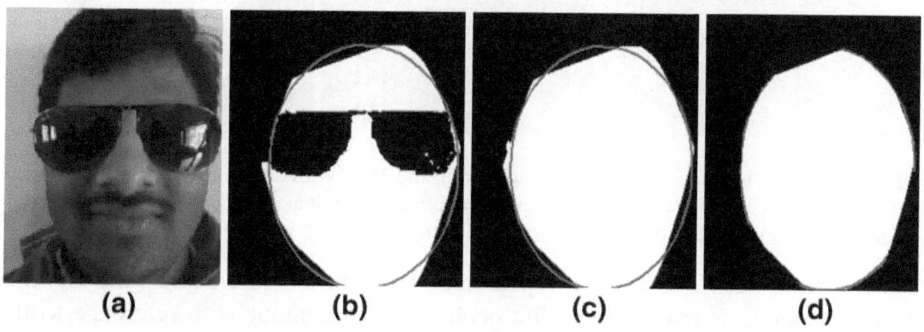


**Fig. 4** Case 4: FIREACH output for a face with glasses/goggles. **a** Pertains to the sample collected with occlusion due to goggles; **b** the output after the application of Convex Hull and Ellipse Fitting with evident occlusions; **c** The output after extending the Convex Hull into the depleted/occluded face regions until the bounds set by ellipse fit; **d** Signifies the output after the application of Step 7 in the FIREACH Algorithm

each of the 50 random image samples (from WWW) inclusive of the ground truth totaling to 500 samples. The Compiled Dataset is composed of a total of 1500 samples (inclusive of the ground truth which is not included in either the Training

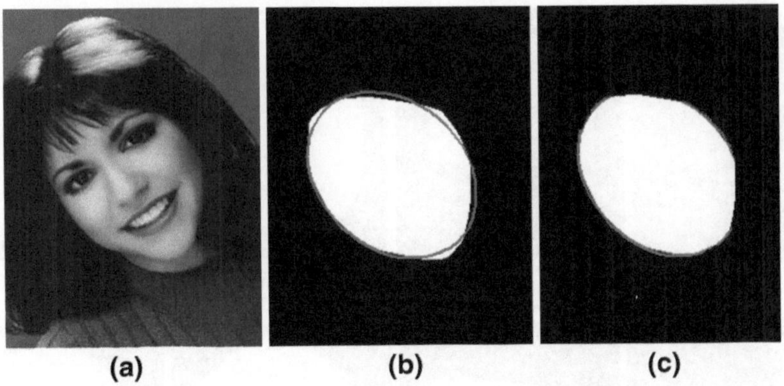


**Fig. 5** Case 5: **a** indicates the image from the WWW; **b** indicates the evolved boundary of the image sample in 'a' after the application of the Convex Hull and Ellipse Fitting; **c** indicates the face contour evolved after the application of Step 7 in the FIREACH Algorithm and is considered as the Ground Truth for all the corresponding occlusions induced on the image sample mentioned in 'a'

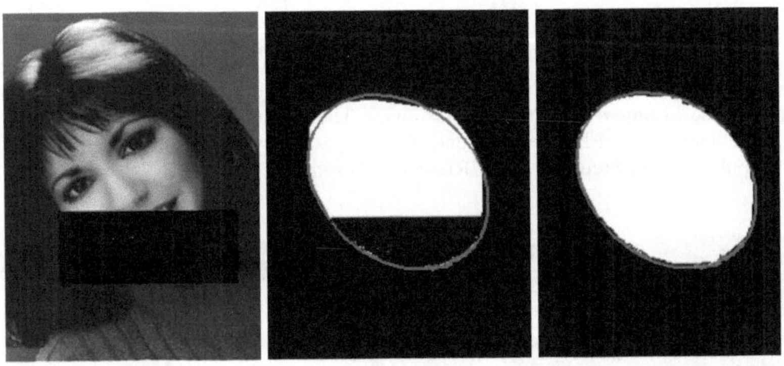

**Fig. 6** Case 6: **a** indicates the image sample from the Web with artificially induced occlusion; **b** indicates the Occluded Face Contour after the application of Step 3 and Step 4 in the FIREACH Algorithm; **c** indicates the output after the application of Step 7 in the FIREACH Algorithm

or the Testing Set). The Tabular representation in the Table 1 depicts the fact that the performance of evolution of the occluded face contour with reference to the closeness of the evolved contour to the corresponding ground truth increases as the number of training images increase.

The results displayed in the Table 1 are the values obtained with reference to the closeness of the evolved face contour of the occluded face regions in comparison with the corresponding evolved ground truth face contour. The comparison of the evolved face contour with that of the ground truth is established with compatibility/acceptable variation of evolved contour with that of the reference contour being $\pm$ 3 %. It is evident that the performance of the algorithm demonstrates a commendable improvement as the number of training samples increases.

**Table 1** The accuracy FIREACH algorithm in evolving occluded face contour

| Degree of occlusion induced | Training data: testing data (wrt., percentage of samples in the data set) | | | | | | | | |
|---|---|---|---|---|---|---|---|---|---|
| | 10:90 % | 20:80 | 30:70 | 40:60 | 50:50 | 60:40 | 70:30 | 80:20 | 90:10 |
| 10 % Occlusion (lower part of the face) | 0.90 | 0.90 | 0.90 | 0.93 | 0.93 | 0.96 | 0.97 | 0.97 | 0.97 |
| 10 % Occlusion (upper part of the face) | 0.87 | 0.89 | 0.90 | 0.90 | 0.92 | 0.92 | 0.92 | 0.93 | 0.93 |
| (20–30) % Occlusion (lower part of the face) | 0.80 | 0.80 | 0.86 | 0.86 | 0.86 | 0.87 | 0.88 | 0.90 | 0.90 |
| (20–30) % Occlusion (upper part of the face) | 0.79 | 0.79 | 0.79 | 0.80 | 0.80 | 0.80 | 0.80 | 0.80 | 0.81 |
| (30–40) % Occlusion (lower part of the face) | 0.74 | 0.75 | 0.75 | 0.77 | 0.76 | 0.77 | 0.77 | 0.79 | 0.80 |
| (30–40) % Occlusion (upper part of the face) | 0.70 | 0.72 | 0.72 | 0.72 | 0.75 | 0.75 | 0.75 | 0.74 | 0.76 |
| (40–50) % Occlusion (around 50 %) (upper part/lower part/side parts of the face) | 0.60 | 0.60 | 0.57 | 0.56 | 0.56 | 0.57 | 0.60 | 0.63 | 0.62 |

It is obvious from some of the tabular entries that the performance drops even with the increase in the training sample number. The reason being that when the system operates by randomly selecting the samples for testing and training during every iteration, if in a particular iteration the presence of web selected face images exceed the manually selected ones the efficiency is found to depreciate as the web images capturing conditions are not very ideal. But, still the depreciation is not very prominent as is evident from Table 1.

The system gives good results with occlusions present on the lower part of the face images when compared to the occlusion present in the upper part of the facial image, which is because of the variations in the hair style (because of which it becomes difficult to match with the exact face contour).

## 6 Conclusion

FIREACH achieves a commendable initiative towards evolution of occluded face region contours which would predominantly aid as a preliminary phase in achieving efficient and reliable face reconstruction for Face Recognition Systems. FIREACH algorithm works well with occlusions varying around 50 %. The Future Enhancement would be directed towards solving the contour evolution lacunas present in FIREACH with respect to occlusions covering the face regions beyond 50 %.

**Acknowledgments** The authors would like to extend their sincere thanks to JSSRF for having provided us with the resources for carrying out our Research Work efficiently.

## References


1. Penn JG (1976) Geographical boundaries of facial reconstruction. S Afr Med J 50:1468
2. Jiang X, Binkert M, Achermann B, Bunke H (1990) Towards detection of glasses in facial images. In: Proceeding Int'l Conference Pattern Recognition, pp 1071–1973
3. Jing Z, Mariani R (2000) Glasses detection and extraction by deformable contour. In: Proceeding Int'l Conference Pattern Recognition, vol 2, pp 933–936
4. Wu C, Liu C, Shum HY, Xu YQ, Zhang Z (2004) Automatic eye glasses removal from face images. IEEE Trans Pattren Anal Mach Intell 26(3): 322–336
5. Tauber Z, Li Z-N, Drew MS (2006) Review and preview: disocclusion by inpainting for image-based rendering, School Computer Science, Simon Fraser University, Burnaby, Canada, Tech Rep
6. Bertalmio M, Sapiro G, Caselles V, Ballester C (2000) Image inpainting. In: Proceeding Computer Graphics (SIGGRAPH 2000) 417–424
7. Bertalmio M, Bertozzi AL, Sapiro G (2001) Navier–Stokes, fluid dynamics, and image and video inpainting. In: Proceedings of IEEE conference on computer visual pattern recognition 1:I-355–I-362
8. Chan TF, Shen J (2001) Non-texture inpainting by curvature driven diffusion. J Vis Commun Image Represent 12(4):436–449
9. Chan TF, Kang SH, Shen J (2002) Euler's elastica and curvature based inpaintings. SIAM J Appl Math 63(2):564–592
10. Kang SH, Chan TF, Soatto S (2002) Landmark based inpainting from multiple views. Univ Caifornia at Los Angeles, Los Angeles, Tech Rep TR-CAM 02–11 Mar 2002
11. Bertalmio M, Vesa L, Sapiro G, Osher S (2003) Simultaneous structure and texture image inpainting. IEEE Trans Image Process 12(8):882–889
12. Walden S (1985) The ravished image. St. Martin's, New York
13. Kumar CNR, Bindu A (2006) An efficient skin illumination compensation model for efficient face detection. IEEE Industrial Electronics, IECON 2006—32nd Annual Conference, ISSN: 1553–572X, 3444–3449
14. King D (1997) The commissar vanishes. Henry Holt, New York
15. Perez P, Criminisi A, Toyama K (2003) Object removal by exemplar based inpainting. Proc IEEE Conf Comput Vis Pattern Recognit 2:721–728
16. Han F, Zhu S-C (2003) Bayesian reconstruction of 3D shapes and scenes from a single image. Proc IEEE Int Conf Comp, Vision
17. Jin H, Soatto S, Yezzi AJ (2003) Multi-view stereo beyond lambert. Proc IEEE Comp Vis Pattern Rec 1:171–178
18. Nitzberg M, Mumford D, Shiota T (1993) Filtering, segmentation and depth. Lecture Notes in Computer Science, Vol 662. Springer, Berlin
19. Masnou S, Morel J-M (1998) Level lines based disoclussion. In Proceeding International Conference Image Process, pp 259–263
20. Ambrosio L, Fusco N, Pallara D (2000) Functions of bounded variationsand free discontinuity problems. Clarendon Oxford, U.K
21. Ballester C, Bertalmio M, Caselles V, Sapiro G, Verdera J (2001) Filling in by join interpolation of vector fields and grey levels. IEEE Trans Image Process 10(8):1200–1211
22. Geman S, Geman D (1984) Stochastic relaxation gibbs distribution and bayesian restoration of images. IEEE Trans Pattern Anal Mach Intell 6:721–741
23. Giustic E (1984) Minimal surfaces and functions of bounded variation. Birkhauser, Boston
24. Mumford D, Nitzberg M, Shiota T (1993) Filtering, Segmentation and Depth. LNCS. Vol 662, Springer, Berlin
25. Mumford D (1994) Geometry driven diffusion in computer vision, Chapter-"The Bayesian Rationale for Energy Functionals". pp 141–153, Kluwer Academic, London
26. Hosoi T, Kobayashi K, Ito K, Aoki T (2011) Fast image inpainting using similarity of subspace method. 18th IEEE International Conference on Image Processing, 2011

27. Mumford D, Shah J (1989) Optimal approximation by piecewise smooth functions and associated variational problems. Comm Pure Appl Math 42:577–685
28. Droske M, Ring W, Rumpf M (2009) Mumford–Shah based registration: a comparison of a level set and a phase field approach. Comput Visual Sci 12:101–114
29. Hwang BW, Lee SW (2003) Reconstruction of partially damaged face images based on a morphable face model. IEEE trans Pattern Anal Mach Intell 25(3):365–372 doi: 10.1109 ITPAMI,2003,1182099]
30. Mo ZY, Lewis JP, Neumann U (2004) Face inpainting with local linear representations. British Machine Vision Conference, London
31. Zhuang Y, Wang Y, Shih TK, Tang NC (2009) Patch-guided facial image inpainting by shape propagation. J Zheijiang Univ Sci A 10(2):232–238
32. Matusik W, Buehler C, Raskar R, Gortler SJ, McMillan L (2000) Image based visual hulls. In Proceeding Computer Graphics (SIGGRAPH 2000) 369–374
33. Chan Timothy M (1996) Optimal output-sensitive convex hull algorithms in two and three dimensions. Discrete and Computational Geometry 16:361–368
34. Graham RL (1972) An efficient algorithm for determining the convex hull of a finite planar set. Inf Process Lett 1(132–133):1972
35. Bindu A, RaviKumar CN (2011) Novel bound setting algorithm for occluded region reconstruction for reducing the inpainting complexity under extreme conditions. Int J Comput Appl 16(5):0975–8887
36. Fitzgibbon AW, Fischer RB (1995) A buyer's guide to conic fitting. In: Proceeding of the British Machine Vision Conference, pp 265–271, Birmingham
37. Fitzgibbon AW (1995) Set of MATLAB files for ellipse fitting. Dept. of Artificial Intelligence, The University of Edingburgh, ftp.dai.ed.ac.uk/pub/vision/src/demofit.tar.gz, Sep 1995
38. Fitzgibbon AW, Pilu M, Fischer RB (1996) Direct least squares fitting of ellipses. Technical Report DAIRP-794, Department of Artificial Intelligence, The University of Edinburgh, Jan 1996

# South Indian Handwritten Script Identification at Block Level from Trilingual Script Document Based on Gabor Features

**Mallikarjun Hangarge, Gururaj Mukarambi and B. V. Dhandra**

**Abstract** The script is a graphical illustration of thinking of a person. Any script can be considered as texture patterns which have linear, oriented and curvilinear sub-pattern primitives. In this paper, the problem of automatic handwritten script identification is considered as texture analysis problem. This paper presents the significance of the traditional Gabor filters in extracting oriented energy distributions. These are tuned efficiently with 24 channels to extract directional energies of text blocks of each script. K nearest neighbor classifier is employed for discriminating six south Indian scripts based on the standard deviations of Gabor filters response. The comprehensive experimentation is conducted on a data set of 600 text block images. Average tri-script classification accuracy with two fold cross validation is 91.99 %.

**Keywords** Monolingual · Bilingual · Trilingual · Multilingual document processing · Script identification · OCR · Gabor feature extraction · KNN classifier

## 1 Introduction

Automatic handwritten script identification has been an active research area in multilingual document image analysis from last three decades. In recent years, advancements in technologies are accomplishing the demands of paperless office. The use of electronic documents in office facilitates easy communication, storage

M. Hangarge
Karnatak Arts, Science and Commerce College, Bidar, Karnataka, India

G. Mukarambi (✉) · B. V. Dhandra
Department of P.G. Studies and Research in Computer Science, Gulbarga University, Gulbarga, Karnataka, India
e-mail: gmukarambi@gmail.com

P. P. Swamy and D. S. Guru (eds.), *Multimedia Processing, Communication and Computing Applications*, Lecture Notes in Electrical Engineering 213,
DOI: 10.1007/978-81-322-1143-3_3, 

of documents, searching, indexing and retrieving of multilingual information. To read multi-script and multi-lingual documents, Optical Character Recognition (OCR) systems necessitate being proficient of recognizing the characters irrespective of the script in which they are written. It is hard to recognize characters of multi-script documents by a single OCR system, because of the shape of the characters differs from one script to another [1]. For example, features used for recognizing Arabic characters are not efficient in recognizing Roman characters. It is also important to note that the Optical Character Recognizer designed to read machine printed characters are not competent in recognizing handwritten characters of the same script. Therefore, to solve this problem, it needs to design a bank of optical character recognizing systems to read the multi-script documents. While reading multi-script documents, it needs to switch over between OCR of one script to another. To do so, automatic script identification is essential. Automatic script identification facilitates sorting, searching, indexing and retrieving of multilingual documents and enhances the efficiency of OCR system. Most of the published work on automatic script identification of Indian scripts, deals with printed documents and few articles were found for handwritten script identification. The problem of script identification may be addressed at bi-script, tri-script and multi-script. However, in India most of the official documents are trilingual in nature (documents with English, Hindi and regional languages) and hence trilingual script identification is the appropriate way to address the problem. Thus, in this paper, we have considered four tri-script documents of South Indian scripts. The shape of the characters of the south Indian scripts has horizontal, vertical, curvilinear, loops and w-formation structures. These insights emphasize the importance of oriented directional energies to characterize the shape of such characters. Hence this paper explores the potentiality of the Gabor filters in capturing oriented energy features of a script.

A detailed review on offline handwritten script identification can be found in [2] and complete survey on script identification can had it from [1]. Ma and Doermann [3] have proposed Gabor filter based multi-class classifier for script identification at word level of scanned document images for Arabic/Roman, Chinese/Roman, Korean/Roman, and Hindi/Roman and obtained the average recognition accuracy of 73.27, 84.08, 83.49, and 92.08 % respectively. Pati et al. [4] have proposed Gabor filter for script identification for Indian bilingual scripts at word level and achieved the recognition accuracy of 99.56, 96.02, and 97.01 % for Hindi, Tamil and Odiya respectively. Chanda et al. [5] have proposed two stage approach for word wise script identification of printed English, Devanagari and Bengali scripts from a single document page and achieved the average recognition accuracy of 98.51 % with SVM classifier. Rajneesh et al. [6] have proposed Gabor features for identification of English numerals at world level from printed Punjabi documents and achieved the average recognition accuracy of above 99 % using fivefold cross validation with SVM classifier. Rajput and Anita [7] have proposed DCT and Wavelet features for trilingual script identification at block level and achieved the average recognition accuracy of 96.4 % with NN classifier. Hangarge and Dhandra [8] have proposed spatial spread features for bi-script and tri-script separation at text lines and text blocks level in document images and achieved the average recognition accuracy of 99.02 and 88.06 % respectively using fivefold cross validation with knn classifier. Padama and Vijaya [9] have proposed profile based features for script identification at text lines from trilingual document images and achieved the average recognition accuracy of 99.5 % with Knn classifier. Swamy et al. [10] have proposed heuristic based algorithms to identify script types from Telugu, Hindi and English text documents and achieved the average recognition accuracy of above 95 %. Abirami and Manjula [11] have mentioned a survey of script identification techniques for multi script document images. From the literature survey it is evident that, still handwritten multilingual script identification is an active filed of research.

In Sect. 2, the overview of data collection and preprocessing is presented. In Sect. 3, feature extraction technique is discussed. The experimental details and results obtained are presented in Sect. 4. Conclusion is given in Sect. 5.

## 2 Data Collection and Preprocessing

The standard database is not available for handwritten Indian scripts, hence collected handwritten documents of different scripts from different professionals belonging to schools, colleges and officials. The documents collected are scanned at 300 DPI and stored as gray scale images. A block of image of size 512 × 512 pixels is extracted manually from different areas of the document image. The handwritten text block region contains only text, and numerals that may appear in the text are not considered. The digitized images are in gray tone and we have used Otsu's global thresholding approach to convert them into two tone images. The two-tone images are then converted into 0−1 label where the label 1 represents the object and 0 represents the background. The small objects (less than are equal to 30 pixels) like, single or double quotation marks, hyphens etc. are removed using morphological opening operations. A total of 600 handwritten image blocks, 100 blocks for each of the six scripts are considered for experimentation. A sample of handwritten text block images representing different scripts is shown in Fig. 1 and preprocessed images is shown in Fig. 2 respectively.

## 3 Feature Extraction

The process of feature extraction is one of the important components in any recognition system. In this paper, features are extracted by transforming the input time domain image into frequency domain. The term frequency refers to variation in brightness or color across the image, i.e. it is a function of spatial coordinates, rather than time. The following is the feature extraction method i.e. Gabor filter Bank.

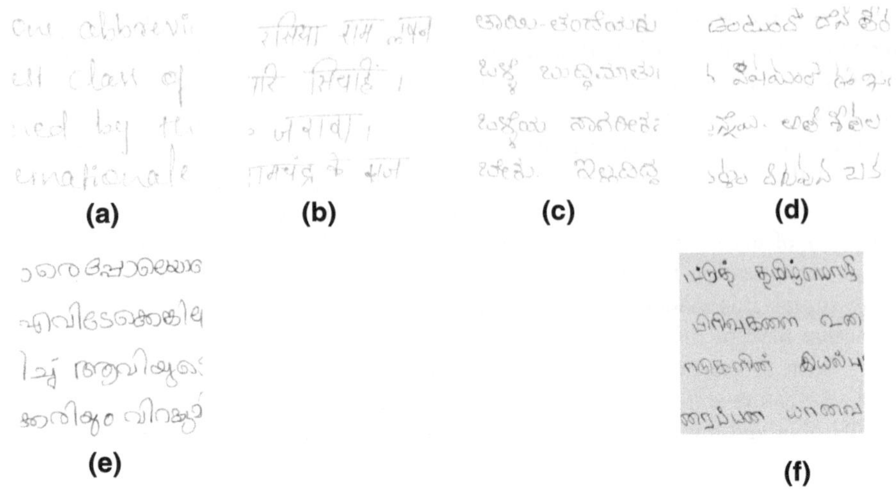

**Fig. 1** Sample image blocks of size 512 × 512 pixels of six scripts. **a** English. **b** Hindi. **c** Kannada. **d** Telugu. **e** Malayalam. **f** Tamil

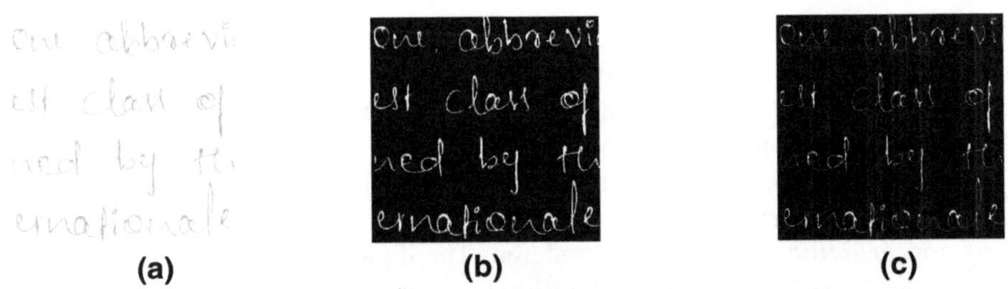

**Fig. 2** Pipeline process of preprocessing operation. **a** *Gray* scale Image. **b** Binary Image. **c** Thinned Image after noise removal

## 3.1 Gabor Filter Bank

The use of Gabor filters in image analysis is biologically motivated as they model the response of the receptive fields of the orientation-selective simple cells in the human visual cortex. Furthermore, they provide the best possible tradeoff between spatial and frequency resolution. Gabor filters are formed by modulating a complex sinusoid by a Gaussian function with different frequencies and orientations. A two dimensional Gabor function consists of a sinusoidal plane wave of some frequency

$$g(x,y) = \left(\frac{1}{2\pi\sigma_x\sigma_y}\right)\exp\left[-\frac{1}{2}\left(\frac{x'^2}{\sigma_x^2}+\frac{y'^2}{\sigma_y^2}\right)2\pi jWx'\right] \tag{1}$$

$$\begin{aligned} x' &= x\cos\theta + y\sin\theta \\ y' &= -x\sin\theta + y\cos\theta \end{aligned} \tag{2}$$

where $\sigma_x^2$ and $\sigma_y^2$ control the spatial extent of the filter, $\theta$ is the orientation of the filter and W is frequency of the filter. Two dimensional Gabor filters are used to extract the features from input text block image. The preprocessed input binary image is convolved with Gabor filters considering six different orientations (0, 30, 60, 90, 120, and 150°) and four different frequencies (a = 0.5, b = 0.125, c = 0.25 and d = 0.0625) with $\sigma_x = 2$ and $\sigma_y = 4$. The values of these parameters are experimentally fixed. Then 24 output images are obtained after convolution with Gabor bank of filters. These output images are used for computing the standard deviation and stored as a feature vector of 24 dimensions. Further, these features are used to train and test the K-NN classifier. Samples of filtered images for 0°, 30° orientations with frequencies are 0.5 are shown in Fig. 3. First line shows for 0° and second for 30°.

Algorithm: Gabor Feature Based Script identification in Trilingual documents
Input: Gray scale images of text blocks of different scripts
Output: Identified script of the text block
Begin

1. Converting gray scale image into binary image using Otsu's method (see Fig. 2b).
2. Noises and small objects around the boundary are removed by using morphological opening operations.
3. Thinning operation is performed (see Fig. 2c).
4. Create Gabor filter Bank by considering six different orientations and four different frequencies to obtain 24 filters.
5. Convolve the input image with created Gabor filter bank (see Fig. 3b and d).
6. For each output image of step 5, compute standard deviation of the entire output images and obtain a feature set of 24 dimensions.
7. Store feature vector of each script of each text block for classification.
8. Initially, KNN classifier is trained with labeled features to provide a knowledge base. Then, features of unknown script is computed as explained in step 5 and 6 and then given to KNN to identify the script of the text block based on its knowledge base.
9. End.

## 4 Experimental Results and Discussions

Experimentations are carried out with KNN classifier. This is a straightforward extension of nearest neighbor. Basically, we try to find the k nearest neighbors and do a majority voting. Typically k is odd when the number of classes is 2. Choosing

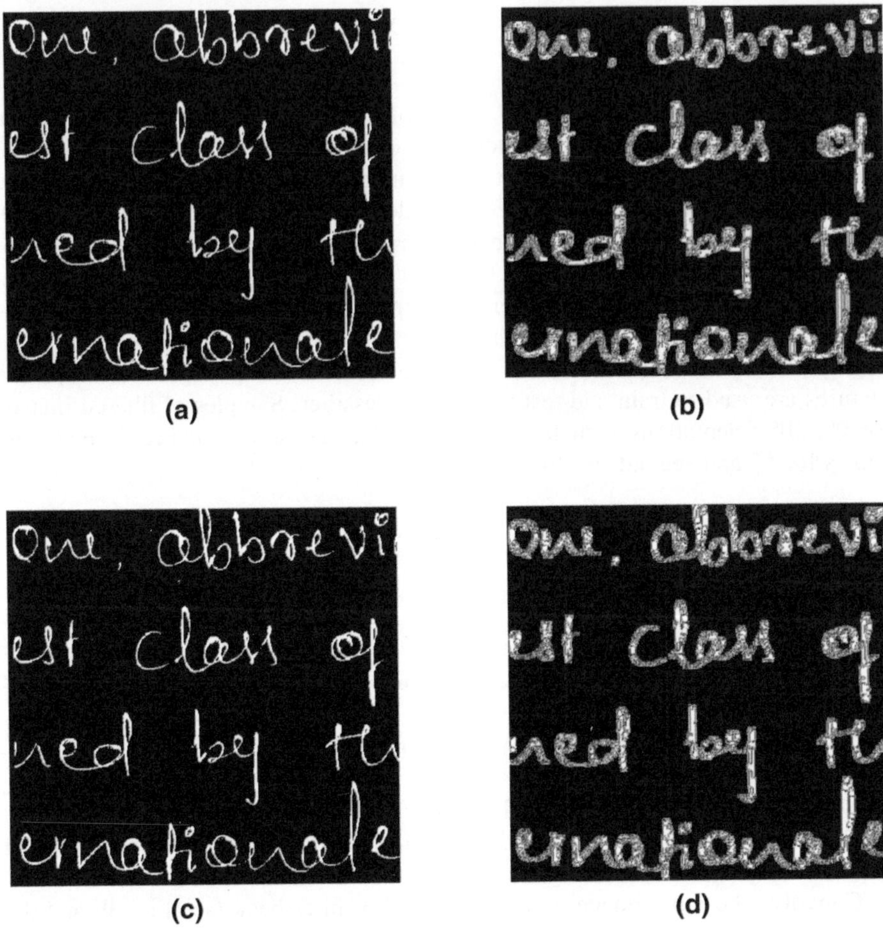

**Fig. 3** Gabor filtered images for 0°, 30° orientations with frequencies 0.5. **a** Binary Image. **b** Gabor output for 0° and frequecny 0.5. **c** Binary Image. **d** Gabor output for 30° and frequecny 0.5

of appropriate values of K is critical. The smaller value of K have high influence of noise on the classification result whereas larger the value of K increases the time complexity of the algorithm. Therefore, to know the optimal performance of the KNN classifier, we have extended our experimentation with varying number of neighbors (K = 1, 3, 5) and found optimal result when K = 1. The proposed script identification problem can be experimented in three ways: (1) bi-script, (2) tri-script and (3) multi-script. However, in this paper we have made an exhaustive experimentation on trilingual documents, because as per the Indian constitution, the rule of trilingual is in governance with each State of the Country. More frequently tri-script documents could be seen in South Indian States. Therefore, identification of scripts of the trilingual documents is the decisive solution. In case of multi-script identification, the recognition accuracy will decrease. Out of

**Table 1** Average recognition accuracy for trilingual script identification using 2 fold cross validation with KNN classifier (k = 1)

| Trilingual script group | Trilingual scripts | Recognition accuracy in (%) |
|---|---|---|
| 1 | English, Hindi, Kannada | 91.33 |
| 2 | English, Hindi, Telugu | 96.00 |
| 3 | English, Hindi, Tamil | 90.33 |
| 4 | English, Hindi, Malayalam | 90.33 |
| Average recognition accuracy | | 91.99 |

curiosity, extended our experimentation on dataset of [7] and noticed the classification average accuracy as 99 % in trilingual case. This result shows how the performance of the algorithm is dependent on the dataset used for experimentation. It is very difficult to justify the performance of the algorithms without testing it on a benchmark dataset with similar experimental conditions. Hence, our claim of average classification accuracy of 91.99 % has the place of justification. A total of 600 text block images (each script 100 text blocks) are considered for experimentation. Half of the text block images are used for training and the remaining for testing. The Table 1 shows the average recognition accuracy for script identification from trilingual document images. The error distribution (Confusion matrix) among different scripts using KNN classifier (i.e. k = 1) with 2 fold cross validation is shown in the Table 2.

**Table 2** Confusion matrix for trilingual script identification using 2 fold cross validation with KNN classifier (k = 1)

| Scripts | English | Hindi | Kannada |
|---|---|---|---|
| English | 87 | 0 | 13 |
| Hindi | 0 | 98 | 2 |
| Kannada | 10 | 1 | 89 |
| Scripts | English | Hindi | Telugu |
| English | 95 | 1 | 4 |
| Hindi | 1 | 97 | 2 |
| Telugu | 0 | 4 | 96 |
| Scripts | English | Hindi | Tamil |
| English | 90 | 1 | 9 |
| Hindi | 1 | 99 | 0 |
| Tamil | 14 | 4 | 82 |
| Scripts | English | Hindi | Malayalam |
| English | 91 | 1 | 8 |
| Hindi | 1 | 93 | 6 |
| Malayalam | 5 | 8 | 87 |

From Table 2 we could observe the confusion of the classifier in discriminating the tri scripts. It can be noticed that considerable confusion has occurred in case of English with regional languages like Kannada, Tamil, Telugu and Malayalam. The reason is straightforward, that is most of the native regional language writers have written English text characters in circular shape which influences the classifier for misclassification. Furthermore, in all the cases Hindi script identification has high accuracy because of Sirorekha (horizontal line at the top of the word) feature which is absent in other scripts. These observations also clarify the importance of directional energies in characterizing the shapes of characters of different scripts.

## 5 Conclusions

This paper presents the potentiality of the traditional Gabor filters in tri-script classification. By its nature, Gabor filters are more efficient in capturing directional energy distributions of the underlying image. The use of Gabor filters is realized based on the clues perceived with the shapes of the characters of different scripts. Features of Gabor filters showed the comparable script recognition accuracy in case of tri-script identification. Though the method is well known and used it in number of cases, still the tuning of the Gabor filters may yield good results. The proposed script identification problem is one of the examples. It is noticed that, when the text block has less coverage of text region the performance of the Gabor filter features decreases. How the coverage of text influence on the performance of the Gabor filter features is under investigation. In future, we would like to extend it for more number of scripts. We would also aim to develop a generalized frame work for handwritten script identification in such a way that it should perform efficiently on inclusion of any unknown script. We are also intended to extend it for word level classification of scripts.

**Acknowledgments** This work is carried out under the UGC sponsored minor research project (ref: MRP(S):661/09-10/KAGU013/UGCSWRO dated, 30/11/2009). Authors are grateful to the reviewers for giving their valuable comments. Authors are also grateful to the UGC for providing financial assistance.

## References

1. Ghosh D, Dube T, Shivaprasad AP (2009) Script recognition—a review. IEEE Trans Pattern Anal Mach Intell 32(12):2142–2161
2. Guru DS, Ravikumar M, Harish BS (2012) A review on offline handwritten script identification. In: Proceedings of international journal of computer applications on national conference on advanced computing and communications
3. Ma H, Doermann D (2003) Gabor filter based multi-class classifier for scanned document images. In: Proceedings of the 7th international conference on document analysis and recognition

4. Pati PB, SabariRaju S, Pati N, Ramakrishnan AG (2004) Gabor filters for document analysis in indian bilingual documents. In: Proceedings of IEEE, ICISIP, pp 123–126
5. Chanda S, Srikanta P, Franke K, Umapada P (2009) Two-stage approach for word-wise script identification. In: Proceedings of IEEE 10th international conference on document analysis and recognition, pp 926–930
6. Rajneesh R, Renu D, Lehal GS (2011) Identification of printed Punjabi words and english numerals using gabor features. World Acad Sci Eng Technol 73:392–395
7. Rajput GG, Anita HB (2010) Handwritten script recognition using dct and wavelet features at block level. IJCA special issue on recent trends in image processing and pattern recognition, pp 158–163
8. Hangarge M, Dhandra BV (2010) Offline handwritten script identification in document images. Int J Comput Appl 4(6):6–10
9. Padama MC, Vijaya PA (2010) Script identification from trilingual documents using profile based features. Int J Comput Sci Appl Techno math Res Found 7(4):16–33
10. Swamy MD, Rani S, Reddy RK, Govardhan A (2011) Script identification from multilingual Telugu, Hindi and English text documents. Int J Wisdom Based Comput 1(3):79–85
11. Abirami S, Manjula D (2009) A survey of script identification techniques for multi-script document images. Int J Recent Trends Eng 1(2):246–249

[illegible]

# Indic Language Machine Translation Tool: English to Kannada/Telugu

**Mallamma V. Reddy and M. Hanumanthappa**

**Abstract** Natural Language Processing is a field of computer science, AI and linguistics concerned with the interactions between computers and human (natural) languages. Specifically, computer extracts meaningful information from natural language input and/or producing natural language output. The major task in NLP is machine translation, which automatically translates text from one human language to another by preserving its meaning. This paper proposes new model for Machine-Translation system in which Rule-Based, Dictionary-Based approaches are applied for English-to-Kannada/Telugu Language-Identification and Machine Translation. The proposed method has four steps: first, Analyze and tokenize an English sentence into a string of grammatical nodes second, Map the input pattern with a table of English–Kannada/Telugu sentence patterns, third, Look-up the bilingual-dictionary for the equivalent Kannada/Telugu words, reorder and then generate output sentences and fourth step is to Display the output sentences. The future work will focus on sentence translation by using semantic features to make a more precise translation.

**Keywords** Natural language processing (NLP) · Language identification · Transliteration · Morphological analyzer · Machine translation (MT)

M. V. Reddy (✉) · M. Hanumanthappa
Department of Computer Science and Applications, Bangalore University, Bangalore, Karnataka, India
e-mail: mallamma_vreddy@yahoo.co.in

M. Hanumanthappa
e-mail: hanu6572@hotmail.com

P. P. Swamy and D. S. Guru (eds.), *Multimedia Processing, Communication and Computing Applications*, Lecture Notes in Electrical Engineering 213,
DOI: 10.1007/978-81-322-1143-3_4, 

## 1 Introduction

India has 18 officially recognized languages: Assamese, Bengali, English, Gujarati, Hindi, Kannada, Kashmiri, Konkani, Malayalam, Manipuri, Marathi, Nepali, Oriya, Punjabi, Sanskrit, Tamil, Telugu, and Urdu. Clearly, India owns the language diversity problem. In the age of Internet, the multiplicity of languages makes it even more necessary to have sophisticated machine translation [1, 2] systems. In this paper we are presenting the Machine translation system particularly from English to Kannada/Telugu and vice versa, Kannada [3] or Canarese is one of the 1,652 mother tongues spoken in India. Forty three million people use it as their mother tongue. Telugu is a Central Dravidian language primarily spoken in the state of Andhra Pradesh, India, where it is an official language. According to the 2001 Census of India, Telugu [4] is the language with the third largest number of native speakers in India (74 million), 13th in the Ethnologies list of most-spoken languages world-wide, and most spoken Dravidian language. As the English Language has ASCII encoding system for identifying the specification of a character, similarly Indian Languages have encoding systems named Unicode [5] such as “UTF-8”, “UTF-16”, “UTF-32”, ISCII. Machine Translation Model broadly classified into three modules.

- **Language Identification Module:** Identifying the Language [6] of the Document(s) by uploading file(s) or by entering the text
- **Transliteration Module:** Transliteration is mapping of pronunciation and articulation of words written in one script into another script preserving the phonetics.
- **Translation Module:** Change in language while preserving meaning.

## 2 Language Identification

The language identification problem refers to the task of deciding in which natural language a given text is written is the major challenge in Natural Language Processing. Several corpora were collected to estimate the parameters of the proposed models to evaluate the performance of the proposed approach. Using the unigram statistical approach for each Language, the proposed model [7, 8] is learnt with a training data set of 100 text lines from each of the three Languages- English, Kannada and Telugu. Language Identification [9] algorithm is described and result is shown in Fig. 1.

```
Algorithm LandId ()
Input: Pre-processed text lines of English, Kannada and
Telugu text Doc(s)
Output: Identify the Language of the document.
```

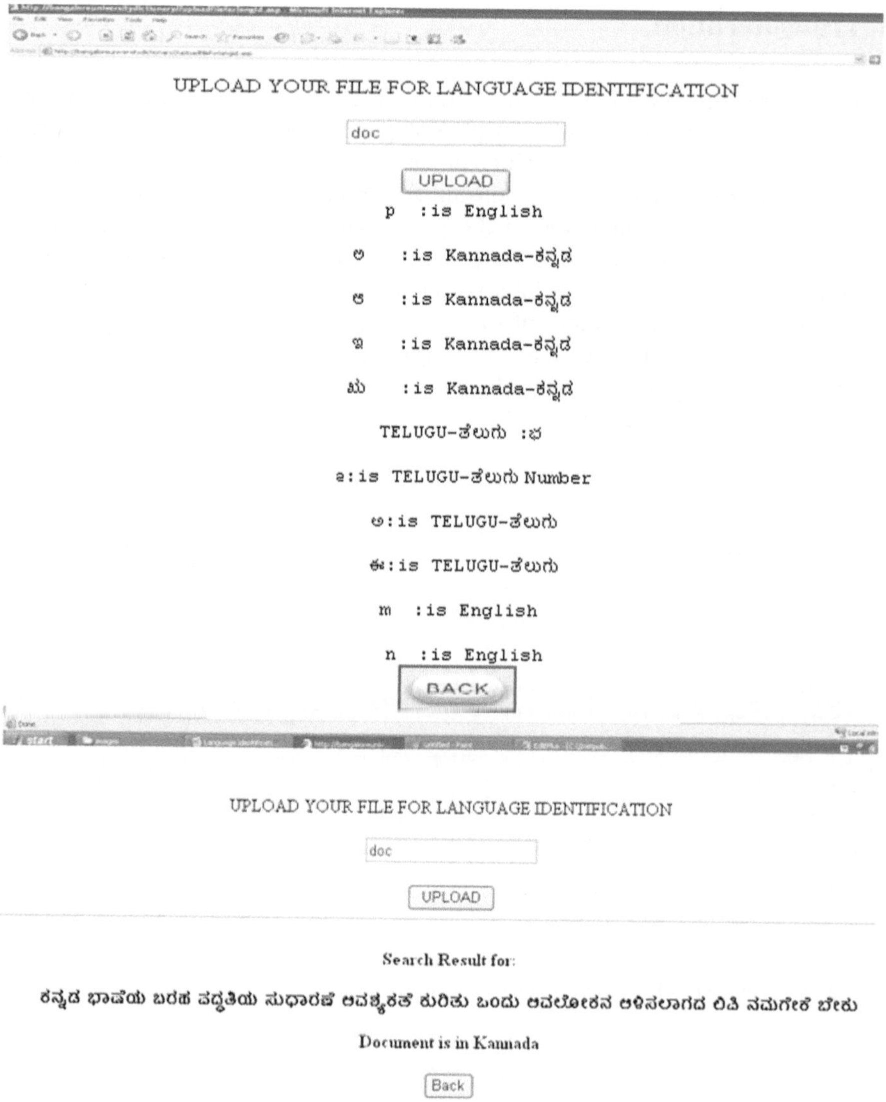


**Fig. 1** Language identification for English, Kannada and Telugu by uploading docs

```
Do for i = 1 to 3 Language document types
Do for k = 1 to 100 text lines of ith document
Compare until i == k if yes display the type of Lang.
Otherwise display the unknown language
```

## 3 Transliteration

Machine Transliteration is the conversion of a character or word from one language to another without losing its phonological characteristics. It is an orthographical and phonetic converting process. Therefore, both grapheme and phoneme information should be considered. Accurate transliteration of named entities plays an important role in the performance of machine translation and cross-language information retrieval process CLIR [10]. Dictionaries have often been used for query translation in cross language information retrieval. However, we are faced with the problem of translating Names and Technical Terms from English to Kannada/Telugu. The most important query words in information retrieval are often proper names. Mapping of characters are used for Transliteration as shown in Figs. 2 and 3.

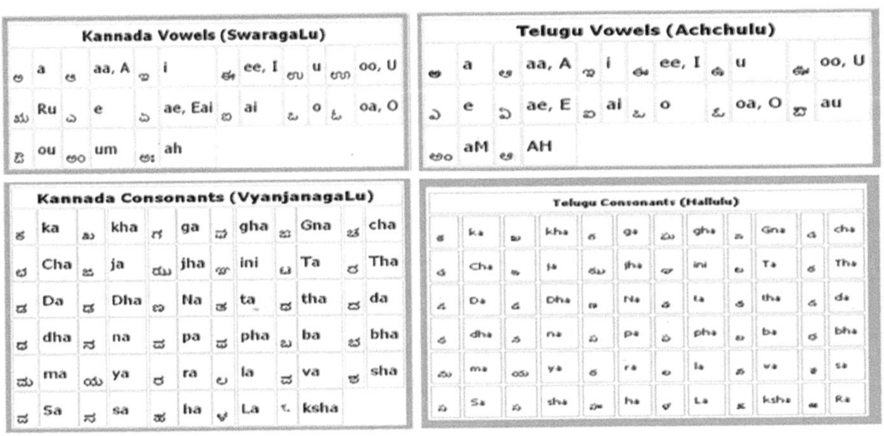


**Fig. 2** English–Kannada/Telugu character mapping

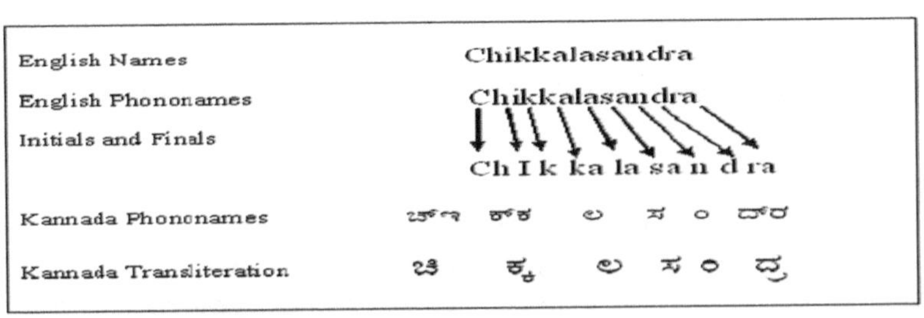


**Fig. 3** English–Kannada name transliteration

### 3.1 Transliteration Standards

- **Complete**: Every well-formed sequence of characters in the source script should transliterate to a sequence of characters from the target script, and vice versa.
- **Predictable**: The letters themselves (without any knowledge of the languages written in that script) should be sufficient for the transliteration, based on a relatively small number of rules.
- **Pronounceable**: The resulting characters have reasonable pronunciations in the target script.
- **Reversible**: It is possible to recover the text in the source script from the transliteration in the target script. That is, someone that knows the transliteration rules would be able to recover the precise spelling of the original source text.

### 3.2 Algorithm

We constructed a Dictionary with the help of training data that stores the possible mappings between English characters and Kannada/Telugu characters. Mapping was created between single English to single Kannada/Telugu character or two English characters to single Kannada/Telugu characters. Algorithm followed for making dictionary is as follows:

```
for each (name_english,name_Kannada) in the training
data:index = 0
while index ! = len (name_english) and index ! =
len(name_Kannada):
map name_english [index] to name_Kannada [index]
if index < len (name_english) − 1
map (name_english [index] + name_english [index + 1]) to
name_Kannada [index];index ++
index_english = len (name_english) − 1
index_Kannada = len (name_Kannada) − 1
while index_Kannada > − 1 and index_english > − 1:
map name_english[index_english] to ame_Kannada[index_
Kannada]
if index_english > 0:map (name_english [index_english-
1] + name_english [index_english])to name_Kannada[index_
Kannada]
index_english;index_Kannada
```

# 4 Translation

Machine translation [11, 12] systems that produce translations between only two particular languages are called bilingual systems and those that produce translations for any given pair of languages are called multilingual systems. Multilingual systems may be either uni-directional or bi-directional. The ideal aim of machine translation systems is to produce the best possible translation without human assistance. Query translation module with Bilingual Dictionary is depicted in Fig. 4.

Kannada and Telugu, like other Indian languages, are morphologically rich. Therefore, we stem the query words before looking up their entries in the bilingual dictionary. In case of a match, all possible translations from the dictionary are returned. In case a match is not found, the word is assumed to be a proper noun and therefore transliterated by the UTF-8 English transliteration module. The above module, based on a simple lookup table and corpus, returns the best three English transliterations for a given query word. Finally, the translation disambiguation module disambiguates the multiple translations/transliterations returned for each word and returns the most probable English translation of the entire query to the monolingual IR engine.

## 4.1 Kannada Morphology

Kannada is a morphologically rich language in which morphemes combine with the root words in the form of suffixes. Kannada grammarians divide the words of the language into three categories namely:

- **Declinable words** (namapada): Morphology of declinable words shown in Fig. 5, as in many Dravidian languages is fairly simple compared to verbs.

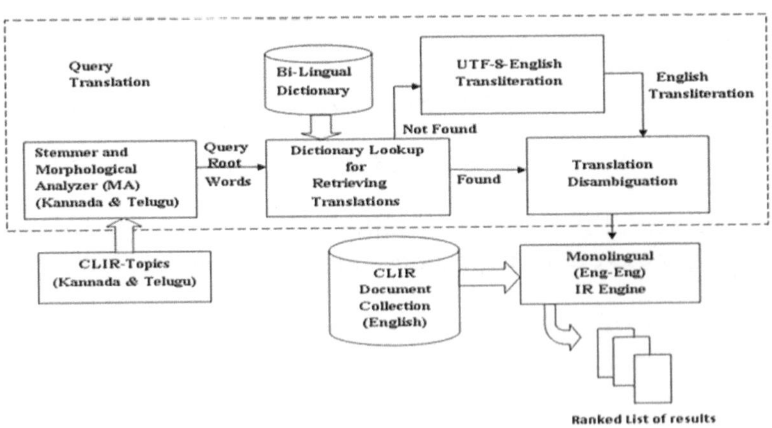


**Fig. 4** Query based translation module

**Fig. 5** Formal grammar for Kannada nouns

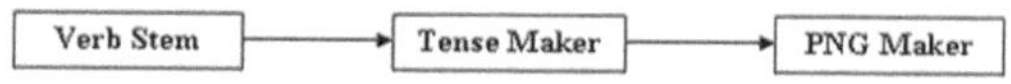


**Fig. 6** A formal grammar for Kannada verbs

Kannada words are of three genders- masculine, feminine and neutral. Declinable and Conjugable words have two numbers- singular and plural.

- **Verbs** (kriyapada) or Conjugable words: The verb is much more complex than the nouns. There are three persons namely first, second and third person. Tense of verbs as shown in Fig. 6. is past, present or future. Aspect may be simple, continuous or perfect. Verbs occur as the last constituent of the sentence. They can be broadly divided into finite or non-finite forms. Finite verbs have nothing added to them and are found in the last position of a sentence. They are marked for tense with Person-Number-Gender (PNG) markers. Non-finite verbs, on the other hand cannot stand alone. They are always marked for tense without PNG marker.
- **Uninflected words (avyaya)**: Uninflected words may be classified as adverbs, postpositions, conjunctions and interjections. Some of the example words of this class are haage, mele, tanaka, alli, bagge, anthu etc.

## *4.2 Morphophonemics*

In Kannada, adjacent words are often joined and pronounced as one word. Such word combinations occur in two ways- Sandhi and Samasa. Sandhi (Morphophonemics) deals with changes that occur when two words or separate morphemes come together to form a new word. Few sandhi types are native to Kannada and few are borrowed from Sanskrit. We in our tool have handled only Kannada sandhi. However we do not handle Samasa. Kannada sandhi is of three types—lopa, agama and adesha sandhi. While lopa and agama take place both in compound words and in the junction of the crude forms of words and suffixes, adesha sandhi occurs only in compound words.

- **Morphological analysis and generation**: Morphological analysis [13] determines the word form such as inflections, tense, number, part of speech, etc. shown in "Table 1" and Fig. 7. Syntactic analysis determines whether the word is subject or object. Semantic and contextual analysis determines a proper interpretation of a sentence from the results produced by the syntactic analysis. Syntactic and semantic analyses are often executed simultaneously and produce syntactic tree structure and semantic network respectively. This results in internal structure of a sentence. The sentence generation phase is just reverse of the process of analysis.

**Table 1** Different cases and their corresponding and few inflections of a verb stem

| Kannada name | English name | Characteristic suffix |
|---|---|---|
| Prathama | Nominative | 0 (nu/ru/vu/yu) |
| Dwitiya | Accusative | annu/vannu/rannu |
| Tritiya | Instrumental | iMda/niMda/riMda |
| Chaturthi | Dative | ge/ige/kke |
| Pachami | Ablative | deseyiMda |
| Shashti | Genitive | a/ra/da/na |
| Saptami | Locative | alli/nalli/dalli/valli |
| Sambhodana | Vocative | ee |

**Fig. 7** Characteristic suffixes for nouns and its corresponding meanings

| Inflected Verb | Meaning in English | Tense | Aspect | PNG |
|---|---|---|---|---|
| ಮಾಡುವನು | He will do. | Future | Simple | 3SM |
| ಮಾಡುತ್ತಿದ್ದಾನೆ | He is doing. | Present | Continuous | 3SM |
| ಮಾಡಿರುವಳು | She has done. | Future | Perfect | 3SF |
| ಮಾಡುತ್ತಿದ್ದಳು | She was doing. | Past | Continuous | 3SF |
| ಮಾಡಿದಿರಿ | You did. | Past | Simple | 2P- |
| ಮಾಡುತ್ತೇನೆ | I will do. | Future | Simple | 1S- |
| ಮಾಡಿದ್ದರು | They did. | Past | Perfect | 3P- |
| ಮಾಡಿರುತ್ತದೆ | It did. | Present | Perfect | 3SN |

Computational morphology deals with recognition, analysis and generation of words. Some of the morphological processes are inflection, derivation, affixes and combining forms as shown in Fig. 8. Inflection is the most regular and productive morphological process across languages. Inflection alters the form of the word in number, gender, mood, tense, aspect, person, and case. Morphological analyzer gives information concerning morphological properties of the words it analyses.

In this section we are going to describe about the new algorithm which is developed for morphological analyzer [13] and generator. The main advantage for this algorithm is simple and accurate.

```
Algorithm
1: Get the word to be analyzed.
2: find entered word is found in the Root Dict.
3: If the word is found in the Dict, stop; Else
```

**Fig. 8** Sandhi types and examples for word combination

| Complex word | Simple/inflected words | Sandhi type |
|---|---|---|
| ಚೆಂಡಾಟ | ಚೆಂಡು + ಆಟ | ಲೋಪ ಸಂಧಿ |
| ಸುಂದರವಾದ | ಸುಂದರ + ಆದ | ಆಗಮ ಸಂಧಿ |
| ಕೈದೋಟ | ಕೈ + ತೋಟ | ಆದೇಶ ಸಂಧಿ |

```
4: Separate any suffix from the right hand side
5: If any suffix is present in the word, then check the availability of the suffix in the dictionary. Then
6: Remove the suffix present, Then re-initialize the word without identified suffix, Go to Step 2.
7: Repeat until the Dictionary finds the root/stem word.
8: Store the English root/stem word in a variable and then get the corresponding Kannada word from the bilingual dictionary
9: Check what all grammatical features does the English word have given and then generate the corresponding features for the Kannada word
10: Exit.
```

- **Dictionary based approach**: Dictionary based translation [14] is basically translation with a help of a bi-lingual dictionary. Only translation words with high coherence scores will be selected for the translation of the query as shown in Fig. 9. Query translation is relatively efficient and can be performed as needed. The principal limitation of query translation is that queries are often short and short queries provide little context for disambiguation.
- **Rule-Based Approach**: This approach consists of (1) a process of analyzing input sentences of a source language morphologically, syntactically and/or semantically and (2) a process of generating output sentences of a target language based on an internal structure. Each process is controlled by the dictionary and the rules.

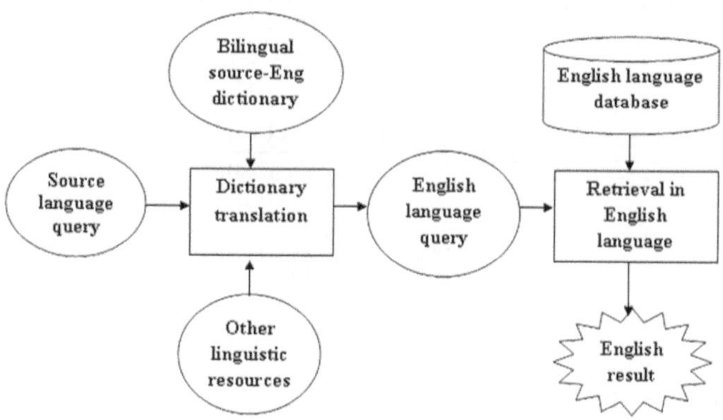


**Fig. 9** Dictionary based method for query translation

## 4.3 The Selection of Word Translation

Normally in CLIR words that are not included in phrases are translated word-by-word shown in Fig. 8. However, this does not mean that they should be translated in isolation from each other. Instead, while translating a word, the other words (or their translations) form a "context" that helps determine the correct translation for the given word.

Working in this principle of translation our assumption is that the correct translations of query words tend to co-occur in target language documents and incorrect translations do not. Therefore, given a set of original source language query words, we select for each of them the best translation word such that it co-occurs most often with other translation words in destination language documents. For example as shown in Fig. 10.

Finding such an optimal set is computationally very costly. Therefore, an approximate greedy algorithm is used. It works as follows: Given a set of m original query terms $\{a_1 \ldots a_n\}$, we first determine a set $T_i$ of translation words for each $a_i$ through the dictionary. Then we try to select the word in each $T_i$ that has the highest degree of cohesion with the other sets of translation words. The set of best words from each translation set forms our query translation.

Cohesion is the study of textual equivalence defining it as the network of lexical, grammatical, and other relations which provide links between various parts of a text and works based on term similarity. The EMMI weighting measure has been successfully used to estimate the term similarity in [7]. We take a similar approach. However, we also observe that EMMI does not take into account the distance between words. In reality, we observe that local context is more important for translation selection. If two words appear in the same document but at two distant places, it is unlikely that they are strongly dependent. Therefore, we add a distance factor in our calculation of word similarity. Formally, the similarity between terms $x$ and $y$ is

$$SIM(x,y) = p(x,y) \times \log_2\left(\frac{p(x,y)}{p(x) \times p(y)}\right) - K \times \log_2 Dis(x,y) \tag{1}$$

where

$$p(x,y) = \frac{c(x,y)}{c(x)} + \frac{c(x,y)}{c(y)} \tag{2}$$

$$p(x) = \frac{c(x)}{\sum_x c(x)} \tag{3}$$

**Fig. 10** Word-by-word translation

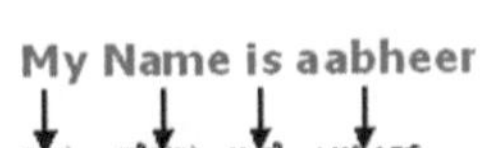

c($x$, $y$) is the frequency that term $x$ and term $y$ co-occur in the same sentences in the collection, c($x$) is the number of occurrence of term $x$ in the collection, Dis($x$, $y$) is the average distance (word count) between terms x and y in a sentence, and $K$ is a constant coefficient, which is chosen empirically ($K = 0.8$ in our experiments).

$$Cohesion(x, X) = Max_{y \in X} SIM(x, y) \tag{4}$$

The cohesion of a term $x$ with a set $X$ of other terms is the maximal similarity of this term with every term in the set, is shown in Eq. 4.

## 5 Experimental Setup

We use machine-readable bi-lingual Kannada → English and Telugu → English dictionaries created by BUBShabdasagar. The Kannada → English bi-lingual dictionary has around 14,000 English entries and 40,000 Kannada entries. The Telugu → English bi-lingual has relatively less coverage and has around 6,110 entries. CLIR Tool [15] is developed by using the ASP.NET as front end and Database as back end. We have trained the systems with corpus size of 200, 500 and 1,000 lexicons and sentences respectively. Performances of the systems were evaluated with the same set of 500 distinguished sentences/Phases that were out of corpus. The experiment results as shown in Figs. 11 and 12. The comparative results are shown in Figs. 13 and 14.

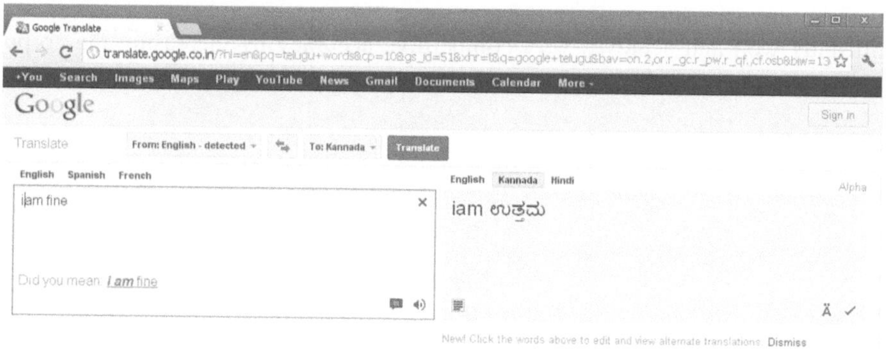


**Fig. 11** Google translation

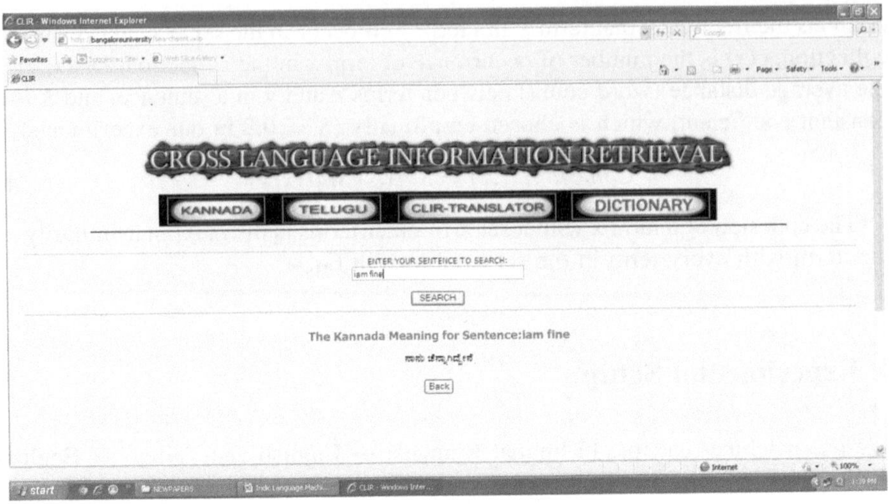


Fig. 12 CLIR translation

## 5.1 Evaluation Metric and Performance

In the experiment, the performance of word translation extraction was evaluated based on precision and recall rates at the word. Since, we considered exactly one word in the source language and one translation in the target language at a time. The word level recall and precision rates were defined as follows:

$$WordPrecision\,(WP) = \frac{number\ of\ correctly\ extracted\ word}{number\ of\ extracted\ words} \tag{5}$$

$$WordRecall\,(WR) = \frac{number\ of\ correctly\ extracted\ Words}{number\ of\ correct\ words} \tag{6}$$

From the experiment we found that the performances of our systems are significantly well and achieves very competitive accuracy by increasing the corpus size as shown in Fig 15.

Fig. 13 Sample results for word

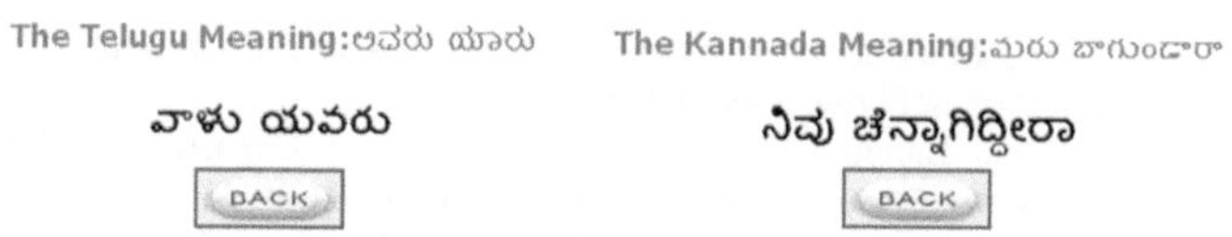

Fig. 14 Sample results for sentences

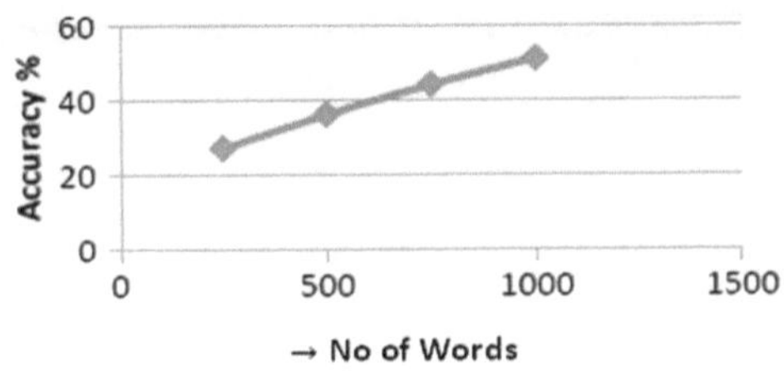


**Fig. 15** Performance graph

# 6 Conclusion and Future Work

In this paper, we presented our Kannada → English and Telugu → English CLIR system developed for the Ad-Hoc bilingual Task. Separate text lines of English, Kannada and Telugu documents from a trilingual document are presented for Natural Language Identification. The approach is based on the analysis of the Unigram statistical approach of individual text lines and hence it requires character or word segmentation. In future we can also use this language identification module for translation with the help of bilingual dictionary. This will be very useful for machine translation from English to Kannada/Telugu language. One of the major challenges in CLIR is that English has Subject Verb Object (SVO) structure while Kannada has Subject Object Verb (SOV) structure in Machine translation will be unraveled by using morphology.

**Acknowledgments** I owe my sincere feelings of gratitude to Dr. M. Hanumanthappa, for his valuable guidance and suggestions which helped me a lot to write this paper. This is the major research project entitled Cross-Language Information Retrieval sanctioned to Dr. M. Hanumanthappa, PI-UGC-MH, Department of computer science and applications by the University grant commission. I thank to the UGC for financial assistance. This paper is in continuation of the project carried out at the Bangalore University, Bangalore, India.

# References

1. Konchady M (2008) Text mining application programming, 3rd edn. Charles River Media, Boston
2. Homiedan AH (2010) Machine translation. http://faculty.ksu.edu.sa/homiedan/Publications/Machine%20Translation.pdf. Accessed 4 Sep 2011
3. The Karnataka Official Language Act. Official website of Department of Parliamentary Affairs and Legislation. Government of Karnataka. Retrieved 29 July 2012
4. http://en.wikipedia.org/wiki/Telugu_language Retrieved 29 July 2012
5. http://www.ssec.wisc.edu/~tomw/java/unicode.html#x0C80. Retrieved 29 July 2012
6. Botha G, Zimu V, Barnard E (2007) Text-based language identification for South African languages. Published in South African Institute Of Electrical Engineers vol 98 (4)
7. Sibun P, Reynar JC (1996) Language identification: examining the issues. http://www.cis.upenn.edu/~nenkova/Courses/cis430/languageIdentification.pdf. Accessed on 10 Jan 2012

8. Vatanen T, Väyrynen JJ, Virpioja S (2010) Language identification of short text segments with N-gram models
9. Ahmed B, Cha SH, Tappert C (2004) Language identification from text using N-gram based cumulative frequency addition. In: Proceedings of Student/Faculty Research Day, CSIS, Pace University
10. Pingali P, Varma V (2006) Hindi and Telugu to English cross language information retrieval at CLEF 2006, 20–22 Sep, Alicante, Spain
11. Kereto S, Wongchaisuwat C, Poovarawan Y (1993) Machine translation research and development. In: Proceedings of the Symposium on Natural Language processing in Thailand, pp. 167–195
12. Knowles F (1982) The pivotal role of the dictionaries in a machine translation system. In: Lawson V (ed) Practical experience of machine translation. North-Holland, Amsterdam
13. Ritchie G, Whitelock (eds) (1985) The lexicon. pp. 225–256
14. Ballesteros L, Croft WB (1997) Phrasal translation and query expansion techniques for cross-language information retrieval. In: Proceedings of ACM SIGIR Conference 20: 84–91
15. Reddy MV, Hanumanthappa M (2009) CLIR Project (English to Kannada and Telugu). Available at http://bangaloreuniversitydictionary//menu.asp

# Real-Time Stereo Camera Calibration Using Stereo Synchronization and Erroneous Input Image Pair Elimination

**M. S. Shashi Kumar and N. Avinash**

**Abstract** With the increasing usage of stereo cameras in electronic gadgets, it becomes necessary to have fast, accurate, and automatic stereo calibration during production. In this paper we propose a novel method to achieve fast, accurate and automatic stereo calibration on the video feed using with chessboard pattern. Stereo calibration requires optimal stereo pair images at various orientations without delay between capturing of left and right cameras when objects are moving. In this paper we have developed a novel software based approach to capture synchronized frames from the stereo camera setup with very minimum delay between left and right camera of stereo setup. An approach to reject unmatched stereo pairs is developed based on z-score method, so that only valid optimal image pairs are used in stereo calibration. The optimal sets of stereo synchronized error free images are used for stereo camera calibration and calibration results are stored. The entire process runs in one shot real time without human intervention thus speeding up the stereo camera calibration process.

**Keywords:** Z-score · Stereo calibration · Stereo synchronization

## 1 Introduction

Stereo calibration is used to estimate geometric relation between cameras in the stereo setup. Stereo calibration requires input image pairs with different views. If we use time variant video streams for calibration then we need to select correct

M. S. Shashi Kumar (✉) · N. Avinash
Wittybot Technologies, Bangalore, India
e-mail: shashikumar@wittybot.com

N. Avinash
e-mail: avinash@wittybot.com

P. P. Swamy and D. S. Guru (eds.), *Multimedia Processing, Communication and Computing Applications*, Lecture Notes in Electrical Engineering 213,
DOI: 10.1007/978-81-322-1143-3_5, 

stereo pairs which are captured at same instance of time. Hence, it requires specialized system architecture to select stereo image pair simultaneously. This process is economically expensive because of additional hardware. In this paper, we handle this issue using software technique where we use timestamps while capturing the images from the stereo camera setup. Stereo image pair selection is based on minimum time difference between two images of stereo cameras. Some of the stereo synchronized image pair may be blurred or may not cover the field of view. Hence these image pairs have to be removed from the input set of images for stereo calibration. Z-score technique is used to eliminate the wrong image pairs and retain appropriate image pairs for stereo calibration. The above process is used to stereo synchronized image pairs from video feed, selecting optimal stereo pairs from accumulated pairs and calibrating these optimal stereo pairs is automated as a one step process for calibrating the stereo setup. These calibration parameters can be further used as needed by application.

Stereo calibration is the process of computing the geometrical relationship between the two cameras in space. Stereo calibration depends on finding the rotation matrix (R) and translation vector (T) between the two cameras. Further Rx, Ry, Rz and Tx, Ty, Tz define the R and T about the x, y, z axes of the 3d Cartesian space. Cipolla [1] proposes a method to that uses rigidity constraints of parallelism and orthogonality and uses vanishing point technique to find intrinsic and extrinsic parameters. Faugeras and Toscani [2] proposes a technique to use least squares method to obtain a transformation matrix which relates 3D points with their 2D projections. The advantage here is the simplicity of the model which consists in a simple and rapid calibration. In this work we use the technique explained in [3, 4, 8] to find R and T between two cameras.

Stereo Rectification is the process of aligning image planes to a common plane so that the epipolar lines become collinear without rotating the actual cameras [5]. The epipolar lines become parallel to the horizontal axis of image after rectification, so that it is possible to scan along corresponding rows from two images. In this paper we use stereo rectification to verify the stereo calibration parameters.

Most of the work carried out on automatic stereo calibration is about detecting the chess board corners. Arturo de la Escalera and Jose María Armingol [6] propose a method to detect number of chess board corners using Hough lines. Very less work has emphasized about stereo synchronization of frames in stereo setup and elimination of error pairs. Stereo synchronization problems arise when the calibration object is moving and there is delay in capturing between left and right image of the stereo setup. Error in calibration can also happen due to blurring of image, calibration object might have moved in one of the images due to image synchronization problems, calibration object may not be visible in one or both images of the stereo setup to achieve optimal calibration parameters. Filtering of all these kinds of images is required to generate optimal set of images for accurate calibration.

In this paper we emphasize on accurate stereo calibration by selecting image pair with minimal delay between capturing of images from left and right cameras of stereo setup. Followed by error pair elimination we find and remove unmatched image pairs by applying z-score technique to find optimal set of image pairs for stereo calibration.

This paper is organized as follows: In Sect. 2, overview of the system is described. In Sect. 3, detailed descriptions of proposed methodology are described. In Sect. 4, experimental results are presented and finally in Sect. 5 we add concluding remarks.

## 2 Overview of the System

The experimental setup consists of two cameras rigged side by side as shown in the Fig. 1. The chess board pattern is held in front of the camera such that entire chess board is inside the field of view of both left and right cameras of stereo setup. Once the program is started the user continuously changes the orientation of the chess board till the calibration is successfully completed. Internally in the program stereo synchronization followed by error pair elimination to generate optimal image pairs for calibration which are calibrated. The calibration results are checked for consistency over 1 sigma standard deviation over few trials. If not inside the 1 sigma limit one more stereo synchronized frame is added to the queue. Stereo calibration is performed including the new frame added and standard deviation is checked for 1 sigma limits. Like this more stereo pairs are added in the queue till stereo calibration parameters are converged to 1 sigma limits. Once calibration results get converged these calibration parameters are stored for further usage.

**Fig. 1** Stereo camera setup

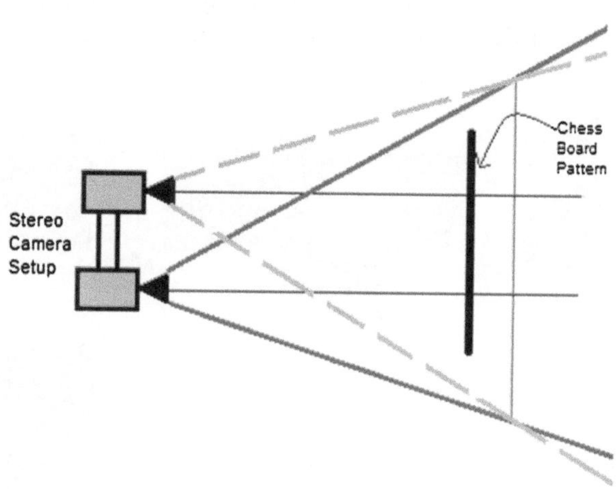

## 3 Proposed Methodology

Stereo setup as shown in Fig. 1 gives the input video feed for stereo calibration. Stereo synchronization selects left and right camera images with minimum time delay. Error pair elimination removes any unmatched pair of images for calibration. Using these above steps sufficient number of optimal stereo pairs are collected from the video feed. From the stereo calibration process we calculate the R and T vectors and further rectification matrix. This rectification matrix is computed which can be used to rectify the images (row aligned images) as shown in Fig. 2.

The proposed system consists of following steps:

- Stereo synchronization
- Error Pair elimination
- Stereo calibration
- Image rectification.

### *3.1 Stereo Synchronization*

Video stream from stereo camera are used as input for calibration, and there is always time delay between capturing images for left and right cameras. We need to choose proper image pair that has very minimal delay (nearly equal to zero) between capturing for right and left cameras. This is achieved by using time stamp assigned for each captured image by camera driver. The images with minimum time stamp difference are only selected for camera calibration.

The stereo synchronization process for pairing of images in Fig. 3 and algorithm follows. Here two left and right queues are maintained to fill images from left and right camera in two different threads.

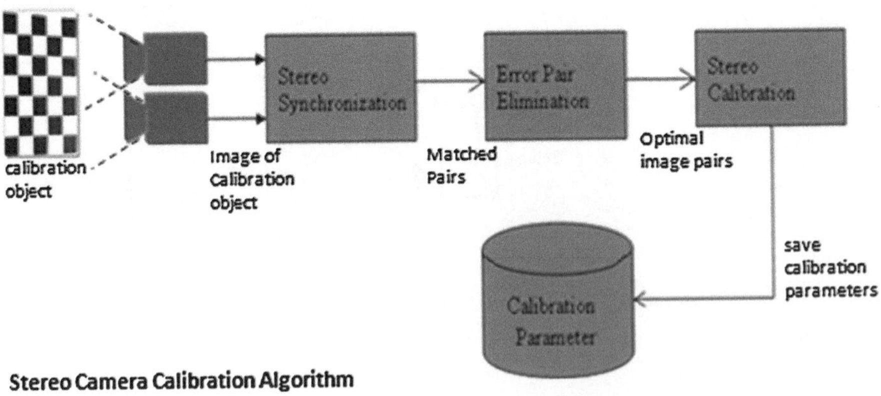


**Fig. 2** System block diagram

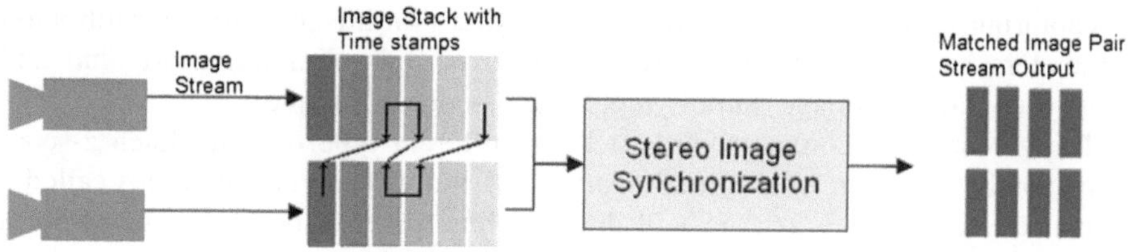


**Fig. 3** Stereo synchronization sample output depiction

Start with Left queue as reference queue and 1st frame of left queue as reference frame. Find the time difference with all the frames in the right queue. Consider 3rd frame of right queue has the least time difference. Image Pair 1st frame of left and 3rd frame of right queue. Remove 1st frame in left and 1st, 2nd, 3rd frame from right queue. Now make right queue as reference queue and 1st frame in the right queue (after removal of frames) as reference frame and find the time difference with all the frames in the right queue. Consider 2nd frame in the left queue matches. Pair 1st frame in right queue and 2nd frame in the left queue. Remove 1st frame in right and 1st and second frame in left queue. Now once again make left queue as reference queue. This procedure continues…

Step 1: Capture images simultaneously from left and right cameras in two different threads into left and right queue in parallel.
Step 2: Assign left queue for left camera images and right queue for right camera images.
Step 3: Start with Left queue as reference queue and 1st Frame as the reference frame.
Step 4: Find the time difference with all the frames in the opposite queue.
Step 5: Make the next frame in the reference queue as reference frame. Repeat step 4 for all the frames in the queue.
Step 6: Find the frame with least time difference.
Step 7: If below threshold, pair these images and remove all the frames above and including paired images in both the queues.
Step 8: Swap the reference queue to the opposite queue (left to right or right to left).
Step 9: Go to step3 till all the frames in a queue are emptied.
Step 10: Go to step 1 till sufficient number of stereo image pairs are generated.
Step 11: Considering an example with respect to Fig. 3.

## 3.2 Error Pair Elimination

All the stereo pair images obtained from stereo synchronization may not be useful for stereo calibration. In few stereo paired images the calibration object may be located in the field of view, some are blurred, object may not be visible in one or both of the images, the object might have moved in one of the images due to delay in capturing. Error is introduced by these kinds of image pairs in calibration parameters and in rectification process. In error pair elimination we find and remove unmatched image pairs by applying z-score technique.

The z-score is a common yard stick for different type of data. Each z-score corresponds to a point in a normal distribution and as such is sometimes called a normal deviate since a z-score will describe how much a point deviates from a median or specification point. The z-score is calculated by subtracting your sample median from a data point and dividing by the standard deviation. This value is a measure of the distance in standard deviations of a sample from the median.

The R and T parameters of individual images will give cue that object is clearly visible in the given image or not. These extrinsic parameters of the camera are obtained by using Zhang's monocular calibration method [3]. The detailed algorithm for error pair elimination is given below.

Following is the algorithm for error pair elimination.

Median of absolute deviation (MAD) can be calculated using the following formula. For a data set $X_1$, $X_2$... $X_n$, the MAD is defined as the median of the absolute deviations from the data's median:

$$MAD = median_i(|X_i - median_j(x_j)|) \quad (1)$$

Z-score can be calculated using the following formula

$$x = \frac{\text{ZSCORE\_CONST}[\text{Xi} - \text{median}(\text{Xj})]}{\text{MAD}(\text{Xj})} \quad (2)$$

where ZSCORE_CONST is a normalizing constant corresponding to the z-score partitioning of the normal distribution.

Step 1: Apply monocular calibration and use the same to find Stereo R and T vector from left & right for the stereo pair.
Step 2: Find Median of Absolute Deviation (MAD) for all the {Rx, Ry, Rz, Tx, Ty, Tz} for each of the image pair.
Step 3: Calculate z-score for all the extrinsic parameters {Rx, Ry, Rz, Tx, Ty, Tz} for each image pair.
Step 4: Calculate cumulative z-score for each image pair.
Step 5: If z-score of given stereo image pairs greater than z-score threshold then eliminate the corresponding image pairs, use the image pair for store it for stereo calibration.

### *3.3 Stereo Calibration*

Stereo image pairs in the image queue are the optimal pairs of images pairs selected by error pair elimination. Stereo calibration is performed on these optimal pairs. Stereo calibration is the process of computing the geometrical relationship between the two cameras in space. Here in this process of calibrating two cameras at the same time and will be looking to relate them together through a rotation matrix and a translation vector. We obtained the intrinsic parameters of two cameras using Zhang's method [3] as represented below.

$$M_{Lold} = \begin{pmatrix} f_l & 0 & C_{lx} \\ 0 & f_l & C_{ly} \\ 0 & 0 & 1 \end{pmatrix}, \quad M_{Rold} = \begin{pmatrix} f_r & 0 & C_{rx} \\ 0 & f_r & C_{ry} \\ 0 & 0 & 1 \end{pmatrix}$$

Let P(x, y, z) be a point in world co-ordinate observed by stereo cameras. The corresponding image points $P_l(u, v, 1)$ and $P_r(u, v, 1)$ in left and right image planes. In stereo calibration we obtain the relation (with respect to left camera) how much right camera is rotated and translated using essential matrix [5].

$$P_l^T \epsilon P_r = 0 \tag{3}$$

where, $\varepsilon = [t_x]R$ is a $3 \times 3$ essential matrix with three degrees of freedom of the rotation matrix R and the three degrees of freedom of the translation vector T.

### 3.4 Image Rectification

Image rectification is the process of "correcting" the individual images so that they appear as if they had been taken by two cameras with row-aligned image planes [3]. After Rectification, the optical axes (or principal rays) of the two cameras are parallel and so we say that they intersect at infinity thus making it stereo disparity calculation between two images simple.

For stereo rectification we used Bouguet's algorithm [7] to minimize the amount of change in reprojection produces for each of the two images while maximizing common viewing area. It rotates each camera half a rotation, so their principal rays each end up parallel. As a result, such a rotation puts the cameras into coplanar alignment as shown in Fig. 4.

## 4 Results and Analysis

Analysis is concentrated on the following experiments.

1. To evaluate the number of frames required and time to converge.
2. To evaluate the correctness by measuring 3D distance.

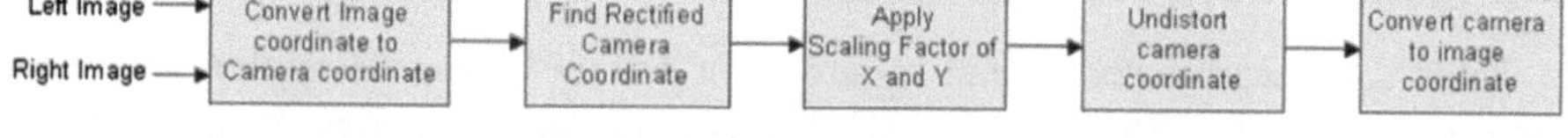


**Fig. 4** Block diagram of image rectification algorithm

The experimental setup consists of two Microsoft Lifecam 5,000 webcams placed side by side as shown in the proposed methodology in Fig. 1. Size of captured Image is set same for both the cameras at 640 × 480 pixels. Here we have used a desktop with configuration–Intel Core 2 Duo Processor 2 GHz; 512 MB DDR2 RAM for calibration of stereo setup.

Experiment 1: Following is the algorithm used for obtaining the calibration parameters of the stereo calibration setup. Here we add one frame each time and check for the standard deviation of each of the calibration parameters. Consistent calibration parameters for 1sigma standard deviation is chosen as it covers 34.1 % from mean value so as to achieve accurate calibration parameters.

Step 1: Start the live feed keep changing the orientation of the chess board continuously. Capture first 3 frames at different instances.
Step 2: Perform stereo calibration and store results.
Step 3: Add one more (next) frame to calibration image set and perform stereo calibration and store results.
Step 4: Check if 5set of readings of stereo calibration results are available. If not, go to step 3.
Step 5: If 5 set of readings is available, check if all the extrinsic parameters (Rotation and translation parameters) of the last 5 readings of stereo calibration results are within 1 sigma limits.
Step 6: If not inside 1 sigma limits. Go to step 3 to add one more pair of image to calibration data set.
Step 7: If under 1 sigma limits save calibration details of the last stereo calibration results.
Step 8: Repeat the experiment few times to verify results.

From the Table 1 and Fig. 5 we infer that on an average our algorithm requires about 11–15 frames to calibrate the stereo setup within a time of 15 s where as the system without stereo synchronization and error pair elimination is inconsistent. The accuracy of the calibration is validated in experiment 2.

Experiment: 2 To evaluate the correctness of the proposed algorithm the following experiment is performed. In this experiment stereo rectified image pair is randomly picked and 3D co-ordinates is calculated for each of the chessboard corners. The root mean square error for all the corners in the chess board id calculated. The experimentation steps are as given below.

Step 1: Select one stereo rectified pair.
Step 2: Find the all the corners of chess board in left and right images.
Step 3: Find 3D points for each of the corners. Find the root mean square error in measurement.
Step 4: Repeat steps 1, 2, 3.

Table 2 shows the root mean square error for the proposed method. Here we can observe that the RMS error of 3D distance measured. A chess board of 5 × 8 is used in this experimentation whose box size is of 45 × 45 mm.

**Table 1** Table showing number of frames required for calibration over 8 iterations

| Trial no | With stereo synchronization and error pair elimination | | Without stereo synchronization and error pair elimination | |
|---|---|---|---|---|
| | Number of frames | Time taken in sec | Number of frames | Time taken in sec |
| 1 | 13 | 15 | 68 | 220 |
| 2 | 12 | 14 | 66 | 186 |
| 3 | 14 | 15 | 60 | 178 |
| 4 | 13 | 15 | 42 | 198 |
| 5 | 11 | 12 | 78 | 196 |
| 6 | 13 | 14 | 92 | 248 |
| 7 | 11 | 12 | 72 | 188 |
| 8 | 11 | 11 | 59 | 172 |

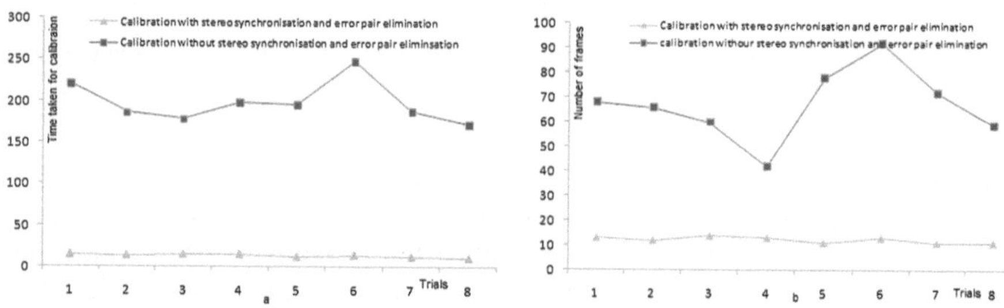


**Fig. 5** Graph plotted for (**a**) Time required for with and without stereo synchronization and error pair elimination (**b**) Number of frames required for with and without stereo synchronization and error pair elimination

**Table 2** RMS error of 3D distance measurement with and without z-score method

| Trial no | RMS error with z-score method in mm | RMS error without z-score method in mm |
|---|---|---|
| 1 | 2.098 | 13.368 |
| 2 | 1.478 | 6.235 |
| 3 | 1.621 | 8.628 |
| 4 | 1.416 | 9.998 |
| 5 | 1.996 | 12.160 |
| 6 | 1.849 | 14.963 |
| 7 | 1.288 | 11.818 |
| 8 | 1.385 | 5.314 |

In this experimentation we are measuring error in 3D distance measurement between successive 40 chess board corners. From Fig. 6 we can observe that the RMS error with stereo synchronization and error pair elimination is at a maximum of 2.1 mm over 8 trials whereas without stereo synchronization and error pair elimination is at a maximum of 15 mm.

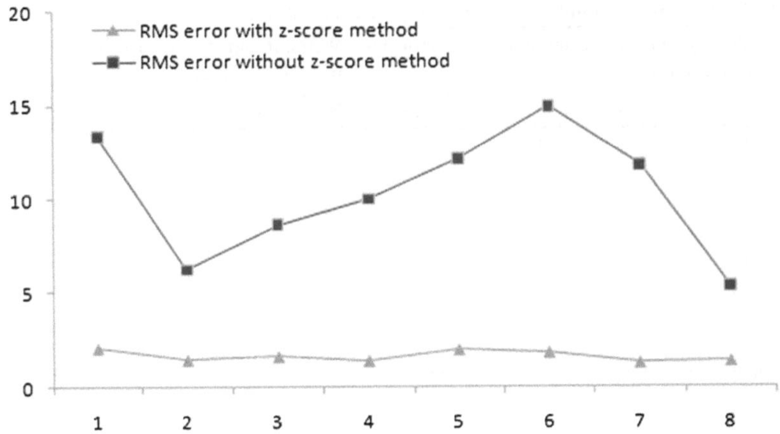


**Fig. 6** Graph plotted for RMS error of 3D distance measured with and without z-score method

The inferences obtained from the experiments carried out in this section are as follows. From experiment 1, with proposed method we can calibrate the stereo setup without human intervention for selecting the image pairs or to remove the error pairs. This proposed method converges within 1 sigma standard deviation with in as little as 15 frames. On our experimental setup it takes a maximum of 15 s. On a faster computer it takes significantly less. Also from the experiment 2 it can be noted that the measurement error with the proposed method is significantly lesser as compared to conventional method without stereo synchronization and error pair elimination. Also we can see that the error in the proposed method is quiet consistent.

## 5 Conclusion

This paper presents a real time efficient and accurate calibration method without human intervention to calibrate a stereo camera setup. The advantage of this method is that it automatically selects stereo synchronized images with very less time difference between image pair, selects optimal pairs for by calibration eliminating bad pairs and calibrates the optimal image pairs selected. All these steps happen in a single step which saves time by fast calibration. Money is saved as extra hardware for stereo synchronization is replaced by stereo approach.

## References

1. Cipolla R, Drummond T, Robertson D (1999) Camera calibration from vanishing points in images of architectural scenes. pp 382–391 BMVC
2. Faugeras OD, Toscani G (1986) The calibration problem for stereo. In: Proceedings of the IEEE computer vision and pattern recognition, pp 15–20
3. Zhang Z (2000) A flexible new technique for camera calibration. IEEE Trans Pattern Anal Mach Intell 22(11):1330–1334
4. Opencv 2.1 Reference manual [18th March 2010]
5. Loop C, Zhang Z (1999) Computing rectifying hymnographies for stereo vision. In: Proceedings of IEEE computer society conference on computer vision and pattern recognition, vol 1, pp 125–131
6. Escalera A de la, Armingol JM (2010) Automatic chessboard detection for intrinsic and extrinsic camera parameter calibration sensors (Basel). 10(3), pp 2027–2044
7. Guerchouche R, Coldefy F (2007) Robust camera calibration and evaluation procedure based on images rectification and 3D reconstruction. In: The 5th international conference on computer vision systems, pp 2967–2977
8. Bradski GR, Kaehler A (2008) Learning openCV: computer vision with the openCV library. O'REILLY

# Topic Modeling for Content Based Image Retrieval

**Hemant Misra, Anuj K. Goyal and Joemon M. Jose**

**Abstract** Latent Dirichlet allocation (LDA) topic model has taken a center stage in multimedia information retrieval, for example, LDA model was used by several participants in the recent TRECVid evaluation "Search" task. One of the common approaches while using LDA is to train the model on a set of test images and obtain their topic distribution. During retrieval, the likelihood of a query image is computed given the topic distribution of the test images, and the test images with the highest likelihood are returned as the most relevant images. In this paper we propose to project the unseen query images also in the topic space, and then estimate the similarity between a query image and the test images in the semantic topic space. The positive results obtained by the proposed method indicate that the semantic matching in topic space leads to a better performance than conventional likelihood based approach; there is an improvement of 25 % absolute in the number of relevant results extracted by the proposed LDA based system over the conventional likelihood based LDA system. Another not-so-obvious benefit of the proposed approach is a significant reduction in computational cost.

**Keywords** Multimedia information retrieval · Topic modeling · Latent Dirichlet allocation · Semantic information

---

H. Misra (✉)
Philips Research Asia, Philips Electronics India Limited, Bangalore, India
e-mail: Hemant.Misra@philips.com

A. K. Goyal
Language Technologies Institute, Carnegie Mellon University, Pittsburgh, USA
e-mail: Anuj@cs.cmu.edu

J. M. Jose
Department of Computing Science, University of Glasgow, Glasgow, UK
e-mail: Joemon.Jose@glasgow.ac.uk

P. P. Swamy and D. S. Guru (eds.), *Multimedia Processing, Communication and Computing Applications*, Lecture Notes in Electrical Engineering 213,
DOI: 10.1007/978-81-322-1143-3_6, 

## 1 Introduction

The amount of personal multimedia data present on the internet is so huge that using traditional methods of annotating the data for retrieval is no more practical. There is a constant search for algorithms which can assist in unsupervised retrieval of multimedia data. The two important requirements of these algorithms are good retrieval performance and low computational cost.

Content based image retrieval (CBIR) involves extracting relevant digital images on the basis of their visual content [1–4]. TRECVid and Image-CLEF evaluations are typical meeting grounds for participants to showcase their image extraction algorithms and compare the performance of their algorithms on a common benchmark. In the recent TRECVid evaluations, quite a few participants have used latent Dirichlet allocation (LDA) [5, 6] topic model for the search task [7, 8]. Notably the top performing team in 2008 evaluations used LDA as one of the components in their system. Apart from TRECVid evaluations, LDA and probabilistic latent semantic analysis (PLSA) [9] have enjoyed prominence in many CBIR publications [1–3].

In most of the LDA based systems, the typical approach for retrieval is to compute the likelihood of a query image given the topic distribution of the test images, and return those test images as relevant which give the highest likelihood [1, 8]. However, as pointed out in [10] for the information retrieval task, likelihood based systems do not perform well on their own.

In this paper, we propose to project bag-of-words (BOW) representation of query images also in the LDA topic space; this has two major advantages: (1) we are able to capture the semantic information present in a query (relation among visual words of the query), (2) in the lower dimension LDA topic space, the cost associated with matching a query image with the test images is significantly lower as compared to the cost associated with a likelihood based approach. A brief description of the previous use of topic models for image retrieval is in Sect. 2.3.

The rest of the paper is organized as follows: In Sect. 2, we give a brief description of LDA model. The description of the proposed system and its main differences from the LDA based systems previously used for image retrieval tasks are provided in Sect. 3. Experimental setup of this paper is described in Sect. 4. In Sect. 5, we compare the performance of the two LDA based methods on TRECVid 2009 benchmark and analyze the results. Conclusions of this study are presented in Sect. 6.

## 2 Latent Dirichlet Allocation

The LDA model for the task of unsupervised topic detection was proposed in [5, 6]. The authors demonstrated the advantages of the LDA model vis-à-vis several other models, including multinomial mixture model [11] and probabilistic latent semantic analysis (PLSA) [9]. Like most models of text, LDA uses the bag-of-words (BOW) representation of documents. The key assumptions of LDA are that *each document is represented by a topic distribution* and *each topic has an underlying word distribution*.

LDA is a generative model and specifies a probabilistic method for generating a new document. Assuming a fixed and known number of topics, $T$, for each topic $t$, a distribution $\phi_t$ is drawn from a Dirichlet distribution of order $V$, where $V$ is the vocabulary size. The first step in generating a document is to choose a topic distribution, $\theta_{dt}, t = 1...T$, for that document from a Dirichlet distribution of order $T$. Next, assuming that the document length is fixed, for each word occurrence in the document, a topic, $z_i$, is chosen from this topic distribution and a word is selected from $\phi_t$, the word distribution of the chosen topic. Given the topic distribution of the document, each word is drawn independently of every other word.

Therefore, the probability of $w_i$, the $i$th word token in document $d$, is:

$$P(w_i|\theta_d, \phi) = \sum_{t=1}^{T} P(z_i = t|\theta_d)P(w_i|z_i = t, \phi) = \sum_{t=1}^{T} \theta_{dt}\phi_{tw_i} \tag{1}$$

where $P(z_i = t|\theta_d)$ is the probability that given the topic distribution $\theta_d$, $t$th topic was chosen for the $i$th word token and $P(w_i|z_i = t, \phi)$ is the probability of word $w_i$ given topic $t$.

The likelihood of document $d$ is a product of terms such as (1), and can be written as:

$$P(C_d|\theta_d, \phi) = \prod_{v=1}^{V} \left[ \sum_{t=1}^{T} (\theta_{dt}\phi_{tv}) \right]^{C_{dv}} \tag{2}$$

where $C_{dv}$ is the count of word $v$ in $d$ and $C_d$ is the word-frequency count in $d$.

## 2.1 LDA: Training

In the LDA training, the following two sets of parameters are estimated from a set of documents (train data): the topic distribution in each document $d$ ($\theta_{dt}, d = 1...D$, $t = 1...T$) and the word distribution in each topic ($\phi_{tv}, t = 1...T, v = 1...V$). In this paper, Gibbs sampling [6] method is used to estimate these two distributions due to its better convergence and it being less sensitivity to initialization. $\alpha$ and $\beta$, two hyper-parameters of the LDA model, define the non-informative Dirichlet priors on $\theta$ and $\phi$ respectively.

The training process for LDA model using Gibbs sampling is explained in [6]. For each word token in the training data, the probability of assigning the current word token to each topic is conditioned on the topic assigned to all other word tokens except the current word token. A topic is sampled from this conditional distribution and assigned to the current one. In every pass of Gibbs sampling, this process of assigning a topic for all the word tokens in the training data constitutes one Gibbs sample. The initial Gibbs samples are discarded as they are not a reliable estimate of the posterior. For a particular Gibbs sample, the estimates for $\theta$ and $\phi$ are derived from the counts of hypothesized topic assignments as:

$$\phi_{tv} = \frac{J_{tv} + \beta}{\sum_{k=1}^{V} J_{tk} + V\beta} \text{ and } \theta_{dt} \frac{K_{dt} + \alpha}{\sum_{k=1}^{T} K_{dk} + T\alpha}$$

where $J_{tv}$ is the number of times word $v$ is assigned to topic $t$ and $K_{dt}$ is the number of times topic $t$ is assigned to some word token in document $d$.

## 2.2 LDA: Testing

In a typical information retrieval (IR) setting, where the main focus is on computing the similarity between a document $d$ and a query $d'$, a natural similarity measure is given by $P(C_{d'}|\theta_d, \phi)$, computed according to (2) [12]. An alternative would be to compute the similarity through measures which are well suited for comparing distributions such as cosine distance, Bhattacharyya distance or KL divergence between $\theta_d$ and $\theta_{d'}$ (the topic distribution in $d$ and $d'$); this however requires to infer the latter quantity. As the topic distribution of a (new) document gives its representation along the latent semantic dimensions, computing this value is helpful for many applications such as language model adaptation [13] and text classification [14]. In [5], an approximate convexity based variational approach was proposed for inference. However, as pointed out in [15], the variational approach for inference has high bias and high computational cost.

In this paper, we use the expectation-maximization (EM) like iterative procedure suggested in [13, 14] for estimating topic distribution. The update rule is given by:

$$\theta_{dt} \leftarrow \frac{1}{l_d} \sum_{v=1}^{V} \frac{C_{dv}\theta_{dt}\phi_{tv}}{\sum_{t'=1}^{T} \theta_{dt'}\phi_{t'v}} \tag{3}$$

where $l_d$ is document length, computed as the number of running words. It was shown in [14] that this update rule converges monotonically towards a local optimum of the likelihood, and the convergence is typically achieved in less than 10 iterations.

## 2.3 LDA: Application in CBIR

As mentioned previously, though the topic models were initially proposed for processing text [5, 6, 9] recently they have gained popularity in many other applications related to text processing [7, 10, 12–14, 16, 17] and image processing [1–3,

18–20]. PLSA based approaches [2, 3, 20] and LDA based approaches [1, 7, 8] have given good performance on very large databases. As was the case in text processing applications, LDA has typically yielded better performance than PLSA in image processing domain as well [1, 5].

The work which is closest in nature to the idea presented in this paper is [1]. In Ref. [1], the authors did users studies to compare the performance of LDA based approach with PLSA based approach, and found LDA to perform better than PLSA. Also, they used the concept of relevance feedback to improve the results. The major differences between [1] and the present work are: (a) in [1], the authors used approximate convexity based variational approach [5] for computing the topic distribution of queries (unseen data). It has been pointed out in [15] that this approach is prone to bias and has high computational cost. In the present paper, we use the method presented in Sect. 2.2 for computing the topic distribution of queries. This method has been shown to have excellent performance on several other tasks [13, 14], and to the best of out knowledge is more accurate and much faster than any other methods proposed in the literature for LDA inference [15], (b) time complexity of the approach proposed in this paper is very low, thus making it possible to use this approach for online applications, (c) the results reported in this paper are on TRECVid 2009 benchmark in terms of standard measures such as mean average precision (MAP) and precision at 10 (P@10) making it possible to compare these results with other algorithms, whereas the results presented in [1] were on user evaluations.

## 3 System Description

Figure 1 shows the major components of the proposed and conventional LDA based methods; the most important difference is projecting the queries into the LDA topic space (LDA Testing block) and then doing the similarity match in the LDA topic space. Both these blocks are shaded in the figure.

### *3.1 Low-Level Features*

Our images are represented in terms of low-level local features obtained by Scale Invariant Feature Transform (SIFT) [21]. SIFT was the preferred choice in our case because we wanted a feature representation which is able to model regions of variable size in an image; also SIFT is invariant to scale, orientation and affine distortion and is partially immune to changes in illumination conditions.

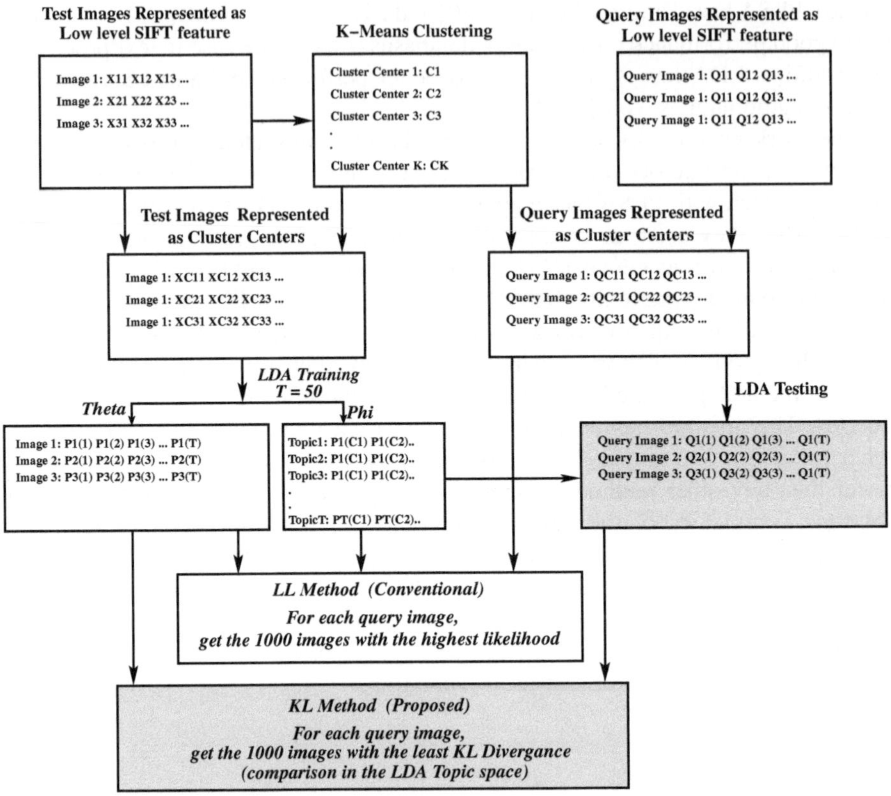


**Fig. 1** Different components of the proposed system (**KL Method**: *shaded blocks*) and the conventional system (**LL Method**)

## *3.2 Clustering*

LDA assumes that the documents to be modeled are represented in terms of a word vector; though the order of the words is not important, the words belong to a fixed vocabulary. Images represented in terms of features obtained by SIFT (henceforth we will call them SIFT features) cannot be used in an LDA model because the number of SIFT features is not limited. In order to limit the vocabulary size, we need to do some quantization of SIFT features and in our experiments we employed simple K-Means clustering to perform this desired quantization. We treated each SIFT feature as independent to obtain the $K$ cluster centers. In the experiments reported in this paper, we have fixed $K$, the number of cluster centers (visual words), to be 10,000, which also becomes the size of our visual vocabulary.

We used the clustering code available in LIBPMK library [22] after making some necessary modifications. The time complexity of the K-Means algorithm increases with an increase in number of data points to be clustered; in order to keep the computational cost under check and also to explore the dependence of the final performance on the amount of data used for clustering, we used 02, 05, 10 and 20 % of the *TRECVid 2008 relevant test data* for obtaining the 10,000 cluster centers. More details about this setup are provided in Sect. 4.

The criterion for stopping the K-Means clustering in LIBPMK is the number of iterations. As the time complexity increases linearly with an increase in the number of iterations, in this paper we have explored the following four settings for the number of iterations: 10, 20, 40 and 80.

$L2 - norm$ was used to compute distances in the K-Means algorithm. The useful output of this clustering procedure is the $K = 10{,}000$ cluster centers (visual words) which are used in the next stage of the proposed system.

### 3.2.1 Cluster Centers for 2009 Test Data

Once the visual words are obtained on a percentage of *TRECVid 2008 relevant test data*, representing the entire *TRECVid 2009 test data* in terms of these visual words is relatively simple: for each SIFT feature in each test image find the cluster center which is closest to it. We have used $L2 - norm$ while estimating the distance between test SIFT features and cluster centers. At the end, each test image is represented in terms of cluster centers.

From a document perspective, each image is a document described by a word vector derived from a fixed vocabulary ($K = 10{,}000$ cluster centers). This representation of images can be used for training the LDA model.

### 3.2.2 Cluster Centers for 2009 Query Data

Similar to the case of representing the *TRECVid 2009 test data* in terms of 10$K$ visual words, one can represent each query image of the *TRECVid 2009 query data* in terms of visual words.

## 3.3 LDA Training on TRECVid 2009 Test Data

In this step, we train an LDA model on the entire TRECVid 2009 test collection represented in terms of $K = 10{,}000$ cluster centers. *The hyper-parameters of the LDA model were* $\alpha = 1$, $\beta = 0.1$ *and number of topics,* $T = 50$. The procedure for training the LDA model was explained in Sect. 2.1.

At the end of the training, the following two parameters of the LDA model are obtained: (1) each test image represented in terms of $T$-dimensional LDA topic distribution ($\theta$), and (2) each LDA topic represented in terms of $K$-dimensional word distribution ($\phi$). The LDA based approaches reported in the past followed all the steps up to this point. In the next section we briefly explain the difference between the previous approaches and the approach proposed in this paper, highlighting the main advantages of the proposed approach.

## 3.4 Image Retrieval

As explained in Sect. 2.2, in response to a query there are two possibilities to retrieve a set of images from a test collection.

1. **LL Method**: Compute the likelihood of the query given the images in the test collection and select those images as relevant that give the highest likelihood. In this case the likelihood is computed using (2) as follows (computation is typically done in log domain to avoid underflow that is why we work with log-likelihood (LL) instead of likelihood): $P(C_q|\theta_d,\phi) = \prod_{v=1}^{V}\left[\sum_{t=1}^{T}(\theta_{dt}\phi_{tv})\right]^{C_{qv}}$.
2. **KL Method**[1]: First estimate the topic distribution of the query image using the iterative procedure given in (3) as follows: $\theta_{qt} \leftarrow \frac{1}{l_q}\sum_{v=1}^{V}\frac{C_{qv}\theta_{qt}\phi_{tv}}{\sum_{t'=1}^{T}\theta_{qt'}\phi_{t'v}}$. This is an extremely fast procedure and convergence is typically reached in less than 10 iterations. Then symmetric KL divergence between $\theta_q$, the topic distribution of the query image, and $\theta_d$, the topic distribution of the test image $d$ is computed. We select those test images as relevant that give the least KL divergence, where $KL(\theta_q,\theta_d) = \sum_{t=1}^{T}[\theta_{dt}\log(\theta_{dt}/\theta_{qt}) + \theta_{qt}\log(\theta_{qt}/\theta_{dt})]$.

By projecting the queries in the LDA topic space we are able to capture the semantics (relationship among words in a query) whereas this information is missing when each word in a query is treated independent of every other word.

The second, but not so obvious, advantage of the **KL Method** is the significant reduction in computational cost that can be realized by projecting the queries onto a lower dimensional LDA topic space and then doing the matching. The computational cost of the **LL Method** is dependent upon the query length which is significantly much higher than the number of LDA topics, specially when the images are represented by SIFT features. The average number of visual word tokens in a query were found to be 795 in the *TRECVid 2009 query data* collection. The high computational cost of **LL Method** was cited as its drawback in [1] as well. We will discuss more about the performance and the computational efficiency of the proposed **KL Method** in Sect. 5.

---

[1] It was reported in [1] that cosine distance performs poorly as compared to KL divergence. In this paper we have considered symmetric KL divergence as the measure to estimate the similarity/distance between two images.

## 4 Experimental Setup

In TRECVid 2009 evaluations, approximately 280 hours of video data was provided as test set for the search task. The information about the shot boundaries was provided by NIST for the test set. In our experiments, we extracted one frame per shot. With this setup, our TRECVid 2009 evaluation test database for the search task consists of 97149 images. The query dataset had 471 images, either extracted from video shots or static images. In case of video, again we extracted only one frame per shot.

The number of search topics in TRECVid 2009 were 24. These topics can be considered as multimedia statement of information need. TRECVid results are typically reported as follows: given the search test collection, a topic (multimedia statement of information need of a user), and the common shot boundary reference for the search test collection, return a ranked list of at most $N$ common reference shots from the test collection which best satisfy the user need. In TRECVid evaluations, $N = 1{,}000$ for the standard search task whereas $N = 10$ for the high-precision search task.

In our experiments, while performing K-Means clustering, we kept the number of cluster centers, $K$, fixed at 10,000. In the clustering algorithm, we studied the effect of the following two variables on the final retrieval performance:

- the number of iterations, and
- the amount of data from *TRECVID 2008 relevant test dataset* that was used for clustering.

We used the number of iterations as 10, 20, 40, and 80. The amount of data used for clustering was either 2, 5, 10 or 20 % of the relevant test dataset. For example, 2 % of the SIFT features from each image in the test collection were pooled together to create 2 % of the relevant test dataset for clustering. *The hyper parameters of the LDA model were kept fixed as* $\alpha = 1$, $\beta = 0.1$ *and* $T = 50$ *in all the experiments.*

For each TRECVid topic a result list containing 1,000 shots was to be generated. When there were several query examples in a topic, each example generated a list of 1,000 most relevant shots. In such a case, each example was given equal importance and a voting was performed to generate the final list of 1,000 most relevant shots from these individual lists.

## 5 Results and Analysis

### *5.1 Retrieval Performance*

In this section, we present the results of the two LDA based systems, one proposed in this paper and the other typically used in the literature. We present the performance of the systems in terms of mean average precision (MAP), precision at R (R-prec), precision at 10 (P@10), precision at 1,000 (P@1,000) and total number of relevant results returned out of 10619 relevant results (Relevant).

Owing to the high computational cost associated with the **LL Method**, we were able to run only a few experiments for this system. In Table 1 we compare the performance of the two systems for two different number of iterations, 10 and 20.

It must be noted that the underlying LDA model for the **KL Method** and **LL Method** is exactly the same (the LDA model changes with the change in amount of data and number of iterations used for K-Means clustering).

Comparing the results of the two methods we observe that the proposed **KL Method** gives a better performance then **LL Method** across all the measures. This result is valid for two different training setups of the LDA model. Projecting the queries into the LDA topic space and then performing the similarity between the queries and the test documents in the LDA topic space not only gives a better performance, it also brings a significant reduction in cost complexity of the system. The time complexities of the two systems are described in the next section.

The results in Table 2 show the performance of **KL Method** for different *number of iterations used for K-Means clustering* when data used for clustering is 2 % of the total *TRECVid 2008 relevant test data*.

The results show a trend, though it is weak, that increasing the number of iterations of the K-means algorithm brings an improvement in the performance of the system by retrieving more relevant documents towards the end of the list (note that though P@1,000 improves, P@10 drops with an increase in number of iterations)

In Table 3, we present the performance of **KL Method** obtained by changing the *amount of data used for K-Means clustering*. The number of iterations were fixed at 10. Again we see a weak trend that increasing the amount of data used for K-Means clustering leads to a small improvement in performance. As in the previous case. though P@1,000 improves slightly, the P@10 drops, indicating that more relevant documents are added towards the end of the list. Also, for 20 % data size, performance drops slightly. The reason for this could be that as we increase

**Table 1** The performance of **KL Method** and **LL Method** for the same LDA model

| Measure | KL Method | | LL Method | |
|---|---|---|---|---|
| | Iterations | | Iterations | |
| | 10 | 20 | 10 | 20 |
| MAP | 0.0174 | **0.0196** | 0.0104 | 0.0110 |
| R-prec | 0.0405 | **0.0433** | 0.0340 | 0.0322 |
| P@10 | 0.0792 | **0.1083** | 0.0375 | 0.0500 |
| P@1,000 | 0.0424 | **0.0456** | 0.0360 | 0.0363 |
| Relevant | 1017 | **1095** | 863 | 872 |

The performance is shown for two different number of iterations, 10 and 20, whereas the data size is kept at 10 %. Performance in **bold** indicates that the imporvement is statistically significant as compared to all the other systems

**Table 2** The performance of **KL Method** for different number of iterations in K-Means clustering

| Measure | Iterations | | | |
|---|---|---|---|---|
| | 10 | 20 | 40 | 80 |
| MAP | 0.0172 | 0.0183 | 0.0178 | 0.0177 |
| R-prec | 0.0388 | 0.0392 | 0.0412 | 0.0427 |
| P@10 | 0.0917 | 0.0917 | 0.0708 | 0.0792 |
| P@1,000 | 0.0408 | 0.0417 | 0.0425 | 0.0430 |
| Relevant | 979 | 1001 | 1019 | 1033 |

Data size = 2 % of the total TRECVid 2008 relevant test data

**Table 3** The performance of **KL Method** for different data sizes from TRECVid 2008 relevant test data used in K-Means clustering

| Measure | % Data | | | |
|---|---|---|---|---|
| | 2 | 5 | 10 | 20 |
| MAP | 0.0172 | 0.0178 | 0.0174 | 0.0166 |
| R-prec | 0.0388 | 0.0410 | 0.0405 | 0.0391 |
| P@10 | 0.0917 | 0.0833 | 0.0792 | 0.0750 |
| P@1,000 | 0.0408 | 0.0420 | 0.0424 | 0.0412 |
| Relevant | 979 | 1008 | 1017 | 990 |

Number of iterations is 10

the data size, we start selecting more data points which are non-relevant (like stop words) and these data points change the final cluster centers.

The encouraging result obtained from these experiments is that the performance of the *KL Method* is stable for different data sizes and different number of iterations which were used for clustering the data.

## 5.2 Time Complexity

In the proposed **KL Method**, the results are obtained in two steps: first we project each query image into the LDA topic space and then we compute the KL-divergence between the topic distribution of a query image and topic distribution of a test image. The test images are ranked based on their similarity to the query image in the topic space. Projecting a query image onto the LDA topic space is a very fast process and takes less than a second (298.7 seconds for 471 query images). LDA topic space has a much lower dimensionality than the test and query images represented in terms of cluster centers; as a consequence, the distance or similarity computation is very fast. Comparing this to the **LL Method** we find that though the test images are projected into the LDA topic space during LDA training, the query images are represented only in terms of cluster centers. Computing the

**Table 4** Time taken, in seconds, by **LL Method** and **KL Method**

| **KL Method** | | **LL Method** |
|---|---|---|
| Topic Estimation (of queries) | KL Divergence | LL Similarity |
| 298.7 s | 4544.16 s | 76368 s |

**KL Method** has two steps; time taken by the two steps is reported separately

likelihood of a query image with respect to all the test images requires that either (1) topic distribution of every test image ($\theta_{dt}$) is multiplied with $\phi_{tv}$ **only for the visual words which are present in the query image** to compute the likelihood of the test images with respect to the query, or (2) multiply $\theta_{dt}$ of all the test images at once with $\phi_{tv}$ and then **depending upon each query just sum the components which are relevant to that query**. (1) has less memory requirements but very high time requirements where as (2) has high memory requirements ([*NumberOfTestImages X SizeOfVisualVocabulary*]) and moderate time requirements. Average time required by (2) and the **KL Method** are shown in Table 4. Memory requirement of (2) is too high to be processed on a simple machine and we used the grid facility provided by IRF, Vienna, to complete this task.

Comparing the results presented in Table 4 we observe that the proposed **KL Method** brings down the computational cost of the LDA approach by a factor of 15.7. The proposed LDA based approach not only gives an improvement of approximately 20 % over the conventional LDA based approach, it also reduces the time complexity by approximately 93.7 %.

It may also be noted that the LDA model was trained on TRECVid 2008 data whereas retrieval was performed on TRECVid 2009 data. Further improvement in the performance may be obtained if the model is trained and then used for retrieval on the same dataset. Moreover, it is possible to use more sophisticated clustering algorithms than K-Means to obtain a better visual vocabulary.

## 6 Conclusions

In this paper we proposed an LDA based system wherein the low-level SIFT features obtained from query images are first projected into the LDA topic space and then the matching between the query images and the test images is done in the LDA topic space. This is a departure from the conventional LDA based systems where typically the likelihood of the query images with respect to the test images is estimated to retrieve the most similar images. The proposed method not only leads to a significant improvement in the performance, it also reduces the computational cost associated with score estimation. In absolute terms, on TRECVid 2009 dataset, the number of relevant images retrieved by the proposed LDA system is approximately 20 % more than that obtained by the conventional likelihood based LDA system. The reduction in computational cost while estimating the matching scores is more than 90 %. This result generalizes across all the training setups used in this study.

## References


1. Hörster E, Lienhart R, Slaney M (2007) Image retrieval on large-scale image databases. In ACM international conference on image and video retrieval, Amsterdam
2. Lienhart R, Slaney M (2007) PLSA on large-scale image databases. In IEEE international conference on acoustics, speech and signal processing, Honolulu, Hawaii
3. Monay F, Gatica-Perez D (2007) Modeling semantic aspects for cross-media image indexing. IEEE Transactions on Pattern Analysis and Machine Intelligence
4. Datta R, Joshi D, Li J, Wang JZ (2008) Image retrieval: ideas, influences, and trends of the new age. ACM computer surveys 40(2):1–60
5. Blei DM, Ng AY, Jordan MI (2003) Latent Dirichlet allocation. Mach Learn Res 3:993–1022
6. Griffiths TL, Steyvers M (2004) Finding scientific topics. Proc Nat Acad Sci 101(supl 1):5228–5235
7. Cao J, Li J, Zhang Y, Tang S (2007) LDA-based retrieval framework for semantic news video retrieval. In IEEE international conference on semantic computing, Irvine, California, pp 155–160
8. Tang S, Li J-T, Li M, Xie C, Liu Y-Z, Tao K, Xu S-X (2008) TRECVid 2008 high-level feature extraction by MCG-ICT-CAS. In TRECVID 2008 Workshop. Gaithersburg, Maryland
9. Hofmann T (2001) Unsupervised learning by probabilistic latent semantic analysis. Mach Learn J 42(1):177–196
10. Wei X, Croft BW (2006) LDA-based document models for ad-hoc retrieval. In Proceedings of the 29th annual international ACM SIGIR conference on Research and development in information retrieval, New York, p 178–185, ACM
11. Nigam K, McCallum AK, Thrun S, Mitchell TM (2000) Text classification from labeled and unlabeled documents using EM. Mach Learn 39(2/3):103–134
12. Buntine W, Löfström J, Perkiö J, Perttu S, Poroshin V, Silander T, Tirri H, Tuominen A, Tuulos V (2004) A scalable topic-based open source search engine. In Proceedings of the IEEE/WIC/ACM international conference on web intelligence, p 228–234, Beijing
13. Heidel A, an Chang H, shan Lee L (2007) Language model adaptation using latent Dirichlet allocation and an efficient topic inference algorithm. In proceedings of EuroSpeech, Antwerp, Belgium
14. Misra H, Cappé O, Yvon F (2008) Using LDA to detect semantically incoherent documents. In Proceedings of CoNLL, Manchester, pp 41–48
15. Yao L, Mimno D, McCallum A (2009) Efficient methods for topic model inference on streaming document collections. In Proceedings of the 15th ACM SIGKDD international conference on Knowledge discovery and data mining, New York, ACM, p 937–946
16. Xing D, Girolami M (2007) Employing latent Dirichlet allocation for fraud detection in telecommunications. Pattern recognition letters 28(13):1727–1734
17. Biró I, Siklósi D, Szabó J, Benczúr AA (2009) Linked latent Dirichlet allocation in web spam filtering. In Adversarial Information Retrieval on the Web, Madrid
18. Barnard K, Duygulu P, de Freitas N, Forsyth D, Blei D, Jordan MI (2003) Matching words and pictures. J Mach Learn Res 3:1107–1135
19. Blei DM, Jordan MI (2003) Modeling annotated data. In Proceedings of the 26th annual international ACM SIGIR conference on Research and development in informaion retrieval, New York, ACM, p 127–134

20. Bosch A, Zisserman A, Muñoz X (2006) Scene classification via pLSA. In European Conference on Computer Vision, p 517–530
21. Lowe DG (2004) Distinctive image features from scale-invariant keypoints. Int J Comput Vis 60(2):91–110
22. Lee JJ (2008) Libpmk: A pyramid match toolkit. Technical, Report MIT-CSAIL-TR-2008-17

# Texture in Classification of Pollen Grain Images

**D. S. Guru, S. Siddesha and S. Manjunath**

**Abstract** In this paper we present a model for classification of pollen grain images based on surface texture. The surface textures of pollens are extracted using different models like Wavelet, Gabor, Local Binary Pattern (LBP), Gray Level Difference Matrix (GLDM) and Gray Level Co-Occurrence Matrix (GLCM) and combination of these features. The Nearest Neighbor (NN) classifier is adapted for classification. Unlike other existing contemporary works which are designed for a specific family or for one or few different families, the proposed model is designed independent of families of pollen grains. Experimentations on a dataset containing pollen grain images of about 50 different families totally 419 images of 18 classes have been conducted to demonstrate the performance of the proposed model. A classification rate up to 91.66 % is achieved when Gabor wavelet features are used.

**Keywords** Pollen grain images · Texture features · Nearest neighbor classifier

## 1 Introduction

Conserving earth's biodiversity for future generations is a fundamental global task. 20 % of the all world's plants are already at the edge of becoming extinct [1] and many methods must be combined to achieve this goal. Saving flora biodiversity

D. S. Guru (✉) · S. Siddesha
Department of Studies in Computer Sicence, Manasagangotri, University of Mysore, Mysore 570006, Karnataka, India
e-mail: dsg@compsci.uni-mysore.ac.in

S. Siddesha
e-mail: siddesh.shiv@gmail.com

S. Manjunath
PG Department of Studies in Computer Sicence, JSS Arts, Commerce and Science College, Mysore 570025, Karnataka, India
e-mail: manju_uom@yahoo.co.in

P. P. Swamy and D. S. Guru (eds.), *Multimedia Processing, Communication and Computing Applications*, Lecture Notes in Electrical Engineering 213,
DOI: 10.1007/978-81-322-1143-3_7, 

involves mapping plant distribution by collecting pollen and later identifying and classifying them in a laboratory environment. Pollen classification is a qualitative process, involving observation and discrimination features [2]. The manual method is depending on experts, but takes large amount of time. Therefore pollen grain classification using computer vision is highly needed in Palynology. Palynology is the study of external morphological features of mature pollen grains [3]. Several characteristic features such as leaf, flower, seed etc., of plants are used to determine the rank of the taxa (a taxonomic unit), of which Palynological evidence has proven useful in verifying relationships in established taxonomic groups. Pollen grains are distinguished primarily by their structure and surface sculpture (Texture) [4]. There are approximately 300,000 species of flowering plants and these are classified under 410 families as per Takthajan system of classification [5]. As per the general study by taxonomists, in each and every family there are plants whose external characteristics looks similar but their identities are under dispute (doubtful of their species).

The main objective of classification of pollen grains is to solve the species and family of the plants which are under dispute in the field of plant taxonomy. Classification of pollen grains also finds its applications in, identifying pollens available in the environment which causes allergy (Aerobiology), to solve legal problems (Forensic palynology), study of pollens in fossils (Quaternary Palaeopalynology) and study of botanical and geographical origin of honey (Melissopalynology) etc.

Classification of pollen grains using image processing techniques focuses on getting maximum quality output. Many attempts have been made to automate identification, classification and recognition of pollen grain by the use of image processing.

In [6] non linear features from pollen grain images extracted using wavelet transforms. The extracted features are used to perform the classification of pollen grain images using self organizing map (SOM) neural network. Also attempts have been made for pollen texture identification using neural network multi layer perceptron (MLP) technique over the statistical classifier methods [7]. In [8] a prototype of a system presented for classifying the two genders of pollen grains of three types of plants of the family Urticaceae. Here the classification is based on shape analysis using area, perimeter and compactness as features.

A work carried out for recognition of pollen grain images. Five types of pollen grains are classified based on surface texture and the geometric shapes. Surface texture extracted using Gabor transform and geometric shapes using moment invariants with artificial neural network as a classifier [9]. Work has been done for investigation of feasibility of the application of computer vision, to determine in a fast and precise way, the flower origin of pollen from honey of northwest Spain [10]. They classified the pollen grain images using support vector machine (SVM) and multi-layer perceptron (MLP), using a minimum distance classifier.

Specifically, several well-known classifiers, k-nearest neighbor (KNN), support vector machine and multi-layer perception are used to increase the classification rate. The method was to identify honeybee pollen. This work mainly focuses on the improvement of the classification stage. The combination of SVM classifier and local linear transformations (LLT) texture vector achieved the best performance to discriminate among the five most abundant plant species from three geographical places in north-west of Spain.

Almost all the works reported in literature for classification of pollen grains are limited to very few families or dependent on specific family or area. No work has been carried out for classification of the pollen grains independent of families. In this work we designed the model for pollen grain classification which is independent of families.

The rest of the work is organized as follows. In Sect. 2, we present a texture based model for classification of pollen grain images. Details of experimentation are discussed in Sect. 3. The paper is concluded along with the scope for future work in Sect. 4.

## 2 Proposed Model

The proposed model has two stages feature extraction and classification. In training phase, from a given set of pollen grain images the texture features (Wavelet/Gabor/LBP/GLDM/GLCM) are extracted and used to train the system. In classification stage, from a given unknown test pollen grain image, texture features are extracted and these features are queried to nearest neighbor (NN) classifier to label the unknown pollen grain. The block diagram of the proposed model is given in Fig. 1.

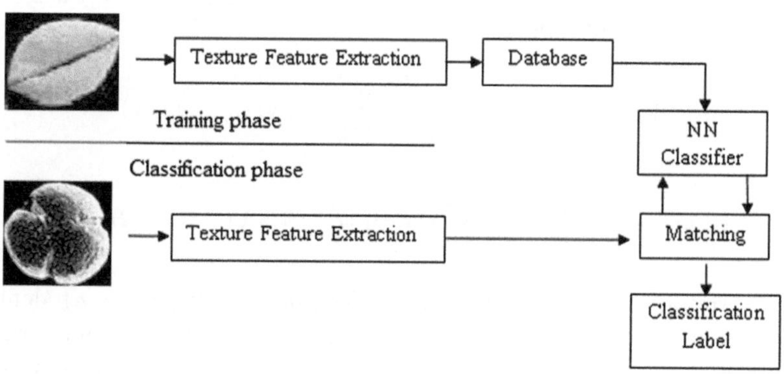


**Fig. 1** Block diagram of proposed model

## 2.1 Feature Extraction

Surface texture of the pollen grain plays very vital role in pollen classification. The outer surface of the pollen grain is covered with sculpture elements and it has different structure of apertures, a thin region through which one pollen grain can be differentiated from another. Hence in this work we recommend to use the texture feature for classification of pollen grains. Different texture features such as Wavelet, Gabor Wavelet, Local Binary Pattern (LBP), Gray Level Co-occurrence Matrix (GLCM), Gray Level Difference method (GLDM) and their combinations are studied here. The following sub sections provide an overview of above mentioned texture features.

### 2.1.1 Wavelet Transformation

Wavelet transforms are an alternative to the short time Fourier to overcome problems related to its frequency and time resolution properties. The basic idea of discrete wavelet transform (DWT) is to provide the time–frequency representation. In two dimension DWT, a two dimensional scaling function $\varphi(x, y)$ and three two dimensional wavelets $\psi^H(x, y)$, $\psi^V(x, y)$, $\psi^D(x, y)$ are required. Each one is the product of two one dimensional functions. Excluding the product which produce one dimensional results, like $\varphi(x)\,\psi(y)$, the four remaining products produce the separable scaling function, $\varphi(x, y) = \varphi(x)\,\varphi(y)$, and separable directly sensitive wavelets $\psi^H(x, y) = \psi(x)\,\varphi(y)$, $\psi^V(x, y) = \varphi(x)\,\psi(y)$, and $\psi^D(x, y) = \psi(x)\,\psi(y)$. These wavelets measure functional variables, intensity variables for images along different directions. $\psi^H$ measures variation along column (horizontal edges), $\psi^V$ responds to variation along rows (vertical edges) and $\psi^D$ corresponds to variation along diagonals. The two dimensional wavelet functions based on scaling and translation are, $\varphi_{j,m,n}(x, y) = 2^{\frac{j}{2}}\,\varphi(2^j x - m, 2^j y - n)$ and $\psi^i_{j,m,n}(x, y) = 2^{\frac{j}{2}}$ $\psi^i(2^j x - m, 2^j y - n)$, $i = \{H, V, D\}$ where index $i$ identifies the directional wavelets $\psi^H(x, y)$, $\psi^V(x, y)$ and $\psi^D(x, y)$. The discrete wavelet transform of image $f(x, y)$ of size $M \times N$ is,

$$W_\varphi(j_0, m, n) = \frac{1}{\sqrt{MN}} \sum_{x=0}^{M-1} \sum_{y=0}^{N-1} f(x, y)\, \varphi_{j_0\, m,n}(x, y) \tag{1}$$

$$W^i_\psi(j, m, n) = \frac{1}{\sqrt{MN}} \sum_{x=0}^{M-1} \sum_{y=0}^{N-1} f(x, y)\, \psi^i_{j,m,n}(x, y), \quad i = \{H, V, D\} \tag{2}$$

$j_0$ is an arbitrary starting scale and the coefficients $W_\varphi(j_0, m, n)$ define an approximation of $f(x, y)$ at scale $j_0$. The $W^i_\psi(j, m, n)$ coefficients add horizontal, vertical and diagonal details for scales $j \geq j_0$ normally $j_0 = 0$ and $N = M = 2$ so that $j = 0, 1, 2, \ldots J - 1$ and $m = n = 0, 1, 2, \ldots 2^j - 1$ [11].

### 2.1.2 Gabor Wavelet transforms

It is similar to the short time Fourier transforms, the Gabor wavelet transforms has been utilized as an effective and powerful time–frequency analysis tool for identifying rapidly varying characteristics of wave signals. The use of Gabor filters in extracting texture features motivated by several factors. These filters are considered as orientation and scale tunable edge and line detectors, and the statistics of these features in a given region are used to characterize the texture information [12]. A two dimensional Gabor function $g(x, y)$ and its Fourier transform $G(u, v)$ is,

$$g(x, y) = \left[\frac{1}{2\pi\sigma_x\sigma_y}\right] \exp\left[-\frac{1}{2}\left[\frac{x^2}{\sigma_x^2} + \frac{y^2}{\sigma_y^2}\right] + 2\pi jWx\right] \tag{3}$$

$$G(u, v) = \exp\left\{-\frac{1}{2}\left[\frac{(u-W)^2}{\sigma_u^2} + \frac{v^2}{\sigma_v^2}\right]\right\} \tag{4}$$

where $\sigma_u = \frac{1}{2\pi\sigma_x}$ and $\sigma_v = \frac{1}{2\pi\sigma_y}$. Gabor function forms a non-orthogonal but a complete set. Let $g(x, y)$ be the mother wavelet by dilation and rotations of $g(x, y)$ through the generating function, $g_{mn}(x, y) = a^{-m} g(x', y')$, $a > 1$, $m, n$ are integers, where $x' = a^{-m}(x\cos\theta + y\sin\theta)$, and $y' = a^{-m}(-x\sin\theta + y\cos\theta)$ and $\theta = \frac{n\pi}{N}$ and $N$ is the total number of orientations and $a^{-m}$ is the scale factor. Gabor wavelet transform of a image $f(x, y)$ is given as,

$$W_{mn}(x, y) = \int f(x_1, y_1) g_{mn} * (x - x_1, y - y_1) dx_1 dy_1 \tag{5}$$

where * is the complex conjugate. By assuming that the local texture regions are spatially homogeneous and $\mu_{mn}$ the mean and $\sigma_{mn}$ the standard deviation of the magnitude of the transform, coefficients are used to represent the region for classification. $\mu_{mn} = \iint |W_{mn}(xy)| dxdy$and $\sigma_{mn} = \sqrt{\iint (|W_{mn}(x, y)| - \mu_{mn})^2 dxdy}$. In our work, we have used four angular orientation and six scale factors with wavelet features.

### 2.1.3 Local Binary Pattern

Is a gray-scale and rotational invariant texture operator which characterizes the spatial structure of the local image texture [13]. In order to achieve gray scale invariance a unique pattern label has to be assigned to every pixel of an image based on the comparison of value of its binary pattern with its neighborhoods. The pattern label can be computed by

$$\boldsymbol{LBP}_{P,R} = \sum_{p=0}^{P-1} s(g_p - g_c)2^P \tag{6}$$

$$\text{where } s(g_p - g_c) = \begin{cases} 1, & (g_p - g_c) > \; = 0 \\ 0, & (g_p - g_c) < 0 \end{cases}$$

$g_c$ is central pixel's gray value having circular symmetric neighborhood $g_p(p = 0, 1, \ldots, P - 1)$, $g_p$ is neighbor's gray value, $P$ is the number of neighbors and $R$ is the neighborhood radius. The $LBP_{P,R}$ operator produces $2^P$ different output values, corresponding to the $2^P$ different binary patterns that can be formed by $P$ pixel in the neighbor. During rotation of image, $g_p$ the gray value will corresponding move along the perimeter of the circle. Since $g_0$ is assigned as the gray value of element (0, $R$) to the right side of rotating a specific binary pattern naturally results in a different $\mathrm{LBP_{P,R}}$ value. Therefore the rotation invariance is achieved by assigning a unique label to each rotation invariant binary pattern that is,

$$LBP^{ri}_{P,R} = \min\{ROR(LBP_{P,R}, i)| \quad i = 0, 1, \ldots, P - 1\} \tag{7}$$

where $ROR(LBP_{P,R}, i)$ performs a bitwise right shift in a circular way on the $P$ bit number $LBP_{P,R}$ i times. The uniform ($U$) value of $LBP^{ri}_{P,R}$ pattern is defined as the number of spatial transitions (bitwise 0/1 changes) in that pattern and is given by,

$$U\left(LBP^{ri}_{P,R}\right) = |S(g_{P-1} - g_c) - S(g_0 - g_c)| + \sum_{p=1}^{P-1} |S(g_P - g_c) - S(g_{p-1} - g_c)| \tag{8}$$

As per the recommendation in [13], Uniformity measure ($U$) of pattern $\leq 2$, it is referred as uniform pattern and assigned with a label in the range 0 to $P$ corresponding to the spatial transition. Other patterns with $U > 2$ are assigned to a label $P + 1$. Then we have

$$LBP^{riu2}_{P,R} = \begin{cases} \sum_{p=0}^{P-1} S(g_p - g_c), & \text{if } U(LBP^{ri}_{P,R}) \leq 2 \\ P + 1, & \text{otherwise} \end{cases}. \tag{9}$$

The texture features from a pollen grain image is extracted using the above LBP operator. The LBP with radius ($R$) and pixel ($P$) are calculated for the entire image of size $M \times N$ is resulted in a labeled image. The labeled image is represented by histogram as,

$$\underset{k \in 0,K}{H(k)} = \sum_{i=i}^{M} \sum_{j=1}^{N} f(LBP^{riu2}_{P,R}(i,j)k) \tag{10}$$

$$f\left(LBP^{riu2}_{P,R}(i, j), k\right) = \begin{cases} 1, & LBP^{riu2}_{P,R}(i, j) = k \\ 0, & \text{otherwise} \end{cases} \tag{11}$$

where $k$ is the maximal LBP pattern label.

### 2.1.4 Gray Level Co-occurrence Matrix

Texture feature calculations use the contents of the GLCM to give a measure of the variation in intensity at a pixel of interest proposed in [14] and they characterize texture using a variety of quantities derived from second order image statistics. Co-occurrence texture features are extracted from an image in two steps. First, the pair wise spatial co-occurrences of pixels separated by a particular angle and distance are tabulated using a gray level co-occurrence matrix (GLCM). Second, the GLCM is used to compute a set of scalar quantities that characterize different aspects of the underlying texture. The GLCM is a tabulation of how often different combinations of gray levels co-occur in an image. The GLCM is a $N \times N$ square matrix, where $N$ is the number of different gray levels in an image. An element $p(i, j, d, \theta)$ of a GLCM of an image represents the relative frequency, where $i$ is the gray level of the pixel $p$ at location $(x, y)$, and $j$ is the gray level of a pixel located at a distance $d$ from $p$ in the orientation $\theta$. While GLCMs provide a quantitative description of a spatial pattern, they are too unwieldy for practical image analysis. In [14] proposed a set of scalar quantities for summarizing the information contained in a GLCM. He originally proposed a total of 14 quantities, or features; however, typically only subsets of these are used [15]. In our work we considered five subset features of GLCM as shown in the Table 1.

### 2.1.5 Gray Level Difference Method

The Gray Level Difference Method (GLDM) is based on the occurrence of two pixels which have a given absolute difference in gray level and which are separated by a specific displacement $\delta$. For any given displacement vector $\delta = (\Delta x, \Delta y)$, Let $S_\delta(x, y) = |S(x, y) - S(x + \Delta x, y + \Delta y)|$ and $D(i|\delta) = \text{Prob}[S_\delta(x, y) = i]$ be the estimated probability-density function. In this work four possible forms of the vector $\delta$ will be considered $(0, d)$, $(-d, d)$, $(d, 0)$ and $(-d, -d)$, where $d$ is the inter sample spacing [16]. In this work we used four probability density functions for four different displacement vectors are obtained and the texture features are calculated for each probability density function.

**Table 1** Five GLCM features

| | |
|---|---|
| Correlation | $\sum_{i,j} \frac{(i-\mu_i)(j-\mu_j)\overrightarrow{p}(i,j)}{\sigma_i \sigma_j}$ |
| Contrast | $\sum_{i,j} \lvert i-j\rvert^2 p(i,j)$ |
| Energy | $\sum_{i,j} p(i,j)^2$ |
| Entropy | $\sum_{i,j=0}^{N-1} -\ln(P_{i,j})\, P_{i,j}$ |
| Homogeneity | $\sum_{i,j} \frac{p(i,j)}{1+\lvert i-j\rvert}$ |

As our interest is to study the statistics of texture features useful for pollen grain classification, from a pollen grain image, we used all the above feature extraction models for extracting the pollen grain surface texture. The extracted features are then classified using NN classifier.

### 2.2 Classification

Classification is to determine to which of a finite number of physically defined classes (such as different classes of pollen grains) of an unknown sample image of pollen grain belongs. In this work we use nearest neighbor (NN) classifier for the purpose of classification. It is a supervised learning method. In this classifier, to decide whether the sample $S_i$ belongs to class $C_j$, the similarity $Sim(S_i, S_j)$ or dissimilarity $Disim(S_i, S_j)$of $S_i$ to all other samples $S_j$ in the training set is determined. The n most similar training samples (neighbors) are selected. The proportion of neighbors having the same class may be taken as an estimator for the probability of that class and the class with the largest proportion is assigned to the sample $S_i$.

## 3 Dataset and Experimentation

We have created our own dataset of 419 pollen grain images. Out of them around 50 image are collected from World Wide Web sources [17–19], 100 images are collected from experts and around 269 images are collected using the standard procedures [4, 20]. The images of dataset are across 18 classes of pollen grains. These 18 classes are irrespective of families. The dataset contain both LM (Light microscopic) and SEM (Scanning electron microscopic) images. The 18 classes of pollen grains are considered based on NPC (Number, Position and Character of aperture) classification system [20]. The 18 classes considered in this work are, (a) MC: Monocolpate, (b) DC: Dicolpate, (c) TC: Tricolpate, (d) TRC: Tetracolpate, (e) PC: Pentacolpate, (f) HC: Hexacolpate, (g) DCP: Dicolporate, (h) TCP: Tricolporate, (i) TRCP: Tetracolporate, (j) PCP: Pentacolporate, (k) MP: Monoporate, (l) DP: Diporate, (m) TP: Triporate, (n) TRP: Tetraporate, (o) PP: Pentaporate, (p) PPP: Polypantaporate, (q) NAP: Nonaperturate and (r) SAP: Spiraperturate. Sample examples of these classes are shown in Fig. 2.

The pollen images of 18 classes are kept in a database. The pollen images from the database are fed into different feature extraction models individually. We used 5th level decomposition of two dimensional discrete Debucies wavelet transform and extracted 15 features, therefore the feature vector comprises of 15 elements. Of the 15 elements the first 14 elements consists of the Average Intensity Value

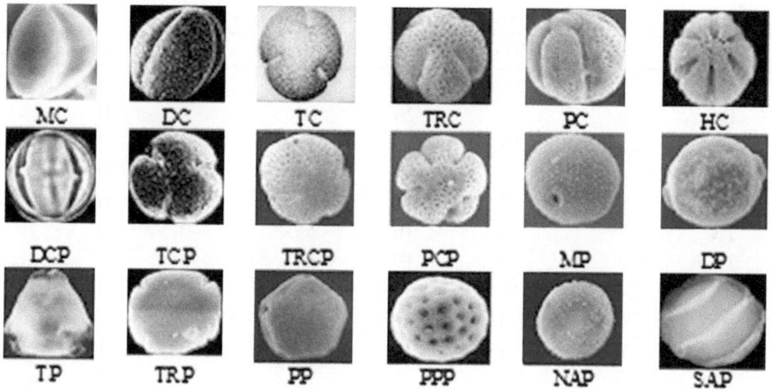


**Fig. 2** 18 classes of pollen grain based on NPC classification

**Table 2** Classification accuracy of different combination of texture features under varying training sets

| Sl. no | Method | Training set | | |
|---|---|---|---|---|
| | | 50 % | 60 % | 70 % |
| 1 | Wavelet (W) | 72.76 | 79.19 | 85.60 |
| 2 | Gabor wavelet(G) | 79.81 | 87.28 | 91.66 |
| 3 | LBP (L) | 74.64 | 79.19 | 86.36 |
| 4 | GLCM (GC) | 67.14 | 73.41 | 84.09 |
| 5 | GLDM (GD) | 74.18 | 78.03 | 84.85 |
| 6 | W + G | 80.74 | 87.28 | 91.66 |
| 7 | W + L | 74.65 | 79.77 | 86.36 |
| 8 | W + GC | 72.77 | 79.19 | 85.61 |
| 9 | W + GD | 74.18 | 78.03 | 84.85 |
| 10 | G + L | 74.65 | 79.77 | 85.61 |
| 11 | G + GC | 79.81 | 87.28 | 91.66 |
| 12 | G + GD | 74.18 | 78.61 | 84.85 |
| 13 | L + GC | 74.65 | 79.19 | 86.36 |
| 14 | L + GD | 75.12 | 79.19 | 85.61 |
| 15 | GC + GD | 74.18 | 78.03 | 84.85 |
| 16 | W + G + L | 74.65 | 80.35 | 86.36 |
| 17 | W + G + GC | 80.75 | 87.28 | 91.66 |
| 18 | W + G + GD | 74.12 | 78.61 | 84.85 |
| 19 | G + L + GC | 74.65 | 79.77 | 85.61 |
| 20 | G + L + GD | 75.11 | 79.19 | 85.61 |
| 21 | L + GC + GD | 75.11 | 79.19 | 85.61 |
| 22 | W + G + L + GC | 74.65 | 80.35 | 86.36 |
| 23 | W + G + L + GD | 75.15 | 79.19 | 85.61 |
| 24 | G + L + GC + GD | 75.12 | 79.19 | 85.61 |
| 25 | W + G + L + GC + GD | 75.12 | 79.19 | 85.61 |

**Table 3** Training, testing, correctly and wrongly classified samples

| Sl. no. | Class | Training | Testing | Total | Correctly classified | Wrongly classified |
|---|---|---|---|---|---|---|
| 1 | MC | 36 | 16 | 52 | 16 | 0 |
| 2 | DC | 18 | 8 | 26 | 8 | 0 |
| 3 | TC | 29 | 13 | 42 | 13 | 0 |
| 4 | TRC | 10 | 5 | 15 | 5 | 0 |
| 5 | PC | 8 | 4 | 12 | 4 | 0 |
| 6 | HC | 11 | 5 | 16 | 5 | 0 |
| 7 | DCP | 13 | 6 | 19 | 5 | 1 |
| 8 | TCP | 24 | 11 | 35 | 8 | 3 |
| 9 | TRCP | 7 | 3 | 10 | 3 | 0 |
| 10 | PCP | 8 | 4 | 12 | 3 | 1 |
| 11 | MP | 13 | 6 | 19 | 5 | 1 |
| 12 | DP | 21 | 10 | 31 | 10 | 0 |
| 13 | TP | 19 | 9 | 28 | 9 | 0 |
| 14 | TRP | 8 | 4 | 12 | 4 | 0 |
| 15 | PP | 10 | 5 | 15 | 3 | 2 |
| 16 | PPP | 23 | 10 | 33 | 8 | 2 |
| 17 | NAP | 21 | 9 | 30 | 9 | 0 |
| 18 | SAP | 8 | 4 | 12 | 3 | 1 |

(AIV) of the matrices that are obtained when passed through high pass filters, which are the horizontal, vertical and diagonal details matrix obtained at each level, whereas the 15th element consists of the AIV of the approximate co-efficient matrix. In case of Gabor wavelet we used four different angular rotations 22.5, 45, 77.5 and 90° with six different scale factors 0, 2, 4, 6, 8, 10 and a total of 15 wavelet features with 4 rotation and 6 scale factors a total of $4 \times 6 \times 15 = 360$ features are extracted. While using LBP we have extracted 256 features. In GLCM we used five subset features as shown in Table 1 and in GLDM we extracted the feature using 4 probability density functions.

All the above experiments are conducted using all 419 pollen grain images of 18 classes for 50, 60 and 70 % of training set. For classification we used five features and their possible combination. The obtained results are shown in Table 2. From Table 2 it is clear that Gabor wavelet perform better than other features and their combinations W + G, G + GC, and W + G + GC having same classification accuracy of 91.66 %. Table 3 shows total number of samples in each class, number of training and testing along with correctly and misclassified samples. The confusion matrix and F-measure graph for 70 % training and 30 % testing of each class are shown in Table 4 and Fig. 3 respectively.

In our experiment Gabor wavelet features gave good result as the Gabor wavelet provides the optimal resolution in both time (spatial) and frequency domains.

**Table 4** Confusion matrix for pollen grain classification

| | MC | DC | TC | TRC | PC | HC | DCP | TCP | TRCP | PCP | MP | DP | TP | TRP | PP | PPP | NAP | SAP |
|---|---|---|---|---|---|---|---|---|---|---|---|---|---|---|---|---|---|---|
| MC | 16 | 0 | 0 | 0 | 0 | 0 | 0 | 0 | 0 | 0 | 0 | 0 | 0 | 0 | 0 | 0 | 0 | 0 |
| DC | 0 | 8 | 0 | 0 | 0 | 0 | 0 | 0 | 0 | 0 | 0 | 0 | 0 | 0 | 0 | 0 | 0 | 0 |
| TC | 0 | 0 | 13 | 0 | 0 | 0 | 0 | 0 | 0 | 0 | 0 | 0 | 0 | 0 | 0 | 0 | 0 | 0 |
| TRC | 0 | 0 | 0 | 5 | 0 | 0 | 0 | 0 | 0 | 0 | 0 | 0 | 0 | 0 | 0 | 0 | 0 | 0 |
| PC | 0 | 0 | 0 | 0 | 4 | 0 | 0 | 0 | 0 | 0 | 0 | 0 | 0 | 0 | 0 | 0 | 0 | 0 |
| HC | 0 | 0 | 0 | 0 | 0 | 5 | 0 | 0 | 0 | 0 | 0 | 0 | 0 | 0 | 0 | 0 | 0 | 0 |
| DCP | 0 | 0 | 0 | 0 | 0 | 0 | 5 | 0 | 0 | 0 | 0 | 1 | 0 | 0 | 0 | 0 | 0 | 0 |
| TCP | 0 | 0 | 0 | 0 | 0 | 1 | 0 | 8 | 0 | 0 | 0 | 0 | 1 | 0 | 1 | 0 | 0 | 0 |
| TRCP | 0 | 0 | 0 | 0 | 0 | 0 | 0 | 0 | 3 | 0 | 0 | 0 | 0 | 0 | 0 | 0 | 0 | 0 |
| PCP | 0 | 0 | 0 | 0 | 0 | 0 | 0 | 0 | 0 | 3 | 0 | 0 | 1 | 0 | 0 | 0 | 0 | 0 |
| MP | 0 | 0 | 0 | 0 | 0 | 1 | 0 | 0 | 0 | 0 | 5 | 0 | 0 | 0 | 0 | 0 | 0 | 0 |
| DP | 0 | 0 | 0 | 0 | 0 | 0 | 0 | 0 | 0 | 0 | 0 | 10 | 0 | 0 | 0 | 0 | 0 | 0 |
| TC | 0 | 0 | 0 | 0 | 0 | 0 | 0 | 0 | 0 | 0 | 0 | 0 | 9 | 0 | 0 | 0 | 0 | 0 |
| TRP | 0 | 0 | 0 | 0 | 0 | 0 | 0 | 0 | 0 | 0 | 0 | 0 | 0 | 4 | 0 | 0 | 0 | 0 |
| PP | 0 | 1 | 0 | 0 | 0 | 0 | 1 | 0 | 0 | 0 | 0 | 0 | 0 | 0 | 3 | 0 | 0 | 0 |
| PPP | 0 | 0 | 0 | 1 | 0 | 0 | 0 | 1 | 0 | 0 | 0 | 0 | 0 | 0 | 0 | 8 | 0 | 0 |
| NAP | 0 | 0 | 0 | 0 | 0 | 0 | 0 | 0 | 0 | 0 | 0 | 0 | 0 | 0 | 0 | 0 | 9 | 0 |
| SAP | 0 | 0 | 0 | 0 | 0 | 0 | 0 | 0 | 0 | 0 | 0 | 0 | 1 | 0 | 0 | 0 | 0 | 3 |

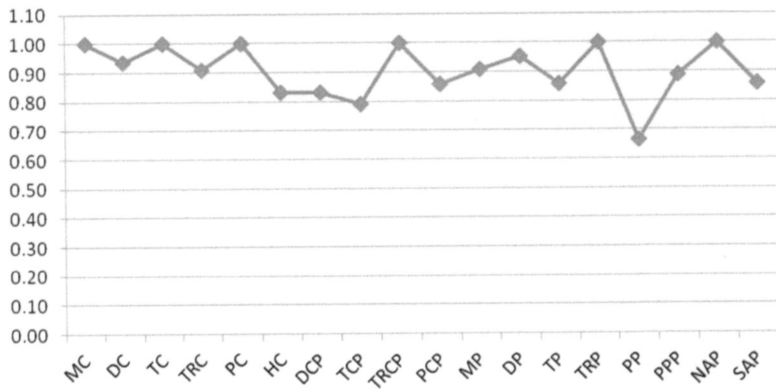


**Fig. 3** F-measure for 18 classes of pollen grain

# 4 Conclusion

The current work deals with the problem of pollen grain classification based on texture retrieval using different models like wavelet, Gabor, LBP, GLCM and GLDM with NN (Nearest Neighbor) classifier. As per the survey, earlier works deals with specific family of pollen grains for specific applications. This work mainly deals with the different classes of pollen grains irrespective of families. Classification of pollengrain using Gabor wavelet with NN Classifier gave better result compared to other models. We can extract other features like shape, contour along texture features using different feature extraction models for better features and use different classifiers other than Nearest Neighbor (NN) for further improvement in classification.

# References

1. Brummitt N, Bachman S (2010) Plants under pressure: a global assessment: first report of the IUCN sampled red list index for plants, Royal Botanic Gardens Kew, UK
2. Travieso MC, Briceno JC, Ticay-Rivas JR, Alonso JB (2011) Pollen classification based on contour features. In: Proceedings of 15th international conference on intelligent engineering systems, IEEE, Poprad, Slovakia
3. Shivanna KR (2003) Pollen biology and biotechnology—special, Indian edn. Oxford and IBH Publishing Co. Pvt. Ltd., New Delhi
4. Kashinath B, Majumdar MR, Bhattacharya SG (2006) A textbook of palynology—(Basic and Applied). New Central Book Agency (P) Ltd., Kolkata
5. Takhtajan AL (1980) Outline of the classification of flowering plants (Magnoiophyta). Bot Rev 46:225–359
6. Araujo A, Perroton L, Oliveira R, Claudino L, Guimaraes S, Bastos, E (2001) Non linear features extraction applied to pollen images. In: Proceedings of nonlinear image processing and pattern analysis, XII, SPIE vol 4303

7. Li P, Flenley JR (1999) Pollen texture identification using neural networks. Int J Grana 38:59–64, ISSN 0017-3134
8. Damian M, Cernadas E, Formilla A, Otero PM (2004) Pollen classification of three types of plants of the family Urticaceae. In: Proceedings of 12th Portuguese conference on pattern recognition, Aveiro
9. Zhang Y, Fountain DW, Hodgson RM, Flenly JR, Gunetileke S (2004) Towards automation of palynology 3: pollen pattern recognition using Gabor transforms and digital moments. J Quat Sci 19:763–768, ISSN 0627-8179
10. Fernandez-Delgado M, Carrion P, Cernadas E, Galvez JF, Otero PM (2003) Improved classification of pollen texture images using SVM and MLP. In: Proceedings of international conference on visualization, imaging, and image processing, vol 2, Benalmadena, ES
11. Gonzalez RC, Woods RE (2009) Digital image processing, 3rd edn. Pearson-Prentice Hall Indian edition. Dorling Kindersley India Pvt.Ltd, New Delhi
12. Manjunath BS, Ma WY (1996) Texture features for browsing and retrieval of image data. IEEE Trans Pattern Anal Mach Intell 18(8):837–842
13. Ojala T, Pietikainen M (2002) Multiresolution gray-scale and rotation invariant texture classification with local binary patterns. IEEE Trans Pattern Anal Mach Intell 24(7):971–987
14. Haralick RM, Shanmugam K, Dinstein I (1973) Textural Features for image classification. IEEE Trans Syst Man Cybern 3(6):610–621
15. Guru DS, Sharath YH, Manjunath S (2010) Texture features and KNN in classification of flower images, IJCA special issue on “recent trends in image processing and pattern recognition”, RTIPPR
16. Kim JK, Park HW (1999) Statistical textural features for detection of micro calcifications in digitized mammograms. IEEE Trans Med Imaging 18(3):231–238
17. Olvera HF, Soriano SF, Hernandez EM (2006) Pollen morphology and systematic of Atripliceae (Chenopodiaceae). Int J Grana 45(3):175–194
18. Harley MM, Paton A, Harley RM, Cade PG (1992) Pollen Morphological studies in tribe Ocimeae (Nepetoideae: Labiatae): I. Ocimum L. Int J Grana 31(3):161–176
19. Remizowa MV, Sokoloff DD, Macfarlane TD, Yadav SR, Prychid CJ, Rudall PJ (2008) Comparative pollen morphology in the early divergent angiosperm family Hydatellaceae reveals variation at the infraspecific level. Int J Grana 47(2):81–100
20. Erdtman G (1966) Pollen morphology and Plant taxonomy in Angiosperms. Hafner Publishing Company, New York and London

# Representation and Classification of Medicinal Plant Leaves: A Symbolic Approach

**Y. G. Naresh and H. S. Nagendraswamy**

**Abstract** In this paper, a method of representing medicinal plant leaves shape in terms of tri-interval valued type symbolic features is proposed. The proposed method exploits the axis of least inertia of a leaf shape to extract invariant features for its characterization. Concept of clustering is used to select class representatives for each medicinal plant species. Multiple class representatives are selected to incorporate intra-class variations in each species to facilitate the task of classification. To study the efficacy of the proposed representation and classification technique, experiments are conducted on Flavia data set. The results obtained are more encouraging and a comparative study with the results of other contemporary methods on the same data set is presented.

**Keywords** Symbolic representation · Shape classification · Shape matching · Axis of least inertia

## 1 Introduction

Ayurveda emphasizes on building better immunity in human body for many ailments. This traditional system of Indian medicine relies largely on Earth's flora. Botanists and the Practitioners of this medicinal system have a great thrive for capturing and unraveling the useful species in the flora. There are many virtual herbaria available online in digitized form. These herbaria provide huge information about the plant species.

Y. G. Naresh (✉) · H. S. Nagendraswamy
DoS in Computer Science, University of Mysore, Mysore, India
e-mail: naresh.yg@gmail.com

H. S. Nagendraswamy
e-mail: swamy_hsn@yahoo.com

P. P. Swamy and D. S. Guru (eds.), *Multimedia Processing, Communication and Computing Applications*, Lecture Notes in Electrical Engineering 213,
DOI: 10.1007/978-81-322-1143-3_8, © Springer India 2013 

Identifying a species from those huge collections is a challenging task unless the name of the species is known. Indeed, it is a challenging task even for an expert, specifically for a common man to have a complete knowledge on the taxonomy of all varieties of species. In this context, computer vision and machine learning algorithms allows identification of species through the algorithms for matching images.

Apart from biological methods to identify the plant species, the characteristics such as shape, texture, color, arrangement and internal vein structure of plant leaves can be used to identify plant species, the shape of the leaf is considered to be the most predominant aspect.

## 2 Related Work

Shape of a plant leaf plays a major role in identifying and classifying the plant species. In the literature, both contour-based and region based methods have been proposed to characterize the shape and experimented on leaf datasets.

In [1], the region based morphological features of a leaf viz., aspect ratio, rectangularity, area ratio of convex hull, perimeter ratio of convex hull, sphericity, circularity, eccentricity, form factor, and invariant moments are used for its description. The concept of physiological width and physiological length of a plant leaf along with other geometric and region based morphological features have been proposed in [2] for characterizing a plant leaf shape. In [3], landmark points are considered for polygon approximation. Given any two points in the set of landmark points, Inner-distance between those two points as a replacement to Euclidean distance is computed to build shape descriptors, for classification. In [4], shape curve is described by using a set of optimal samples, which are extracted using both geometric and topological features to capture symmetric characteristics and spatial relationships of shape structures. In [5], Contour based symbolic representation method has been proposed to describe the shape curve using string of symbols. In [6], a new descriptor called as contour points distribution histogram has been introduced to characterize the shape which is similar to that of shape context [7]. In [8], symbolic representation in terms of multi-interval valued type features has been proposed to characterize two dimensional shapes. In [9], Histogram of gradients is used as shape descriptors for characterizing leaf shape.

From the literature survey, it has been observed that though several methods have been proposed to effectively characterize a shape, the methods cannot be generalized in the sense that a method may show good performance for one class of datasets and may perform poor for another. Thus coming out with an efficient representation technique for characterizing a shape is still a research problem.

In this work, a method of representing shape of a plant leaf in terms of tri-interval valued symbolic features by adopting axis of least inertia of a shape as a unique reference. The data which contain internal variations but structured are referred to as symbolic data [10]. Symbolic data appears in the form of continuous ratio, discrete absolute interval and multi-valued [8]. Efficacy of the proposed method has been corroborated by conducting experiments on the standard leaf database used for the purpose of classification of plant species. The results are more encouraging and comparable to the contemporary works in this direction.

# 3 Proposed Method

## 3.1 Feature Extraction

Extracting relevant and discriminative features for characterizing a leaf shape is an important step in any recognition or classification task. In this section, a method of extracting tri-interval valued features adopting axis of least inertia of a shape is explained.

The color images of plant leaves are first converted into binary images. Then the contour of binary images is obtained by using a suitable contour extraction algorithm. The extracted closed boundary serves as a shape curve of a plant leaf image.

The proposed feature extraction technique is based on the axis of least inertia of a shape which preserves the orientation of a shape curve. It is very important to preserve the orientation of the shape curve to extract features which are invariant to geometric transformations (Rotation, translation and scaling). The details regarding the computation of the axis of least inertia of a shape curve can be found in [11].

Once the axis of least inertia of a shape is computed, all the points of shape boundary are projected onto the axis and the two farthest points are obtained. Figure 1 shows a shape with axis of least inertia and two extreme points. The Euclidean distance $D$ between these two points defines the length of axis of least inertia of a shape. The features are extracted by traversing the shape contour in clock wise direction keeping either $E_1$ or $E_2$ as the starting point. In order to identify this starting point, the distance between $E_1$and the shape centroid and the distance between $E2$ and the shape centroid are computed. The shortest distance among the two is considered as a starting point. In some cases, there is a possibility that these two distances may be same and leads to ambiguity in selecting the starting point. In such case, we resolve the conflict by considering the horizontal width of the shape at subsequent points on the axis starting from the two extreme points.

The two extreme points obtained may not correspond to the points on the shape boundary as shown in the Fig. 1. So, to obtain the starting boundary points for feature extraction, we traverse the axis from the two extreme points of the axis of least inertia till the shape boundary points are found. Now, the axis of least inertia is rotated about an angle $\theta_1$ and the next two extreme points on the shape boundary are obtained as described earlier. Let $C_{11}$ and $C_{12}$ be the two extreme boundary points of a shape before rotating the axis of least inertia. Let $C_{21}$ and $C_{22}$ be the two extreme boundary points of a shape after rotating the axis of least inertia by an angle $\theta_1$. The lengths of the curve segment say $l_1$and the Euclidean distance say $d_1$

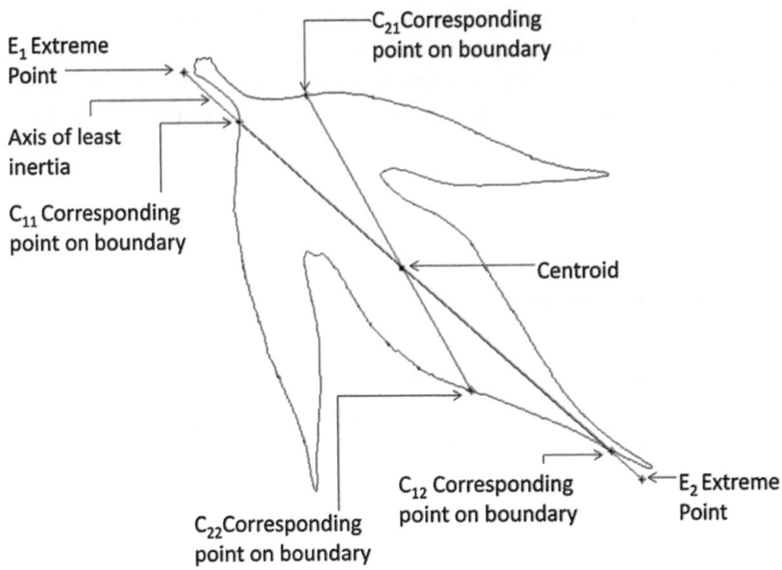


**Fig. 1** After first rotation of axis of least inertia by an angle $\theta_1$

between the boundary points $C_{11}$ and $C_{21}$ are computed and the ratio $R_1$ between $d_1$ and $l_1$ is computed and considered as a feature value. Similarly the lengths of the curve segment say $l_2$ and the Euclidean distance say $d_2$ between the boundary points $C_{12}$ and $C_{22}$ are computed and the ratio $R_2$ between $d_2$ and $l_2$ is computed and considered as another feature value. Also the Euclidean distance $L$ between the two extreme boundary points $C_{21}$ and $C_{22}$ is computed and considered as another feature value. To make the feature value ($L$) invariant to scaling, it is divided by the length of the axis of least inertia ($D$).

$$L = \text{ Euclidean distance between } C_{21} \text{ and } C_{22} \tag{1}$$

$$R_1 = d_1/l_1 \tag{2}$$

$$R_2 = d_2/l_2 \tag{3}$$

Thus the three feature values $L$, $R_1$, $R_2$ are extracted from the shape boundary at an angle $\theta_1$. The process is repeated by rotating the axis of least inertia in clockwise direction for various values of $\theta_1$ with equal intervals. The features thus extracted are used to characterize a shape (Fig. 2).

## 3.2 *Feature Representation*

Let [$LS_1$, $LS_2$, $LS_3$, …, $LS_n$] be the '*n*' number of training shapes in the class of a medicinal plant species $C_i$ where $i = \{1, 2, 3, \ldots n\}$. Since the shape of leaves in a

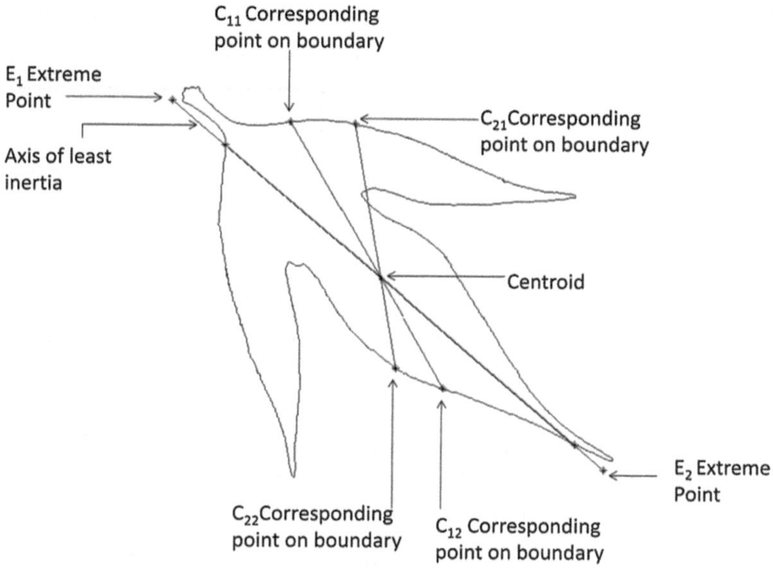


**Fig. 2** After second rotation of axis of least inertia by an angle $2\theta_1$

class may vary due to size, maturity and other biological facts, there will be significant intra-class variations among the shapes. To handle this case, we propose to have multiple representatives for a class by grouping similar shapes into one group and choose a representative for that group within the class. We have used hierarchical clustering technique to group similar shapes in a class by utilizing inbuilt Matlab function available in Matlab R2011a. The natural groups in a class are identified using the inconsistency coefficient for each link of the hierarchical cluster tree yielded by the respective linkage. A representative feature vector for a particular group within the class is obtained by aggregating the corresponding features of the shapes in the group to form an interval valued type feature. Thus, the feature vector representing a group within a class is of interval-valued type rather than crisp. Thus in the proposed approach, the shape of plant leaves in the knowledge base are represented in the form of tri-interval valued symbolic feature vector.

Lower the intra-class variations in the class $C_i$, fewer the number of groups obtained and higher the intra-class variations, more the number of groups are obtained. Hence the number of groups in a class $C_i$ is directly proportional to its intra class variations.

Let $[CL_1, CL_2, CL_3, \ldots, CL_m]$ be the '*m*' number of clusters formed in a class. Let $[S_1, S_2, S_3, \ldots, S_P]$ be the leaf shapes in the cluster $CL_r$ for $r = \{1, 2, 3,..,m\}$.

Each shape in $Si$ for $i = \{1, 2, 3, \ldots, P\}$ in a cluster $CL_r$ is represented by a tri-valued feature vector of dimension $K$ i.e., $Si = \{F_{i1}, F_{i2}, F_{i3}, \ldots, F_{iK}\}$ where $F_{ij} = [L_i\ R_{1j}\ R_{2j}]$ for $j = \{1, 2, 3, \ldots, K\}$. To form a cluster representative vector, we aggregate the corresponding features of the shape $S_i$ to form an interval by

**Table 1** Symbolic representation of a cluster representative using tri-interval valued feature vector

| Shapes in a cluster | 1 | 2 | 3 | … | … | K |
|---|---|---|---|---|---|---|
| $S_1$ | $F_{11}$ | $F_{12}$ | $F_{13}$ | … | … | $F_{1K}$ |
| $S_2$ | $F_{12}$ | $F_{22}$ | $F_{23}$ | … | … | $F_{2K}$ |
| $S_3$ | $F_{13}$ | $F_{32}$ | $F_{33}$ | … | … | $F_{3K}$ |
| .. | … | … | … | … | … | … |
| .. | … | … | … | … | … | … |
| $S_P$ | $F_{P1}$ | $F_{P2}$ | $F_{P3}$ | … | … | $F_{PK}$ |
| Interval form of cluster | $\{[L_1^-, L_1^+]\}$ $\{[R_{11}^-, R_{11}^+]\}$ $\{[R_{21}^-, R_{21}^+]\}$ | $\{[L_2^-, L_2^+]\}$ $\{[R_{12}^-, R_{12}^+]\}$ $\{[R_{22}^-, R_{22}^+]\}$ | $\{[L_3^-, L_3^+]\}$ $\{[R_{13}^-, R_{13}^+]\}$ $\{[R_{23}^-, R_{23}^+]\}$ | … | … | $\{[L_k^-, L_k^+]\}$ $\{[R_{1k}^-, R_{1k}^+]\}$ $\{[R_{2k}^-, R_{2k}^+]\}$ |

choosing the minimum and maximum feature values. Thus, the symbolic feature vector of type tri-interval valued is used to represent the model shape in the knowledge base as shown in the Table 1.

$$K = 180/(\theta_1) \tag{4}$$

Table 2 presents an example of class representative described by twelve tri-interval valued type features with 15° equal intervals of rotation.

At the time of classification, we will encounter only one instance of a leaf shape belonging to a particular class. Thus the test leaf shape is represented in the form of tri-valued crisp type data. For an example, the features for a test leaf shape with 15° of rotation of Axis of least inertia with equal intervals for every increment yields nine tri-valued features comprising of crisp values. The sequence of features representing an example leaf shape is shown in the Table 3.

## 3.3 Matching and Classification

Let $Q_S = \{[L_i\, R_{1i}\, R_{2i}\,]\}$ where $i = \{1, 2, 3, \ldots, k\}$ be the $k$-dimensional tri-valued feature vector representing the shape of a plant species to be classified. Let the model shape representative $M_S = \{[L_i^-, L_i^+]\ [R_{1i}^-, R_{1i}^+]\ [R_{2i}^-, R_{2i}^+]\}$ for $i = \{1, 2, 3, \ldots, k\}$ be the $K$-dimension tri-valued feature vector representing model class representative in the knowledge base of a particular plant species.

We use the similarity measure proposed in [8] to find the similarity score between the test shape and the class representatives as

$$\mathrm{Sim}(Q_S, M_S) = \frac{1}{K}\sum_{i=1}^{K} SL_i + SR_{1i} + SR_{2i} \tag{5}$$

**Table 2** An example of a class representative in tri-interval valued form

| Feature | Rotation | L | $R_1$ | $R_2$ |
|---|---|---|---|---|
| 1 | $\theta + 15°$ | [0.5318, 0.5811] | [0.9957, 0.9997] | [0.9143, 0.9203] |
| 2 | $\theta + 30°$ | [0.4500, 0.4879] | [0.9931, 0.9994] | [0.9121, 0.9426] |
| 3 | $\theta + 45°$ | [0.3747, 0.4190] | [0.9624, 0.9688] | [0.9200, 0.9923] |
| 4 | $\theta + 60°$ | [0.3446, 0.3797] | [0.9179, 0.9838] | [0.9091, 0.9153] |
| 5 | $\theta + 75°$ | [0.5621, 0.6063] | [0.9332, 0.9845] | [0.3772, 0.3947] |
| 6 | $\theta + 90°$ | [0.5460, 0.5529] | [0.9045, 0.9397] | [0.9705, 0.9907] |
| 7 | $\theta + 105°$ | [0.5608, 0.5884] | [0.2858, 0.5890] | [0.9067, 0.9919] |
| 8 | $\theta + 120°$ | [0.3469, 0.3757] | [0.8581, 0.9565] | [0.9558, 0.9996] |
| 9 | $\theta + 135°$ | [0.3942, 0.4037] | [0.8786, 0.9261] | [0.9737, 0.9803] |
| 10 | $\theta + 150°$ | [0.4383, 0.4832] | [0.9063, 0.9823] | [0.9886, 0.9939] |
| 11 | $\theta + 165°$ | [0.5299, 0.5854] | [0.3198, 0.6927] | [0.9756, 0.9812] |
| 12 | $\theta + 180°$ | [0.6893, 0.8099] | [0.3024, 0.9825] | [0.2355, 0.9810] |

**Table 3** An example tri-valued crisp features representing a test leaf shape

| Feature | Rotation |  |
|---|---|---|
| 1 | $\theta + 15°$ | [0.5811, 0.9997, 0.9203] |
| 2 | $\theta + 30°$ | [0.4879, 0.9994, 0.9426] |
| 3 | $\theta + 45°$ | [0.4190, 0.9688, 0.9923] |
| 4 | $\theta + 60°$ | [0.3797, 0.9838, 0.9153] |
| 5 | $\theta + 75°$ | [0.6063, 0.9845, 0.3947] |
| 6 | $\theta + 90°$ | [0.5529, 0.9397, 0.9907] |
| 7 | $\theta + 105°$ | [0.5884, 0.5890, 0.9919] |
| 8 | $\theta + 120°$ | [0.3757, 0.9565, 0.9996] |
| 9 | $\theta + 135°$ | [0.4037, 0.9261, 0.9803] |
| 10 | $\theta + 150°$ | [0.4832, 0.9823, 0.9939] |
| 11 | $\theta + 165°$ | [0.5854, 0.6927, 0.9812] |
| 12 | $\theta + 180°$ | [0.8099, 0.9825, 0.9810] |

where

$$SL_i = \left\{ \begin{array}{l} 1 \quad \text{if}(L_i^- \leq L_i \leq L_i^+) \\ \text{else} \\ \frac{1}{2}\left[\frac{1}{1+abs(L_i^- - L_i)} + \frac{1}{1+abs(L_i^+ - L_i)}\right] \end{array} \right\}$$

$$SR_{1i} = \left\{ \begin{array}{l} 1 \quad \text{if}(R_{1i}^- \leq R_{1i} \leq R_{1i}^+) \\ \text{else} \\ \frac{1}{2}\left[\frac{1}{1+abs(R_{1i}^- - R_{1i})} + \frac{1}{1+abs(R_{1i}^+ - R_{1i})}\right] \end{array} \right\}$$

$$SR_{2i} = \left\{ \begin{array}{l} 1 \quad \text{if}(R_{2i}^- \leq R_{2i} \leq R_{2i}^+) \\ \text{else} \\ \frac{1}{2}\left[\frac{1}{1+abs(R_{2i}^- - R_{2i})} + \frac{1}{1+abs(R_{2i}^+ - R_{2i})}\right] \end{array} \right\}$$

From the Eq. (5), it is evident that the similarity is 1 when $\{[L_i\ R_{1i}\ R_{2i}]\}$ lies between the interval $\{[L_i^-,\ L_i^+]\ [\ R_{1i}^-,\ R_{1i}^+]\ [\ R_{2i}^-,\ R_{2i}^+]\}$. The test leaf shape is compared with all the class representatives of the plant species in the knowledge base. The label of class representative which possess the highest similarity value is assigned to the test leaf shape.

## 4 Experimental Results

We have implemented our method using Matlab R2011a. We have conducted experiments on the data set provided by Wu et al. [2]. All the plant species in the dataset have medicinal values. Thus we have called them as medicinal plant species. The dataset consists of 32 plant species each with more than 40 samples. We considered 30 plant species for our experiments with 30 samples randomly from each species to create knowledge base as discussed earlier and remaining samples are considered for the purpose of testing.

The Fig. 3 the samples of the model plant leaves used in the experiments.

We have extracted 12 tri-valued features from each shape as discussed earlier by rotating the axis of least inertia of a shape with 15° of equal intervals. Once the features are extracted for all the 30 leaf images of each class in the dataset, the class representatives for every class are obtained and stored in the knowledge base as explained in the Sect. 3.2. For our experiments, the complete linkage is chosen empirically for finding the compact clusters in every class of leaf shape.

The performance of any classification system is measured in terms of its accuracy, precision, recall and f-measure and are defined as follows

$$\text{Accuracy} = (\text{TP} + \text{TN}) / \text{N} \tag{6}$$

$$\text{Precision} = \text{TP} / (\text{TP} + \text{FP}) \tag{7}$$

**Fig. 3** Samples of the model plant leaves in the data set

$$\text{Recall} = \text{TP} / (\text{TP} + \text{TN}) \quad (8)$$

$$\text{F-measure} = (2^{*}\text{precision}^{*}\text{Recall}) / (\text{Precision} + \text{Recall}) \quad (9)$$

These measures are defined on the basis of true positive (TP), true negative (TN), false positive (FP) and false negative (FN) for the overall test samples N.

In our experiments, these factors are evaluated for every class. The class wise performance of the proposed scheme is as shown in the Table 4. From Table 4 it has been observed that for most of the classes, all the performance measures are found to be greater than 95 %. But the recall and the f-measure for the class index '21' is less when compared to the other classes. From Fig. 4, we can observe that both the species belonging to class indices '1' and '21' have similar shape structures.

As we have conducted experiments on the dataset used by Wu et al. [2], we have compared the performance of the proposed method with that of [2]. Table 5 shows the average performance of the proposed methodology and the work reported in [2].

Figures 5, 6, 7, 8 show the graphs of the respective performance measures for all the class.

**Table 4** Class wise performance of the proposed scheme

| Class Label | TP | FP | FN | TN | Accuracy | Precision | Recall | F-Measure |
|---|---|---|---|---|---|---|---|---|
| 1 | 10 | 4 | 0 | 286 | 0.9867 | 0.7143 | 1.0000 | 0.8333 |
| 2 | 9 | 1 | 1 | 289 | 0.9933 | 0.9000 | 0.9000 | 0.9000 |
| 3 | 10 | 0 | 0 | 290 | 1.0000 | 1.0000 | 1.0000 | 1.0000 |
| 4 | 10 | 2 | 0 | 288 | 0.9933 | 0.8333 | 1.0000 | 0.9091 |
| 5 | 9 | 0 | 1 | 290 | 0.9967 | 1.0000 | 0.9000 | 0.9474 |
| 6 | 10 | 1 | 0 | 289 | 0.9967 | 0.9091 | 1.0000 | 0.9524 |
| 7 | 9 | 0 | 1 | 290 | 0.9967 | 1.0000 | 0.9000 | 0.9474 |
| 8 | 9 | 0 | 1 | 290 | 0.9967 | 1.0000 | 0.9000 | 0.9474 |
| 9 | 9 | 3 | 1 | 287 | 0.9867 | 0.7500 | 0.9000 | 0.8182 |
| 10 | 10 | 0 | 0 | 290 | 1.0000 | 1.0000 | 1.0000 | 1.0000 |
| 11 | 9 | 2 | 1 | 288 | 0.9900 | 0.8182 | 0.9000 | 0.8571 |
| 12 | 9 | 0 | 1 | 290 | 0.9967 | 1.0000 | 0.9000 | 0.9474 |
| 13 | 9 | 1 | 1 | 289 | 0.9933 | 0.9000 | 0.9000 | 0.9000 |
| 14 | 10 | 1 | 0 | 289 | 0.9967 | 0.9091 | 1.0000 | 0.9524 |
| 15 | 9 | 0 | 1 | 290 | 0.9967 | 1.0000 | 0.9000 | 0.9474 |
| 16 | 10 | 2 | 0 | 288 | 0.9933 | 0.8333 | 1.0000 | 0.9091 |
| 17 | 9 | 0 | 1 | 290 | 0.9967 | 1.0000 | 0.9000 | 0.9474 |
| 18 | 9 | 0 | 1 | 290 | 0.9967 | 1.0000 | 0.9000 | 0.9474 |
| 19 | 9 | 0 | 1 | 290 | 0.9967 | 1.0000 | 0.9000 | 0.9474 |
| 20 | 10 | 0 | 0 | 290 | 1.0000 | 1.0000 | 1.0000 | 1.0000 |
| 21 | 6 | 0 | 4 | 290 | 0.9867 | 1.0000 | 0.6000 | 0.7500 |
| 22 | 10 | 0 | 0 | 290 | 1.0000 | 1.0000 | 1.0000 | 1.0000 |
| 23 | 10 | 0 | 0 | 290 | 1.0000 | 1.0000 | 1.0000 | 1.0000 |
| 24 | 10 | 2 | 0 | 288 | 0.9933 | 0.8333 | 1.0000 | 0.9091 |
| 25 | 9 | 0 | 1 | 290 | 0.9967 | 1.0000 | 0.9000 | 0.9474 |
| 26 | 8 | 0 | 2 | 290 | 0.9933 | 1.0000 | 0.8000 | 0.8889 |
| 27 | 10 | 2 | 0 | 288 | 0.9933 | 0.8333 | 1.0000 | 0.9091 |
| 28 | 10 | 0 | 0 | 290 | 1.0000 | 1.0000 | 1.0000 | 1.0000 |
| 29 | 9 | 0 | 1 | 290 | 0.9967 | 1.0000 | 0.9000 | 0.9474 |
| 30 | 9 | 0 | 1 | 290 | 0.9967 | 1.0000 | 0.9000 | 0.9474 |

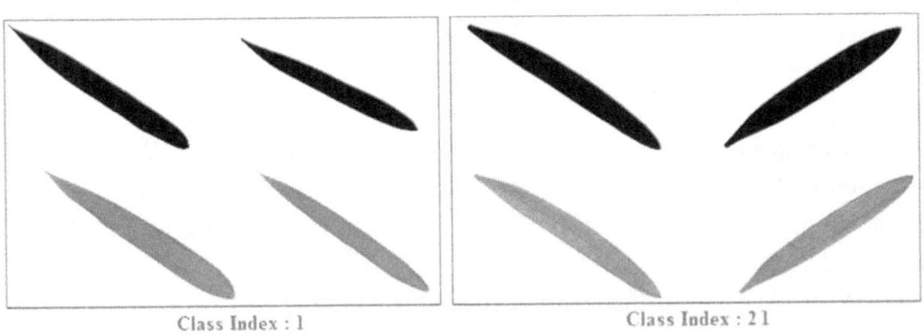


**Fig. 4** Two different Species with similar shape structure

**Table 5** Comparison for the method proposed in [2]

| Scheme | Avg. classification performance (%) |
|---|---|
| Proposed scheme | 93 |
| Stephen et al. [2] | 90.13 |

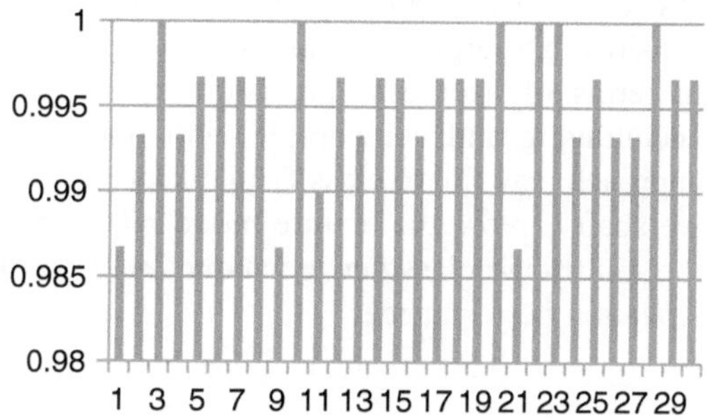


**Fig. 5** Class wise Accuracy obtained with the proposed scheme

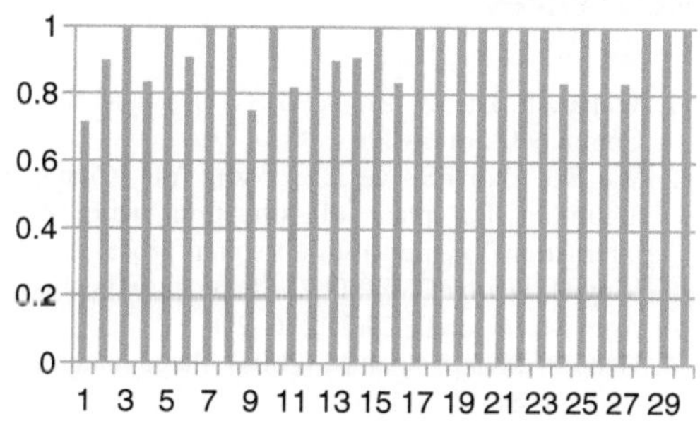


**Fig. 6** Class wise Precision obtained with the proposed scheme

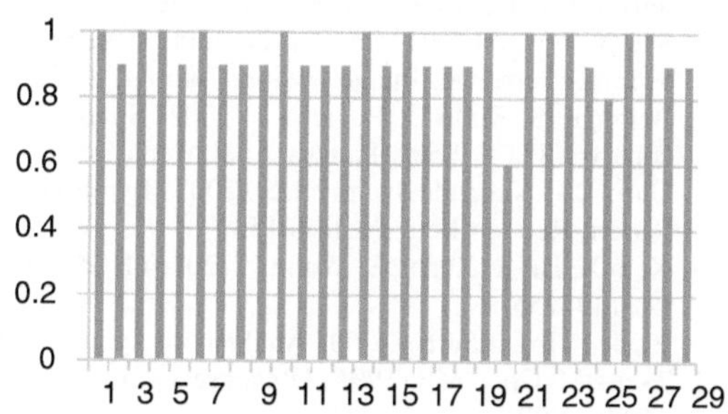


**Fig. 7** Class wise Recall obtained with the proposed scheme

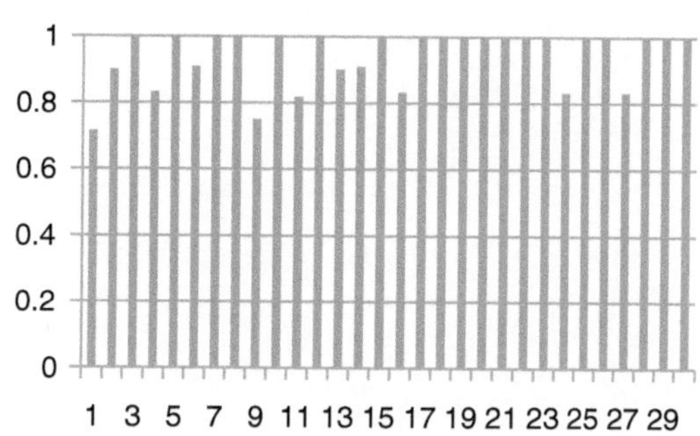


**Fig. 8** Class wise F-Measure obtained with the proposed scheme

## 5 Conclusion

In this work, we proposed novel method of representing medicinal plant leaves for classification. The method exploits the concept of symbolic data analysis for effective representation. Experiments are conducted on the standard database of considerably large size and the performance of the proposed method is evaluated in terms of accuracy, precision, recall and F-measure. The results are more encouraging and comparable with that of state of the art work. However the approach may fail to classify plant species if their leaf shape structure is same. So, we shall explore the texture based representation techniques in such cases. Also, the multistage classifier techniques will be explored to further improve the classification performance.

## References

1. Wang XF, Du JX, Zhang GJ (2007) Leaf shape based plant species recognition. Appl Math Comput, 185:883–893
2. Wu SG, Bao FS, Xu EY, Wang YU, Chang Y-F, Shiang C-L (2007) A leaf recognition algorithm for plant classification using probabilistic neural network. In: IEEE 7th international symposium on signal processing and information technology, Cario, Egypt
3. Ling H, Jacobs DW (2007) Shape classification using the inner distance. IEEE Trans Pattern Anal Mach Intell, 29(2):286–299
4. Heng PA, Xie J, Shah M (2008) Shape matching and modeling using skeletal context. Pattern Recogn, 41:1756–1767
5. MDaliri MR, Torre V (2008) Robust symbolic representation for shape recognition and retrieval. Pattern Recogn, 41:1782–1798
6. Shu X, Wu X-J (2011) A novel contour descriptor for 2D shape matching and its application to image retrieval. Image Vision Comput, 29:286–294
7. Belongie S, Malik J, Puzicha J (2002) Shape matching and object recognition using shape contexts. IEEE Trans Pattern Anal Mach Intell 24(4):509–522
8. Guru DS, Nagendraswamy HS (2007) Symbolic representation of two dimensional shapes. Pattern Recogn Lett, 28:144–155
9. Xiao X, Hu R, Zhang S, Wang X (2010) HOG-based approach for leaf classification. In ICIC (2) 149–155
10. Bock HH, Diday E (2000) Analysis of symbolic data. Springer-Verlag
11. Tsai DM, Chen MF (1995) Object recognition by linear weight classifier. Pattern Recogn Lett 16:591–600

# Linear Discriminant Analysis for 3D Face Recognition Using Radon Transform

**P. S. Hiremath and Manjunath Hiremath**

**Abstract** Face recognition research started in the 70s and a number of algorithms/ systems have been developed in the last decade. Three Dimensional (3D) human face recognition is emerging as a significant biometric technology. Research interest into 3D face recognition has increased during recent years due to the availability of improved 3D acquisition devices and processing algorithms. Three Dimensional face recognition also helps to resolve some of the issues associated with two dimensional (2D) face recognition. Since 2D systems employ intensity images, their performance is reported to degrade significantly with variations in facial pose and ambient illumination. The 3D face recognition systems, on the other hand, have been reported to be less sensitive to the changes in the ambient illumination during image capture that the 2D systems. In the previous works, there are several methods for face recognition using range images that are limited to the data acquisition and preprocessing stage only. In the present paper, we have proposed a 3D face recognition algorithm which is based on Radon transform, principal component analysis (PCA) and linear discriminant analysis (LDA). The radon transform (RT) is a fundamental tool to normalize 3D range data. The PCA is used to reduce the dimensionality of feature space, and the LDA is used to optimize the features, which are finally used to recognize the faces. The experimentation has been done using Texas 3D face database. The experimental results show that the proposed algorithm is efficient in terms of accuracy and detection time, in comparison with other methods based on PCA only and RT + PCA. It is observed that 40 eigenfaces of PCA and 5 LDA components lead to an average recognition rate of 99.16 %.

---

P. S. Hiremath (✉) · M. Hiremath
Department of Computer Science, Gulbarga University, Gulbarga 585106 KA, India
e-mail: hiremathps53@yahoo.com

M. Hiremath
e-mail: manju.gmtl@gmail.com

P. P. Swamy and D. S. Guru (eds.), *Multimedia Processing, Communication and Computing Applications*, Lecture Notes in Electrical Engineering 213,
DOI: 10.1007/978-81-322-1143-3_9, 

**Keywords** 3D face recognition · Range images · Radon transform · Principal component analysis · Linear discriminant analysis

## 1 Introduction

Face Recognition and verification have been at the top of the research agenda of the computer vision community for more than a decade. The scientific interest in this research topic has been motivated by several factors. The main attractor is the inherent challenge that the problem of face image processing, face detection and recognition. However, the impetus for better understanding of the issues raised by automatic face recognition is also fuelled by the immense commercial significance that robust and reliable face recognition technology would entail. Its applications are envisaged in physical and logical access control, security, man–machine interfaces and low bitrate communication.

To date, most of the research efforts, as well as commercial developments, have focused on two dimensional (2D) approaches. This focus on monocular imaging has partly been motivated by costs but to a certain extent also by the need to retrieve faces from existing 2D image and video database. With recent advances in image capture techniques and devices, various types of face-image data have been utilized and various algorithms have been developed for each type of image data. Among various types of face images, a 2D intensity image has been the most popular and common image data used for face recognition because it is easy to acquire and utilize. It, however, has the intrinsic problem that it is vulnerable to the change of illumination. Sometimes the change of illumination gives more difference than the change of people, which severely degrades the recognition performance. Therefore, illumination-controlled images are required to avoid such an undesirable situation when 2D intensity images are used. To overcome the limitation of 2D intensity images, Three Dimensional (3D) images are being used, such as 3D meshes and range images. A 3D mesh image is the best 2D representation of 3D objects. It contains 3D structural information of the surface as well as the intensity information of each point. By utilizing the 3D structural information, the problem of vulnerability to the change of illumination can be solved. A 3D mesh image is suitable image data for face recognition, but it is complex and difficult to handle.

A range image is simply an image with depth information. In other words, a range image is an array of numbers where the numbers quantify the distances from the focal plane of the sensor to the surfaces of objects within the field of view along rays emanating from a regularly spaced grid. Range images have some advantages over 2D intensity images and 3D mesh images. First, range images are robust to the change of illumination and color because the value on each point represents the depth value which does not depend on illumination or color. Also, range images are simple representations of 3D information. The 3D information in 3D mesh images is useful in face recognition, but it is difficult to handle. Different from 3D mesh images, it is easy to utilize the 3D information of range images because the 3D information of each point is explicit on a regularly spaced grid. Due to these advantages, range images are very promising in face recognition.

The majority of the 3D face recognition studies have focused on developing holistic statistical techniques based on the appearance of face range images or on techniques that employ 3D surface matching. A survey of literature on the research work focusing on various potential problems and challenges in the 3D face recognition can be found in the survey [1–5]. Gupta et al. [6] presented a novel anthropometric 3D face recognition algorithm. This approach employs 3D Euclidean and Geodesic distances between 10 automatically located anthropometric facial fiducial points and a linear discriminant classifier with 96.8 % recognition rate. Lu et al. [7] constructed many 3D models as registered templates, then they matched 2.5D images (original 3D data) to these models using iterative closest point (ICP). Chang et al. [8] describe a "multi-region" approach to 3D face recognition. It is a type of classifier ensemble approach in which multiple overlapping sub regions around the nose are independently matched using ICP and the results of the 3D matching are fused. Jahanbim et al. [9] presented an approach of verification system based on Gabor features extracted from range images. In this approach, multiple landmarks (fiducials) on face are automatically detected, and also the Gabor features on all fiducials are concatenated, to form a feature vector to collect all the face features. Hiremath et al. [10] have discussed the 3D face recognition by using Radon Transform and PCA with recognition accuracy of 95.30 %. Tang et al. [11] presented a 3D face recognition algorithm based on sparse representation. In this method they used geometrical features, namely, triangle area, triangle normal and geodesic distance.

In this proposed method, our objective is to propose Discriminant Analysis method for face recognition based on Radon Transformation, principal component analysis (PCA) and linear discriminant analysis (LDA) which are applied on 3D facial range images. The experimentation is done using the Texas 3D face database [12].

## 2 Materials and Methods

For experimentation, we consider the Texas 3D Face Database [12]. The 3D models in the Texas 3D Face recognition Database were acquired using an MU-2 stereo imaging system. All subjects were requested to stand at a known distance from the camera system. The stereo system was calibrated against a target image containing a known pattern of dots on a white background. The database contains 1,149 3D models of 118 adult human subjects. The number of images of each subject varies from 2 per subject to 89 per subject. The subjects age ranges from minimum 22 to maximum 77 years. The database includes images of both males and females from the major ethnic groups of Caucasians, Africans, Asians, East-Indians, and Hispanics. The facial expressions present are smiling or talking faces with open/closed mouths and/or closed eyes. The neutral faces are emotionless.

# 3 Proposed Method

## 3.1 Radon Transform

The Radon transform (RT) is a fundamental tool in many areas. The 3D radon Transform is defined using 1D projections of a 3D object $f(\mathrm{x,y,z})$ where these projections are obtained by integrating $f(\mathrm{x,y,z})$on a plane, whose orientation can be described by a unit vector $\vec{\alpha}$. Geometrically, the continuous 3D Radon transform maps a function $\mathbb{R}^3$ into the set of its plane integrals in $\mathbb{R}^3$. Given a 3D function $f(\vec{x}) \triangleq f(x,y,z)$ and a plane whose representation is given using the normal $\vec{\alpha}$ and the distance s of the plane from the origin, the 3D continuous Radon Transform of '*f*' for this plane is defined by

$$\begin{aligned}\Re f(\vec{a},s) &= \int_{-\infty}^{\infty}\int_{-\infty}^{\infty}\int_{-\infty}^{\infty} f(\vec{x})\,\delta(\vec{x}^T\alpha - s)\,d\vec{x} \\ &= \int_{-\infty}^{\infty}\int_{-\infty}^{\infty}\int_{-\infty}^{\infty} f(x,y,z)\,\delta(x\sin\theta\cos\phi + y\sin\theta\sin\phi + z\cos\theta - s)\,dxdydz\end{aligned}$$

where $\vec{x}$ = [x, y, z]T, $\vec{\alpha}$ = [$\sin\theta$ $\cos\phi$, $\sin\theta$ $\sin\phi$, $\cos\theta$] T, and $\delta$ is Dirac's delta function defined by $\delta(x) = 0,\ x \neq 0,\ \int_{-\infty}^{\infty}\delta(x)\,dx = 1$. The Radon transform maps the spatial domain (x, y, z) to the domain ($\vec{\alpha}$, $s$) are not the polar coordinates of (x, y, z). The 3D continuous Radon Transform satisfies the 3D Fourier slice theorem [10, 13].

## 3.2 Linear Discriminant Analysis

The principal component analysis (PCA) is a standard technique used to approximate the original data with lower dimensional feature vectors [14, 15]. The basic approach is to compute the eigenvectors of the covariance matrix, and approximate the original data by a linear combination of the leading eigenvectors. The mean square error (MSE) in reconstruction is equal to the sum of the remaining eigenvalues. The coefficients of projection of an arbitrary data vector along the principal components (eigenvectors) form the feature vector. In PCA, since no class membership information is used, the data vectors of the same class and of different classes are treated similarly. In the linear discriminant analysis (LDA), the class membership information is used to emphasize the variation of data vectors belonging to different classes and to deemphasize the variations of data vectors within a class. The LDA produces an optimal linear discriminant function $f(x) = W^T x$ which maps the input into the classification space in which the class identification of this sample is decided based on some metric such as

Euclidian distance [16, 17]. A typical LDA implementation is carried out via scatter matrices analysis. The within and between-class scatter matrices as follows:

$$S_w = \frac{1}{M}\sum_{i=1}^{M} \Pr(C_i)\sum_i$$

$$S_b = \frac{1}{M}\sum_{i=1}^{M} \Pr(C_i)(m_i - m)(m_i - m)^T$$

Here $S_w$ is the within-class scatter matrix showing the average scatter $\sum_i$ of the sample vectors x of different class $C_i$ around their respective mean $m_i$:

$$\sum_i = E\left[(x - m_i)\ (x - m_i)^T | C = C_i\right]$$

Similarly, $S_b$ is the between-class scatter matrix, representing the deviation of the conditional mean vectors $m_i$'s from the overall mean vector m. Various measures are available for quantifying the discriminatory power, the commonly used one being,

$$J(W) = \frac{\|W^T S_w W\|}{\|W^T S_b W\|},$$

where $W$ is the optimal discrimination projection and can be obtained via solving the generalized eigenvalues problem $S_b W = \lambda S_w W$. The distance measure used in the matching could be a simple Euclidian distance. Thus, the fundamental difference between the PCA and LDA approaches is that, while PCA performs eigenvalues analysis an covariance matrix, the LDA does it on scatter matrices.

## 3.3 Proposed Methodology

The Radon transform is applied to an input facial range image $I_1$ in steps of h from 0° to 180° orientations; where h may be 1°–3° or any convenient value. It yields a binary image $I_2$ with facial area being segmented. After superposing $I_2$ and $I_1$, the cropped facial range image $I_3$ is obtained. Next, the principal component analysis (PCA) technique is applied to the complete set of such cropped facial range images corresponding to the face images in the face database. It yields the set of eigenfaces. After yielding the eigenfaces we perform linear discriminant analysis (LDA) to these eigenfaces which are used for face recognition in a given test face range image. The Figs. 1 and 2 shows the overview of proposed framework and intermediate results of the Radon transformation of an input face image respectively. The algorithms of the training phase and the testing phase of the proposed method are given below:

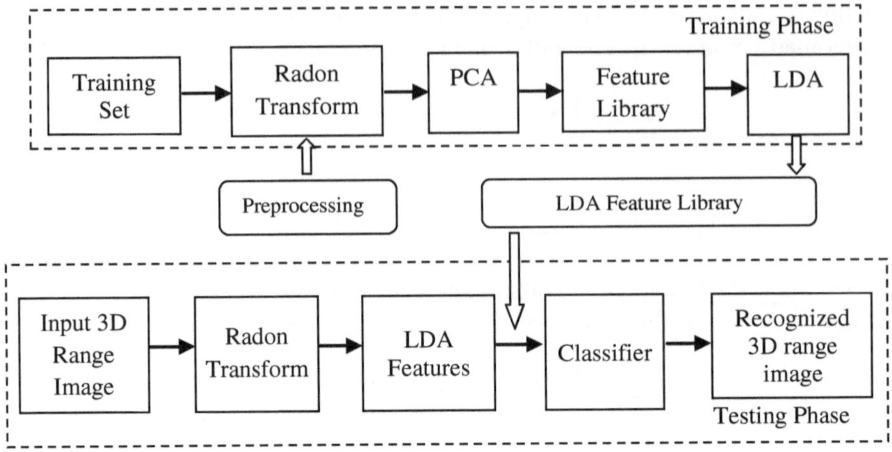


**Fig. 1** Overview of proposed framework

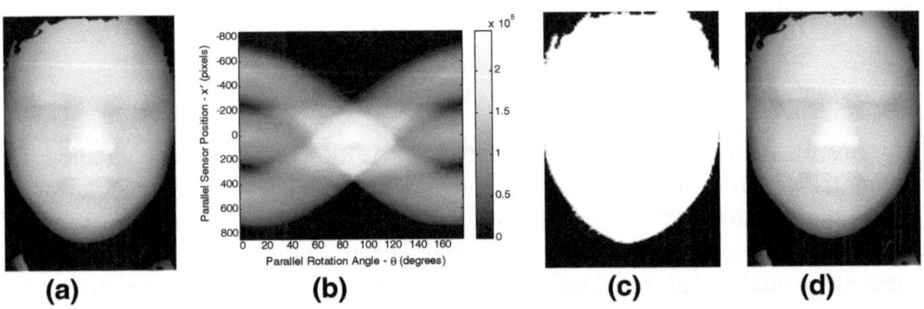


**Fig. 2** **a** The range image $I_1$. **b** Radon transform of $I_1$ in 0°–180° orientation. **c** Binary image $I_2$ obtained after radon transformation. **d** Cropped facial range image $I_3$ after superposing $I_2$ and $I_1$

Algorithm 1: Training Phase

1. Input the range image $I_1$ from the training set containing M images (Fig. 2a).
2. Apply Radon transform, from 0° to 180° orientations (in steps of $h$), to the input range image $I_1$ yielding a binary image $I_2$ (Fig. 2c).
3. Superpose the binary image $I_2$ obtained in the Step 2 on the input range image $I_1$ to obtain the cropped facial range image $I_3$ (Fig. 2d).
4. Repeat the Steps 1–3 for all the M facial range images in the training set.
5. Apply PCA to the set of cropped facial range images obtained in the Step 4 and obtain M eigenfaces.
6. Compute the weights $w_1, w_2, \ldots, w_p$ for each training face image, where $p < M$ is the dimension of Eigen subspace on which the training face image is projected.
7. Store the weights $w_1, w_2, \ldots, w_p$ for each training image as its facial features in the PCA feature library of the face database.

8. Perform LDA on the feature subspace (i.e., weight vectors).
9. Store the LDA components (feature vectors) in the LDA feature library of the face database.

   Algorithm 2: Testing Phase

1. Input the test range image $Z_1$.
2. Apply Radon transform, from 0° to 180° orientations (in steps of h), to the input range image $Z_1$ yielding a binary image $Z_2$.
3. Superimpose the binary image $Z_2$ on $Z_1$ to obtain the cropped facial image $Z_3$.
4. Compute the weights $w_i^{test}, i = 1, 2, \ldots, p$, for the test image $Z_1$ by projecting the test image on the LDA feature subspace of dimension $p$.
5. Compute the Euclidian distance D between the feature vector $w_i^{test}$ and the feature vectors stored in the LDA feature library.
6. The face image in the face database, corresponding to the minimum distance D computed in the Step 5, is the recognized face.
7. Output the texture face image corresponding to the recognized facial range image of the Step 6.

## 4 Results and Discussion

For experimentation, we consider the Texas 3D face database [12]. The proposed method is implemented using Intel Core 2 Quad processor @ 2.66 GHz machine and MATLAB 7.9. In the training phase, 2 frontal face images with neutral expression of each 100 subjects are selected as training data set. In the testing phase, randomly chosen 200 face images of the Texas 3D face database with variations in facial expressions are used. The sample training images which are used for our experimentation are shown in the Fig. 3, and their corresponding texture images are shown in the Fig. 4. The eigenfaces and mean facial range image computed for PCA during the training phase are shown in the Figs. 5 and 6, respectively.

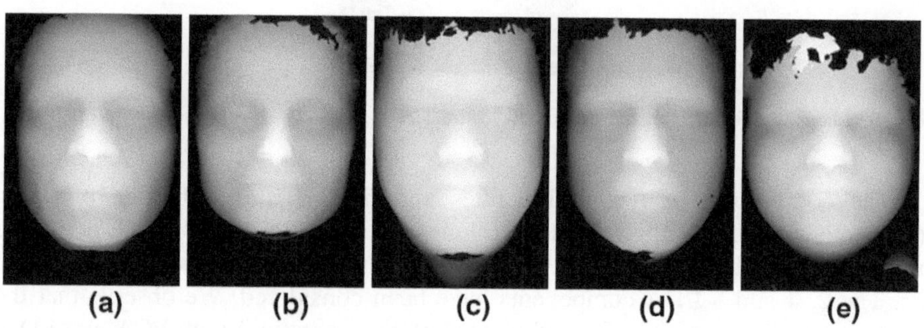


**Fig. 3** The first five range images of the training dataset

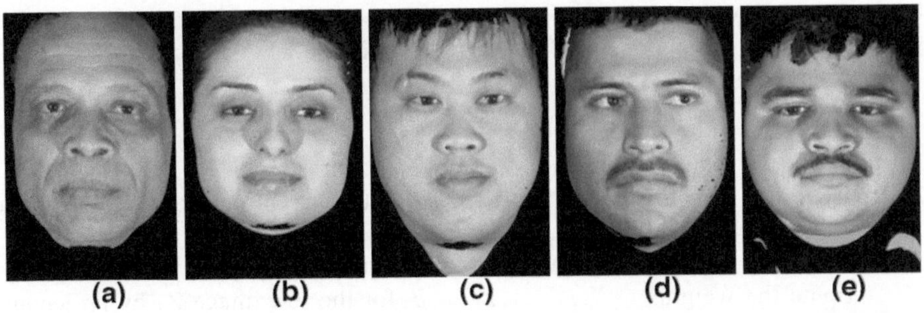


**Fig. 4** The facial texture images corresponding to the training range images of the Fig. 3

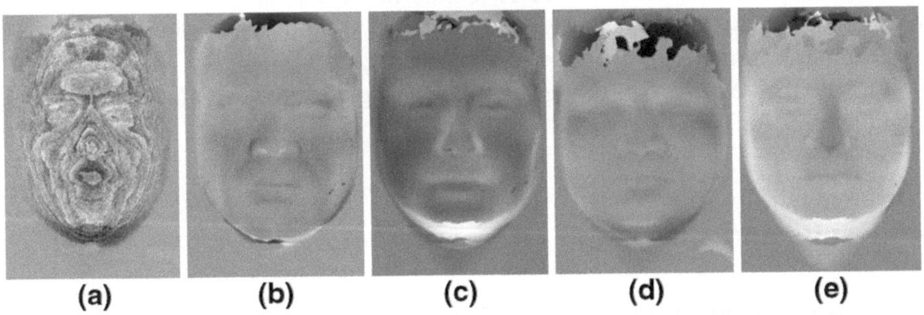


**Fig. 5** The first five eigenfaces obtained by using the PCA in the training phase

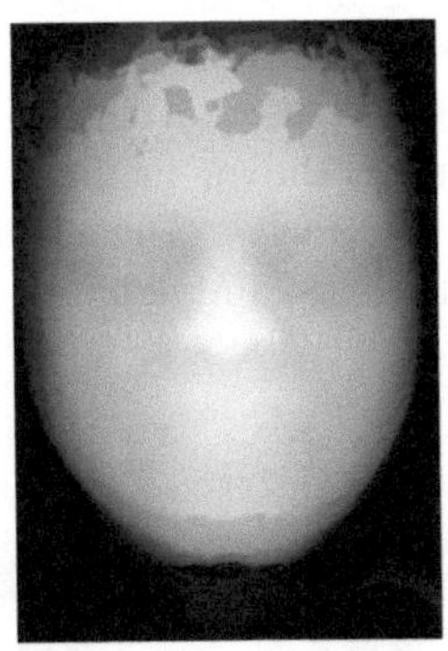

**Fig. 6** Mean facial range image computed in the Step 5 for PCA during the training phase

The comparison of recognition rates and times obtained by the proposed (RT + PCA + LDA) approach, PCA (alone) and RT + PCA approach is presented in the Table 1. The projection orientation of Radon transform is in steps of 1°, 2° and 5°. The 2, 4 and 5 LDA components have been considered. We observe that the proposed method, namely, RT (with steps of 1° orientation) with PCA and LDA, yields better results as compared to the PCA (alone) and RT + PCA method.

**Table 1** The face recognition accuracy (%) obtained by the proposed method using different number of eigenfaces and LDA components

| No.of eigen faces | PCA [10] | | RT + PCA [10] | | RT + PCA + LDA | | | |
|---|---|---|---|---|---|---|---|---|
| | Accuracy (%) | Time (in seconds) | Accuracy (%) | Time (in seconds) | Accuracy (%) | | | Time (in seconds) 5 LDA |
| | | | | | 2 LDA Components | 4 LDA Components | 5 LDA Components | |
| 5 | 58.5 | 9.813 | 60.1 | 9.941 | 60.80 | 61.00 | 61.16 | 9.942 |
| 10 | 76.1 | 9.822 | 77.5 | 9.950 | 77.60 | 77.80 | 77.9 | 9.942 |
| 15 | 81.5 | 9.824 | 84.36 | 9.950 | 84.90 | 85.00 | 85.1 | 9.950 |
| 20 | 87.1 | 9.828 | 90.19 | 9.953 | 90.50 | 91.00 | 91.2 | 9.956 |
| 25 | 87.61 | 9.834 | 94.1 | 9.953 | 94.10 | 94.15 | 94.2 | 9.992 |
| 30 | 88.5 | 9.845 | 94.16 | 9.953 | 94.30 | 95.00 | 95.91 | 10.131 |
| 35 | 89.11 | 9.861 | 95.2 | 10.170 | 96.00 | 97.10 | 97.9 | 10.381 |
| 40 | 89.47 | 10.161 | 95.3 | 10.172 | 98.50 | 99.00 | 99.16 | 11.001 |

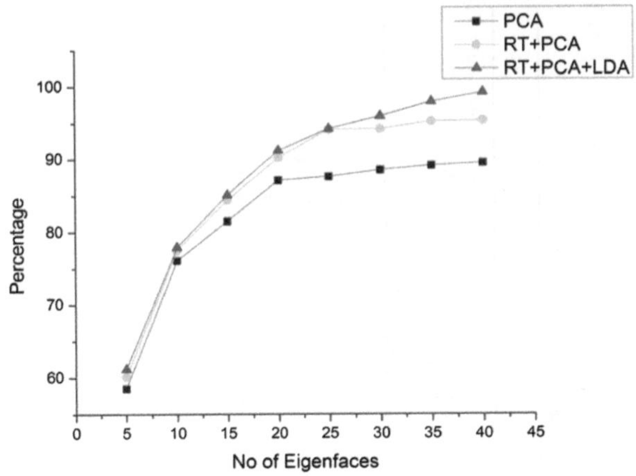


**Fig. 7** The recognition accuracy (%) versus the number of eigenfaces of the proposed method and the other methods

**Table 2** Comparison of the proposed method with the state-of-the-art 3D face recognition algorithms

| Method | Recognition accuracy (%) | Dataset used |
|---|---|---|
| Proposed method (RT + PCA + LDA) | 99.16 | Texas 3D face database |
| 3D face recognition using RT + PCA [10] | 95.30 | Texas 3D face database |
| LDA [8] (Gabor features around facial fiducial points) | 96.80 | Texas 3D face database |
| Sparse representation [11] (triangular area/normal, geodesic distance) | 95.30 | BJUT-3D and FRGC v2 |

The graph of recognition rates versus the number of eigenfaces is shown in the Fig. 7 for the proposed method (RT + PCA + LDA) with other PCA and RT + PCA methods. It is observed that the recognition rate improves as the number of eigenfaces is increased. It is 99.16 % for 40 eigenfaces in case of proposed method. Further, the proposed method based on RT, PCA and LDA outperforms the PCA method.

We compare the rank-one recognition rates of the proposed method to the state-of-the-art 3D face recognition methods, namely, 3D face recognition using RT + PCA [10], LDA [8] and sparse representation [11] in the Table 2.

## 5 Conclusion

In this paper, we have proposed a novel hybrid method for Three Dimensional (3D) face recognition using Radon transform with PCA and LDA based features on face range images. In this method the LDA based feature computation can be done at high speeds, since only few LDA components are adequate to yield good classification results. Our experimental results yield 99.16 % recognition performance with a small number of features, which compares well with other state-of-the-art methods. The experimental results demonstrate the efficacy and the robustness of the method to illumination and pose variations. The recognition accuracy can be further improved by considering a larger training set and a better classifier.

**Acknowledgments** The authors are grateful to the referees for their helpful comments and suggestions. Also, the authors are indebted to the University Grants Commission, New Delhi, for the financial support for this research work under UGC-MRP F.No.39-124/2010 (SR).

## References

1. Chellappa R, Wilson C, Sirohey S (1995) Human and machine recognition of faces: a survey. Proc IEEE 83(5):704–740
2. Zhao W, Chellappa R, Phillip PJ, Rosenfeld A (2003) Face recognition: a literature survey. ACM Comput Surv 35(4):399–458
3. Patil AM, Kolhe SR, Patil PM (2010) 2D face recognition techniques: a survey, Int J Mach Intell, ISSN: 0975–2927, 2(1):74–83
4. Abate AF, Nappi M, Riccio D, Sabatino G (2007) 2D and 3D face recognition: a survey. Patt Recog Lett 28:885–1906
5. Bowyer KW, Chang K, Flynn P (2006) A survey of approaches and challenges in 3D and multi-modal 3D + 2D face recognition. Comput Vis Image Underst 101:1–15
6. Gupta S, Markey MK, Bovik AC (2010) Anthropometric 3D face recognition. Int J Comput Vis. Springer Science + Business Media, LLC
7. Lu X, Colbry D, Jain AK (2004) Matching 2.5D scans for face recognition, Int Conf Pattern Recog, 362–366
8. Chang KI, Bowyer KW, Flynn PJ (2005) Adaptive rigid multi-region selection for handling expression variation in 3D face recognition. Comput Vision Pattern Recogn—Workshops
9. Jahanbim S, Choi H, Jahanbin R, Bovik AC (2008) Automated facial feature detection and face recognition using Gabor features on range and portrait images, 15th IEEE international conference on image processing
10. Hiremath PS, Hiremath M (2012) 3D face recognition using radon transform and PCA. Int J Graph Image Process 2(2):123–128
11. Tang H, Sun Y, Yin B, Ge Y (2011) 3D face recognition based on sparse representation, J Supercomputing, 58(1):84–95
12. Gupta S, Castleman KR, Markey MK, Bovik AC (2010) Texas 3D face recognition database. IEEE southwest symposium on image analysis and interpretation, 97–100, Austin. URL: http://live.ece.utexas.edu/research/texas3dfr/index.htm
13. Averbuch A, Shkolnisky Y (2003) 3D fourier based discrete radon transform, Applied and Computational Harmonic Analysis 15, Elsevier, Amsterdam, 33–69
14. Turk M, Pentland A (1991) Eigenfaces for recognition. J Cognitive Neurosci 3(1):71–86
15. Moon H, Phillips PJ (2001) Computational and performance aspects of PCA-based face recognition algorithms. Perception 30:303–321
16. Etemad K, Chellappa R (1997) Discriminant analysis for recognition of human face images. J Opt Soc Am A 14(8):1724–1733
17. Zhao W, Chellappa R, Krishnaswamy A (1998) Discriminant analysis of principal components for face recognition. Proceedings of the 3rd IEEE international conference on face and gesture recognition, FG'98, 14–16 April 1998, Nara, Japan, pp 336–341

# Fusion of Texture Features and SBS Method for Classification of Tobacco Leaves for Automatic Harvesting

**P. B. Mallikarjuna and D. S. Guru**

**Abstract** In this paper we propose a new model to classify tobacco leaves for automatic harvesting using feature level fusion. The CIELAB color space model is used to segment leaves from their background. Texture features are extracted from segmented leaves using Haar wavelets and gray level local texture pattern (GLTP) separately. These extracted features are fused using the concatenation rule. Discriminative texture features are then selected using the sequential backward selection (SBS) method. The k-NN classifier is designed to classify tobacco leaves into three classes viz., unripe, ripe and over-ripe. In order to corroborate the efficacy of the proposed model, we have conducted an experimentation on our own dataset consisting of 1,300 images of tobacco leaves captured in sunny and cloudy lighting conditions in a real tobacco field.

**Keywords** Tobacco leaves · CIELAB color space model · Wavelets · Gray level local texture pattern · Feature level fusion · Sequential backward selection · k-NN classifier

## 1 Introduction

Precision agriculture is an integrated crop management system that attempts to match the type and quantity of inputs with the actual crop needs for small areas within a farm field. The potential of precision agriculture in terms of economical and

P. B. Mallikarjuna (✉)
Department of Information Science and Engineering,
Bapuji Institute of Engineering and Technology, Davangere,
Karnataka 577004, India
e-mail: pbmalli@yahoo.com

D. S. Guru
Department of Studies in Computer Science, Manasagangothri,
University of Mysore, Mysore, Karnataka 570006, India
e-mail: dsg@compsci.uni-mysore.ac.in

P. P. Swamy and D. S. Guru (eds.), *Multimedia Processing, Communication and Computing Applications*, Lecture Notes in Electrical Engineering 213,
DOI: 10.1007/978-81-322-1143-3_10, © Springer India 2013 

environmental benefits could be visualized through reduced use of water, fertilizers, herbicides and pesticides besides the farm equipments. Instead of managing an entire field based upon some hypothetical average condition, which may not exist anywhere in the field, a precision agriculture approach recognizes site-specific differences within fields and adjusts management actions accordingly [1]. The objectives of precision agriculture are profit maximization, agriculture input rationalization and environmental damage reduction, by adjusting the agriculture practices to the site demands. To achieve these objectives some practices which are site specific application of agrochemicals to remove diseases at seedling (nursery) level and plant level, right time harvesting of crops and grading (quality inspection) of crops are to be adopted. Human intervention in these practices raises many disadvantages such as wrong diagnosis of diseases in crops, wrong quality analysis of crops, man power, labor cost and time consuming. Therefore, we need to automate these practices to increase efficiency and speed using computer vision (CV) algorithmic models.

Harvesting is an important stage in tobacco crop. Tobacco crop is grown for production of quality leaves which largely depends upon the ripeness of a leaf during harvesting. Therefore, while harvesting, farmers should look into factors such as unripe or ripe or over-ripe leaves based on ripeness of a leaf. Ripeness of a leaf begins after 50 days of plantation of tobacco seedlings. Harvesting usually begins after 60 days of plantation of tobacco seedlings. Leaves are removed at intervals as they ripe. Manual classification of unripe, ripe and over-ripe leaves is a laborious, a time consuming, an inefficient and a costly process. Automation of this process helps the tobacco farmers to gain more profit. Computer vision and image processing techniques can be exploited for classification of tobacco leaves for automatic harvesting, thereby increasing the speed and accuracy of harvesting in addition to reducing the number of human labors and hence cost also.

## 2 Related Work

In our recent work [2] we proposed a model for classification of tobacco leaves for automatic harvesting using texture models such as LBP, LBPV and GLTP and k-NN Classifier using Euclidean distance as a similarity measure. We also proposed a model for classification of tobacco leaves for automatic harvesting based on spots density and color [3]. Apart from our works no other attempt can be traced on classification of tobacco leaves for automatic harvesting to the best of our knowledge. However, few attempts could be traced on grading of flue-cured tobacco leaves using machine vision techniques. A 2D feature space was proposed to express feature distribution of tobacco leaves and 'nearest-neighbor' method was used to classify tobacco leaves [4]. A transformation technique from RGB signals to the Munsell system for color analysis of tobacco leaves was proposed [5]. A fuzzy classification model was explored to grade flue-cured tobacco leaves [6]. Recognition of the part of growth of flue-cured tobacco leaves based on support vector machine was recommended [6]. Machine vision techniques have been used to solve problems of feature extraction and analysis of Flue-cured tobacco leaves, which include features of color, size, shape and surface texture [7]. The Barrel theory decision-making algorithm was used in auto-grading of cured tobacco leaves [8]. More than 1,000 Chinese flue-cured tobacco leaf samples, which have 12 genotypes and cultivated from 5 to 10 regions of China in 2003 and 2004, have been discriminated by means of an improved and simplified k-NN classification algorithm based on near infrared spectra [9].

In our previous work [2], we had achieved classification accuracy of 80 % for classification of tobacco leaves for automatic harvesting and experimentation was conducted on our own dataset of 274 tobacco leaves captured in a complex agricultural environment. In this work, to improve classification accuracy we performed a feature level fusion of GLTP and wavelet texture features and exploited wrapper method of feature selection technique such as sequential backward selection (SBS). Further experimentation is conducted on a relatively large database.

# 3 Proposed Model

The proposed model consists of four stages—segmentation, feature extraction, feature level fusion and classification. The color space model CIELAB is used to segment leaf area from the background. Texture features are extracted from segmented image of tobacco leaf using GLTP texture model and wavelet decomposition separately. Texture Features of GLTP and wavelet are fused using concatenation rule and normalize using min–max rule. Discriminative texture features are then selected using Sequential Backward Selection algorithm. For the purpose of classification the k-NN classifier with Euclidean distance as a similarity measure has been exploited in this work.

## *3.1 Segmentation*

We have selected CIELab [10] color model to segment a leaf area from background (soil, stones and noise). CIELab is an approximately uniform color system. Its values are calculated by non-linear transformations of CIE XYZ. In this system, Y represents the brightness (or luminance) of the color, while X and Z are virtual (or not physically realizable) components of the primary spectra. The CIE XYZ tristimuli are standardized with values corresponding to the D65 white point: $X_0 = 95.047, Y_0 = 100$ and $Z_0 = 108.883$. It is then transformed into the standardized tristimuli to the CIELAB Cartesian coordinate system using the following metric lightness function.

$$L = \begin{cases} 166 \times (Y/Y_0)^{1/3} - 16 & \text{for } (Y/Y_0)^{1/3} > 0.00856 \\ 903.3 \times (Y/Y_0) & \text{otherwise} \end{cases} \tag{1}$$

The chromacity coordinates a and b are derived using:

$$\begin{aligned} a &= 500 \times \left[ (X/X_0)^{1/3} - (Y/Y_0)^{1/3} \right] \\ b &= 200 \times \left[ (Y/Y_0)^{1/3} - (Z/Z_0)^{1/3} \right] \end{aligned} \tag{2}$$

The chromacity coordinates represent opponent red-green scales (+a red, −a greens) and opponent blue-yellow scales (+b yellows, −b blues). Since color of the leaf varies from green (unripe) to yellow (over-ripe), the chromacity coordinate a is used segment leaf from the background (soil, stone and noise). The a value is calculated for each image pixel. If the value of a is less than a predefined threshold then the corresponding pixel is considered as a leaf pixel else it is considered as a background pixel.

## *3.2 Feature Extraction*

Top surface of an unripe tobacco leaf is smoother and its roughness increases as ripeness progresses as shown in Fig. 1. This roughness is reflected by transitions in intensity levels on the surface of leaves in the form of uniform and non uniform patterns. To exploit this, we recommend to extract texture features from gray scale images of tobacco leaves using the GLTP texture feature based model and wavelet decomposition which are explained in the following sections.

**Gray Level Local Texture Pattern (GLTP)**. The LBP model is computationally efficient but inadequate to represent a local region whereas TS (Texture spectrum) model will reveal more textural information but it is computationally burden. It was developed by combining the advantages of Texture spectrum (TS) [11] and Local binary patterns (LBP) [12]. The GLTP is computationally acceptable and is robust against variations in the appearance of the texture to meet the real world applications [13]. These variations may be caused by uneven

(a) (b) (c)

**Fig. 1** Images of **a** Unripe, **b** ripe, **c** over-ripe tobacco leaves illumination, different viewing angles and resolving power of the sensor system. Since it is a rotational invariant and gray-scale shift invariant, it is robust against variable illumination. The GLTP model detects the number of transitions or discontinuities in the circular presentation of texture patterns in a small region, thus it suits to our dataset. When such transitions are found to follow a rhythmic pattern, they are recorded as uniform patterns and are assigned with unique labels. All other non-uniform patterns are grouped under single category. The GLTP operator is applied to the resultant intensity image. It assigns a GLTP label to every pixel in an image depending on uniformity of pattern around the pixel. This labeled image is represented using one dimensional histogram with abscissa indicating the GLTP label and ordinate representing its occurrence frequency.

The following rotational and gray-scale invariant GLTP operator is used for describing a local image texture.

$$\mathrm{GLTP}_{P,R}^{riu3} = \begin{cases} \sum_{p=1}^{P} s(g_p, g_c) & \text{if } U \leq 3 \\ P \times 9 + 1 & \text{otherwise} \end{cases} \tag{3}$$

where, *P* is the number of neighbors and *R* is the radius of neighborhood.

$$s(g_p, g_c) = \left\{ \begin{array}{lll} 0 & \text{if} & g_p < (g_c - \Delta g) \\ 1 & \text{if} & (g_c - \Delta g) \leq g_p \leq (g_c + \Delta g) \\ 9 & \text{if} & g_p > (g_c + \Delta g) \end{array} \right\} \tag{4}$$

$$\text{where,} \quad p = 1, 2, \ldots, P$$

Here, $\Delta g$ is a small positive value that represents a desirable gray value and has its importance in forming the uniform patterns. Uniform measure (U) corresponds to the number of spatial transitions in a circular direction to form a pattern string and is defined as

$$U = f(s(g_P, g_c), s(g_1, g_c)) + \sum_{p=2}^{P} f\Big(s(g_1, g_c), s(g_{p-1}, g_c)\Big)$$

$$\text{where, } f(X,Y) = \begin{cases} 1 & \text{if } |X - Y| > 0 \\ 0 & \text{otherwise} \end{cases} \tag{5}$$

For example, GLTP with $P = 1$ and $R = 8$ could generate 46 unique labels. For $U = 0$ there exist 3 GLTP labels (0, 8 and 72), for $U = 2$ there are 21 GLTP labels (1–7, 9, 16, 18, 24, 27, 32, 36, 40, 45, 48, 54, 56, 63 and 64) and for $U = 3$ there exist another 21 GLTP labels (10–15, 19–23, 28–31, 37–39, 46, 47 and 55). All other non uniform patterns ($U > 3$) are grouped under one label 73. Since there are few holes in the GLTP labeling scheme, they are relabeled to form continuous numbering from 1 to 46 using a lookup table. The GLTP labels for some pattern strings are shown in Table 1.

**Table 1** Examples for U and GLTP of uniform and non uniform patterns

| Pattern string | U | GLTP label | Relabeled GLTP | Uniform |
|---|---|---|---|---|
| 00000000 | 0 | 0 | 1 | Yes |
| 11111111 | 0 | 8 | 9 | Yes |
| 99999999 | 0 | 72 | 45 | Yes |
| 00100000 | 2 | 1 | 2 | Yes |
| 00009999 | 2 | 36 | 31 | Yes |
| 99991111 | 2 | 40 | 35 | Yes |
| 91000000 | 3 | 10 | 11 | Yes |
| 11190000 | 3 | 12 | 12 | Yes |
| 09991110 | 3 | 30 | 28 | Yes |
| 01110110 | 4 | 73 | 46 | No |
| 01990100 | 5 | 73 | 46 | No |
| 10011010 | 6 | 73 | 46 | No |

**Wavelet Decomposition**. We performed single level wavelet decomposition on matrix of each channels of original RGB input image of tobacco leaf using Haar wavelet. It results into four coefficients matrix for each channel. They are approximation matrix coefficients (cA) and three details coefficients matrices horizontal (cH), vertical (cV) and diagonal (cD). To analysis the function of wavelet, we compute the reconstructed coefficients matrix for each details coefficients matrices horizontal (cH), vertical (cV) and diagonal (cD) of each channel. Then we extract features by calculating energy for each obtained reconstructed coefficients matrix. Thus we have totally 9 features.

## 3.3 Feature Level Fusion

Let $G = \{g_1, g_2, g_3, \ldots\ldots g_n\}$ and $W = \{w_1, w_2, w_3, \ldots\ldots w_m\}$ represent the feature vector of GLTP and wavelet extracted from a tobacco leaf respectively. The fused feature vector $X_i = \{g_1, g_2, g_3, \ldots\ldots g_n, \ w_1, w_2, w_3, \ldots\ldots w_m\}$ is obtained by concatenating the feature vectors *G* and *W*. The obtained fused feature vector is normalized for storage and classification.

## 3.4 Feature Selection

Feature selection is the process of selecting a subset of relevant features for building robust learning models. There are two types of feature selection models. They are filter model and wrapper model. The filter model relies on general characteristics of the training data to select some features without involving any learning algorithm. The wrapper model requires one predetermined learning algorithm in feature selection and uses its performance to evaluate and determine which features are selected.

In filter model a well known method called relief [14] that relies on relevance evaluation. Time Complexity of relief for a dataset with $P$ instances and $Q$ features is O(tPQ). With $t$ being a constant, the time complexity becomes O(PQ), which makes it very scalable to datasets with both a huge number of instances and a very high dimensionality. However, relief does not help to eliminate redundant features. Empirical evidence from feature selection literature shows that, along with irrelevant features, redundant features also affect the speed and accuracy of learning algorithms and thus should be eliminated as well [15]. Therefore, we have exploited feature selection method based on wrapper model such as sequential backward selection (SBS) [16]. The criterion employed in this method is the correct classification rate of the Bayes classifier assuming that the features obey the multivariate Gaussian distribution. The SBS consists of a backward step which starts from a set of all features $Z_0$. At each backward step at level $l$, it removes the feature $X^+ \in (X - Z_{l-1})$ such that for $Z_l = Z_{l-1} - \{X^+\}$ the probability of correct classification achieved by the Bayes classifier is maximized. This method eliminates irrelevant features as well as redundant features but this method is slightly expensive than filter method Relief. We have used standard cross-validation technique to estimate probability of correct classification rate (CCR) [17, 18].

### *3.5 Classification*

In this work, a k-nearest neighbor classifier based on the Euclidean distance measure has been used to classify tobacco leaves into unripe, ripe and over-ripe for automatic harvesting purpose. Similarity between two samples is computed by comparing their histograms generated by texture feature based models.

## 4 Experimental Results

### *4.1 Dataset*

Color images of tobacco leaves in real tobacco field are acquired using a Sony digital color camera. The leaves used for imaging are randomly selected from the tobacco field at Central Tobacco Research Institute (CTRI), Hunsur, Karnataka, India. Images are acquired at variable illumination conditions (sunny and cloudy). A total of 1,300 sample images of size 250 × 250 are used for evaluating the proposed texture based models.

### *4.2 Results*

Tobacco leaves are classified into three classes: unripe (C1), ripe (C2) and over-ripe (C3). A k-NN classifier based on Euclidean distance has been used for classification. In k-NN classification, k is varied from 2 to 40 with a step size two. Finally the best k value is selected to classify tobacco leaves. Among 1,300 sample images of tobacco leaves, there are 323 samples of type C1, 667 samples of type C2 and 310 samples of type C3. In order to corroborate the efficacy of the proposed model we have studied the effect of classification accuracy under varying size of database. We have varied the training set by 30, 40, 50 and 60 % and remaining is used as testing. Experiment has been conducted 20 times (20 trails) each time selecting a specified number of training and testing samples randomly. Average classification accuracy, minimum classification accuracy, maximum classification accuracy and standard deviation of 20 trails are calculated for proposed model. In order to show the superiority of the proposed model with SBS, we have also calculated average classification accuracy, minimum classification accuracy, maximum classification accuracy and standard deviation for GLTP alone, wavelet alone and feature level fusion of GLTP and wavelet.

The classification results using GLTP alone, wavelet alone and fusion of GLTP and wavelet are tabulated in Tables 2, 3, and 4 respectively. The classification results using the proposed model with SBS are tabulated in Table 5. From Table 5 it is clear that for 60 % training we have achieved a best classification accuracy up to 89 %. The corresponding confusion matrix is tabulated in Table 6. The best average classification accuracy of GLTP alone, wavelet alone, fusion and proposed model are represented in Fig. 2. From Fig. 2 it is clear that proposed model achieves best classification than individual and fusion. In proposed model, classification accuracy is improved by 9 % against to our previous model [2]. The qualitative comparative analysis of the proposed work with previous work [2] is given in Table 7.

**Table 2** Classification results using GLTP

| Training Samples (%) | Minimum accuracy | Maximum accuracy | Average accuracy | Standard deviation |
|---|---|---|---|---|
| 30 | 77.0906 | 82.5419 | 80.1946 | 1.6464 |
| 40 | 80.1983 | 84.4947 | 82.6596 | 1.1769 |
| 50 | 78.7321 | 86.3336 | 82.4641 | 2.0180 |
| 60 | 75.6234 | 86.3903 | 82.6738 | 2.9476 |

**Table 3** Classification results using wavelet

| Training samples | Minimum accuracy | Maximum accuracy | Average accuracy | Standard deviation |
|---|---|---|---|---|
| 30 | 64.0355 | 70.9858 | 68.0577 | 1.5728 |
| 40 | 65.9427 | 71.4480 | 68.3056 | 1.3889 |
| 50 | 67.0934 | 72.5033 | 69.0196 | 1.3347 |
| 60 | 65.8379 | 74.4059 | 70.8379 | 2.0588 |

**Table 4** Classification results using fusion of GLTP and Wavelet

| Training Samples (%) | Minimum accuracy | Maximum accuracy | Average accuracy | Standard deviation |
|---|---|---|---|---|
| 30 | 81.3503 | 85.5002 | 83.4217 | 1.2124 |
| 40 | 82.5901 | 87.0211 | 84.3167 | 1.4200 |
| 50 | 82.9824 | 88.4326 | 85.6530 | 1.3938 |
| 60 | 83.2469 | 89.2456 | 86.4908 | 1.6119 |

**Table 5** Classification results using proposed model

| Training samples (%) | Minimum accuracy | Maximum accuracy | Average accuracy | Standard deviation |
|---|---|---|---|---|
| 30 | 84.2750 | 88.2162 | 86.6704 | 0.9021 |
| 40 | 85.6324 | 89.5413 | 87.8802 | 1.1501 |
| 50 | 85.9982 | 89.9910 | 87.7825 | 1.2500 |
| 60 | 86.1605 | 90.3761 | 88.8085 | 1.2225 |

**Table 6** Confusion matrix obtained for the trail with best classification accuracy

| | Unripe | Ripe | Over-ripe |
|---|---|---|---|
| Unripe | 109 | 20 | 00 |
| Ripe | 19 | 231 | 17 |
| Over-ripe | 00 | 10 | 114 |

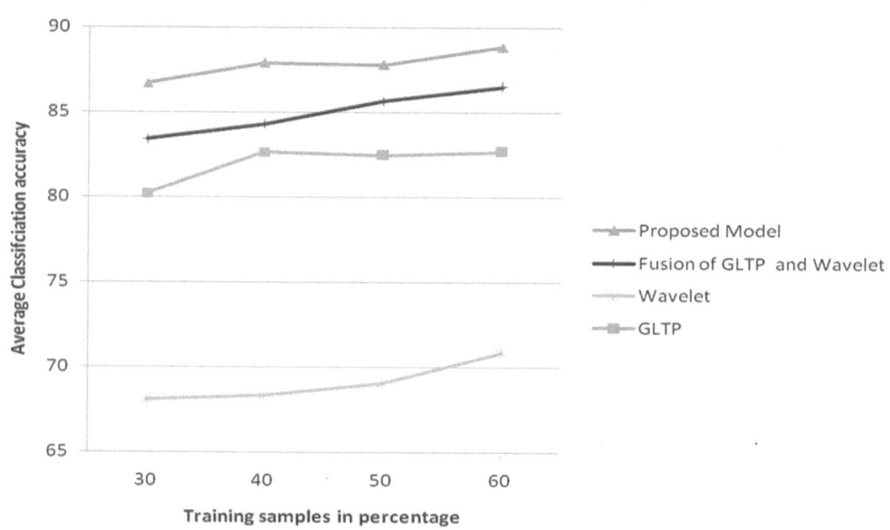


**Fig. 2** Comparison of average classification accuracy of proposed model with fusion and individual texture features (GLTP alone and wavelet alone)

**Table 7** Qualitative comparison with our previous work [2]

| Title | Dataset size | Classifier | Features | Accuracy |
|---|---|---|---|---|
| Machine vision based classification of tobacco leaves for automatic harvesting [2] | 224 | k-nearest neighbor | Local binary pattern alone | 75.9355 |
| | | | Local binary pattern variance alone | 74.1409 |
| | | | Gray level local texture pattern alone | 80.9174 |
| Proposed work | 1,300 | k-nearest neighbor | Gray level local texture pattern alone | 82.6596 |
| | | | Wavelet decomposition | 70.8379 |
| | | | Fusion of gray level local texture pattern and wavelet decomposition | 86.4908 |
| | | | Fusion of gray level local texture pattern and wavelet decomposition and SBS method | 88.8085 |

## 5 Conclusion

In this paper we have explored the fusion of GLTP and wavelet texture features, and SBS feature selection to classify tobacco leaves on a plant for automatic harvesting in a complex agricultural environment. Proposed model has superior performance when compared to GLTP alone, wavelet alone and fusion of GLTP and wavelet. In future, we try to study the classification of tobacco leaves using fusion of different texture features and different feature selection methods, which may improve the classification accuracy.

## References


1. Goovaerts P (2000) Estimation or simulation of soil properties? An optimization problem with conflicting criteria. Geoderma, 165–186
2. Guru DS, Mallikarjuna PB, Manjunath S, Shenoi MM (2012) Machine vision based classification of tobacco leaves for automatic harvesting. Intelligent automation and soft computing. Auto Soft Publisher 18(5):577–586
3. Guru DS, Mallikarjuna PB (2010) Spots and color based ripeness evaluation of tobacco leaves for automatic harvesting. First international conference on intelligent interactive technologies and multimedia (IITM 2010), ACM IIIT Allahabad chapter, pp 198–202
4. Zhang J, Sokhansanj S, Wu S, Fang R, Yang W, Winter P (1997) A trainable grading system for tobacco leaves. Comput Electron Agri 16(3):231–244
5. Zhang J, Sokhansanj S, Wu S, Fang R, Yang W, Winter P (1998) A transformation technique from RGB signals to the Munsell system for color analysis of tobacco leaves. Comput Electron Agri 19:155–166
6. Zhang H, Han L, Wang Z (2003) A fuzzy classification system and its application. International conference on machine learning and cybernetics, pp 2–5
7. Zhang X, Zhang F (2008) Images features extraction of tobacco leaves. Congress Image Signal Process, pp 773–776
8. Huabo L, Liyuan H, Tao L (2009) The barrel theory based decision-making algorithm and its application. Int Conf Comput Intell Nat Comput 1:11–14
9. Ni L, Zhang L, Xie J, Luo J (2008) Pattern recognition of Chinese flue-cured tobaccos by an improved and simplified K-nearest neighbor's classification algorithm on near infrared spectra. Anal Chim Acta 633:43–50
10. Viscarra RA, Minasny B, Roudier P, McBratney AB (2006) Colour space models for soil science. Geoderma 133:320–337
11. He D, Wang L (1990) Texture unit texture spectrum and texture analysis. IEEE Trans Geosci Remote Sens 28(4):509–512
12. Ojala T, Pietikainen M, Maenapaa T (2008) Multi resolution gray-scale and rotation invariant texture classification with Local binary patterns. IEEE Trans Pattern Anal Mach Intell 24(7):971–987
13. Surliandi A, Ramar K (2008) Local texture patterns—a univariate texture model for classification of images. In: Proceedings of the 2008 16th international conference on advanced computing and communications (ADCOM08), pp 32–39
14. Kira K, Rendel L (1992) Feature selection problem: traditional methods and new algorithm. Proceedings of the 10th national conference on artificial intelligence, pp 129–134

15. Hall M (2000) Correlation-based feature selection for discrete and numeric class machine learning. In: Proceedings of the seventeenth international conference on machine learning, pp 359–366
16. Acun E, Coaquira F, Gonzalez M (2003) A comparison of feature selection procedures for classifiers based on kernel density estimation. In: Proceedings of the international conference on computer, communication and control technologies, pp 468–472
17. Ververidis D, Kotropoulos C (2005) Emotional speech classification using gaussian mixture models and the sequential floating forward selection algorithm. In: Proceedings of the 2005 IEEE international conference on multimedia and expo, pp 1500–1503
18. Ververidis D, Kotropoulos C (2008) Fast and accurate sequential floating forward feature selection with the Bayes classifier applied to speech emotion recognition. Elsevier Signal Process 88(12):2956–2970

# Stacked Classifier Model with Prior Resampling for Lung Nodule Rating Prediction

**Vinay Kumar, Ashok Rao and G. Hemanthakumar**

**Abstract** In this work, we are proposing a new machine learning strategy for classification task for imbalanced data. We are using lung image data by Lung Image Database Consortium (LIDC), since LIDC data is a better example for imbalanced dataset. In this work we are using sufficiently large dataset which contains 4,532 nodules extracted from CT images. Later we consider 55 low level nodule image features and radiologists ratings for experiments. This work is being dealt in two stages. (1) data level learning and (2) algorithm level learning. In first stage, we are balancing the dataset prior to classification process. We are using resampling approach for this task. In second stage, we are using ensemble of classifiers to predict lung nodule rating. We are using wide range of classifier models for constructing an ensemble. We use Bagged Decision Tree, naïve Bayes, Boosted Decision Trees, and Support Vector Machine (SVM) in a classifier library. Stacking algorithm is used to combine the different classifier models in library to construct higher level ensemble. We are evaluating the performance of our model on five metrics: Accuracy, precision, recall, F-score and Kappa statistics. Results show that our method yields much improved scores as we are refining at both, data level and algorithm level.

**Keywords** Stacking · Ensemble of classifier · Resampling · Kappa statistics

V. Kumar (✉) · G. Hemanthakumar
DoS in Computer Science, University of Mysore, Mysore, India
e-mail: gotovinni@gmail.com

G. Hemanthakumar
e-mail: ghk.2007@yahoo.com

A. Rao
Freelance Academician, 165, 11th main, S.Puram, Mysore, India
e-mail: ashokrao.mys@gmail.com

P. P. Swamy and D. S. Guru (eds.), *Multimedia Processing, Communication and Computing Applications*, Lecture Notes in Electrical Engineering 213,
DOI: 10.1007/978-81-322-1143-3_11, © Springer India 2013

# 1 Introduction

Lung cancer is one of the major medical challenges that the world is facing today. Recent survey shows that mortality rate of people dying because of lung cancer tend to increase year by year. Computer Aided Diagnostics (CAD) is one such system in medical field which are built using computer programs to effectively assist in diagnosis of the diseases. Many such systems are built using image processing as well as pattern recognition techniques. Even though there are a good numbers of CAD systems that are available in the field, still there is lack of intelligent systems which can adopt themselves to change with respect to variation in input environment. These changes refer to imbalance in input data, uncertainty in domain expert prediction and problem in deciding marginal cases. Hence there is still lot of research required to develop such intelligence into systems and schemes. Machine learning technique is one such approach to solve such issues and recently many efforts have been carried out using various machine learning techniques to address above. We have discussed some of state-of-art work in this domain which has been carried out recently in the Sect. 2.

# 2 Literature Review

In this paper, we brief the literature in two sections. In the first section we discus about some work on CT scan image feature extraction and classification in the perspective of image processing and pattern recognition. In second section we present recent works on machine learning and ensemble of classifier approaches for classification problems. Ekrain et al. [1] investigated several approaches to combine delineated boundaries and ratings from multiple observer and they have used p-map analysis with union, intersection and threshold probability to combine the boundary reading and claimed that threshold probability approach provides good level of agreement. Ebadollahi et al. [2] proposed a framework that uses semantic methods to describe visual abnormalities and exchange knowledge with medical domain.

Nakumura et al. [3] worked on simulating the radiologists perception of diagnostic characteristic rating such as shape, margin, irregularity, Spiculation, Lobulation, texture etc. on a scale of 1–6 and they extracted various statistical and geometric image features including fourier and radiant gradient indices and correlated these features with the radiologists ratings. Significant work towards designing panel of expert machine learning classifier which automates the radiologist work of predicting nodule ratings is done by Dmitriy and Raicu [4]. They have proposed active decorate, a new meta-learning strategy for ensemble of classifier domain. Oza and Tumer [5], presented a survey on applications of ensemble methods covering different fields such as remote sensing, person recognition, one versus all recognition in medicine. In their work they have summarized the most frequently used classifier ensemble methods including averaging, bagging, boosting and order statistics classifiers. They have made a statement that each ensemble method has different properties that make it better suited to particular types of classifiers and applications.

Reid [6] has presented a survey work on several ensemble methods that can accommodate different classifiers for base models types. He has given useful review on heterogeneous ensemble methods with supporting theoretical motivation, empirical results and relationship to other techniques. Kuncheva et al. [7] have mentioned searching for a best classifier is an ill- posed problem because there is no one classifier that is best for all types of data. They have used variety of machine learning techniques to compare the performance of classifier ensembles for fMRI data analysis. Caurana [8] has identified that ensemble method can optimize the performance of the model for classification task. He has experimented with seven test problems and ten performances metric and claimed that ensemble techniques outperformed in all the scenarios.

Datta and Datta el al. [9] have presented their work on adaptive optimal ensemble classifier via bagging and rank aggregation with applications to high dimensional data. In their work they have considered three norm data and simulated microarray dataset. Based on their observation on obtained experimental results they have claimed ensemble classifier performs at the level of best individual classifier or better than individual classifier. They have concluded that for a complex high-dimensional datasets it is wise to consider a number of classification algorithms combined with dimension reduction techniques rather than a fixed standard algorithm. Dzeroski and Zenko et al. [10] have empirically evaluated several state-of-art methods for construction of ensembles of classifiers with stacking and they claimed that stacking method performs at best, comparable to selecting the best classifier from the ensemble by cross validation [5, 11]. Ting and Witten [12] recommended Multi-response Regression (MLR) as suitable for meta-level learning and showed other learning algorithms are not up to the mark as compared to MLR. In this work we are using linear regression as a meta-learner in our stacking model.

## 3 LIDC Dataset

Lung Image Database Consortium (LIDC) provides lung CT image data which is publically available through National Cancer Institute's Imaging Archive (web site—http://ncia.nci.nih.gov) [13]. Dataset consists of image data, radiologist's nodule outline details and radiologist subjective characteristic ratings. The LIDC dataset currently contains complete thoracic CT scans of 399 patients acquired over different periods of time. LIDC data download comes with DICOM image and the nodules information in the XML file. This has information regarding the spatial location information about three types of lesions, they are nodules <3 mm; nodules >3 mm and non-nodules >3 mm in maximum diameter as marked by

**Table 1** Overview of the LIDC data subset considered

| LIDC data subset | |
|---|---|
| Number of cases considered | 124 |
| Number of instances | 14,956 |
| Number of nodules | 4,532 |

panel of 4 expert radiologists. For any lesion greater than 3 mm in diameter XML file contains spatial coordinates of the pixel of nodule outline. Since the number of radiologist in LIDC panel is 4 it is obvious that each nodule >3 mm has 4 nodule outlines. Moreover, any radiologist who identifies the nodule >3 mm also provides subjective ratings for 9 nodule characteristics: Lobulation, internal structure, calcification, subtlety, spiculation, margin, sphericity, texture and malignancy.

LIDC data collection process is in two fold, blinded and unblinded reading session and LIDC did not impose any forced consensus on radiologists, all the lesions indicated by the radiologists at the conclusion of the unblinded reading sessions are recorded and available to the public. With this no consensus on radiologist, lesion >3 mm is marked by a single a radiologist, by two radiologists, by three radiologists or by all four radiologists. The overview of the LIDC data subset we have used in this work has been shown in Table 1.

In our earlier work on lung images [14], we have considered 4,532 nodules that were extracted from LIDC lung image dataset. In this work our objective is to identify the significance of stacking ensemble method on large dataset. Hence we have collected sufficiently large dataset from LIDC CT images. As in literature [4] we are not only concentrating on those nodules which were agreed by all four radiologists, and also it has to be larger in CT scan series to be in dataset. Instead of following same method as in [4] we are considering the entire nodules which may appear in consecutive images in the CT series irrespective of sizes. This is because we have to notice the effect of resampling prior to classification. Therefore, at the end of our dataset preparation we have 4,532 nodules and the details about their distribution in original dataset and resampled dataset are given in Table 2. We have used SMOTE method technique for resampling dataset to make it balanced. The working principle of SMOTE [15, 16] technique will be discussed in Sect. 5.

Table 3 gives the instance distribution for malignancy case. The rating for the malignancy is further divided into multiclass such as Highly Unlikely, Moderately Unlikely, Indeterminate, Moderate and Suspicious cases. As we can see in Table 3 that the number of samples for highly likely cases is 572 where as for the cases Moderately Unlikely and Suspicious are 1,285 and 1,146 respectively. It means the number of cases for Moderately Unlikely and Suspicious is almost the double the number of samples in Highly Unlikely cases. In such scenario when we classify such imbalanced dataset, though using good performing classifier, the result will still be biased. This is because the classifier will get more number of samples of some classes and it will get fewer numbers of samples of other classes. Hence the classifier tends to get biased towards the case which has more number of samples.

**Table 2** Samples distribution across the class in original dataset and resampled dataset

| Class label for malignancy case | Number of samples | |
|---|---|---|
| | Original dataset | Resampled dataset |
| Highly unlikely | 572 | 575 |
| Moderately unlikely | 733 | 742 |
| Indeterminate | 1,285 | 1,219 |
| Moderately | 796 | 854 |
| Suspicious | 1,146 | 1,142 |
| Total number of samples | 4,352 | 4,352 |

**Table 3** Malignancy sample distributions in dataset

| Class label | Number of samples |
|---|---|
| Highly unlikely | 572 |
| Moderately unlikely | 733 |
| Indeterminate | 1,285 |
| Moderate | 796 |
| Suspicious | 1,146 |

It is to be noted that majority of real life medical data is indeed imbalanced. This reflects the distribution of such issues across the general population. Thus, working on such data is important since it captures realistic situation much more effectively. Hence we regard the dataset which we are considering in this work as class imbalanced data.

## 4 Image Feature Extraction

In this work we are using the same set of features which has been used in earlier work [16, 11]. Our image feature set consists of 55 low level image features. In addition to image features we have also used 7 radiologist characteristic ratings making the size of feature set to 62. The details about the image features we have used in this work are given in Table 4.

## 5 Methodology

In our previous work [11] we have investigated the role of single classifier versus panel of classifiers on LIDC data. In this work we consider smaller subset of data consisting of 212 nodules which are extracted from 50 cases. In [16] we have attempted to notice the significance of homogenous ensemble of classifier and heterogeneous ensemble of classifiers on LIDC data. There we have used

**Table 4** Details of low level image features considered

| Size features | Shape features | Intensity features |
|---|---|---|
| Area | Circularity | Min intensity |
| Convex area | Roughness | Max intensity |
| Perimeter | Elongation | Mean intensity |
| Convex perimeter | Compactness | SD intensity |
| Equiv diameter | Eccentricity | |
| Major axis length | Solidity | |
| Minor axis length | Extent | |
| *Texture features* | | |
| 24 Gabor features are mean and standard deviation of 12 different gabor response images at orientation = 0, 45, 90, 135 and time frequency = 0.3, 0.4, 0.5 | | |
| 13 Haralick features calculated from co-occurrence matrices. Energy, correlation, inertia, entropy, inverse difference moment, sum average, sum variance, sum entropy, difference, average, difference variance, difference entropy, information measure of correlation 1, information measure of correlation 2 | | |

DECORATE and stacking ensemble method to construct ensembles. In [14] we had used class imbalanced dataset and single classifier model. Various resampling approaches prior to classification were used and we noticed significant improvement in the results. In this current paper we are using a large dataset, it consists of 4,532 nodules from 124 cases which is a relatively larger dataset compared to the dataset which we have used in our previous works. Here we are considering resampling approach as well as panel of classifiers using stacking method to investigate the performance of classifiers on class imbalanced data. Data resampling and stacking methods are discussed in detail in Sects. 6 and 7.

# 6 Dataset Resample

## 6.1 SMOTE

Synthetic Minority over-sampling Technique (SMOTE) generates synthetic examples by operating in the feature space rather than in the data space [15, 16]. The synthetic examples cause the classifier to create larger and less specific decision regions, rather than smaller and more specific regions. The minority class is over-sampled by taking each minority class sample and introducing synthetic examples along the line segments joining any/all of the k minority class nearest neighbors. The steps involved in SMOTE method is as follows: For each minority observation:- (1) Find its k-nearest minority neighbors (2) Randomly select 'n' of these neighbors (3) Randomly generate synthetic samples along the lines joining the minority sample and its 'n' selected neighbors ('n' depends on the amount of oversampling which is pre defined). The flow diagram of SMOTE method is as shown in Fig. 1.

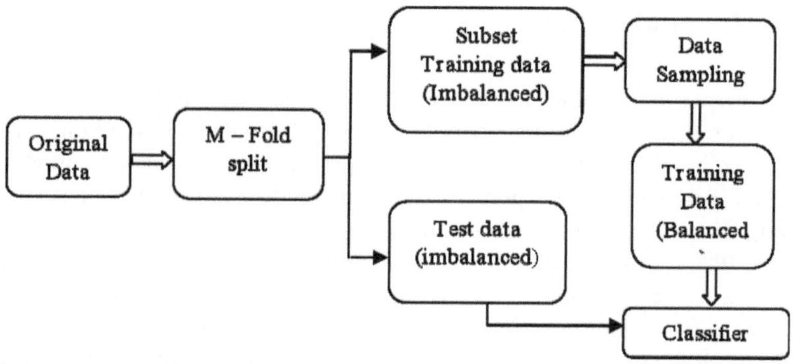


**Fig. 1** SMOTE resample work flow diagram

## 7 Stacking Methodology

In machine learning, ensemble methods use multiple models to obtain better predictive performance than that could be obtained from any of the constituent models [17]. Stacked generalization (or stacking) was first proposed by Wolpert in 1992 [18] is a way of combining multiple models that introduces the concept of a meta-learner. Although it is an attractive idea, it is less used than bagging and boosting in literature. Stacking is a machine learning technique and it is a variant in ensemble literature in that, it actively seeks to improve the performance of the ensemble by correcting the errors. It addresses the issue of classifier bias with respect to training data and aims at learning and using these biases to improve classification and it regarded as stacked generalization. It is concerned with combining multiple classifiers generated using different base classifiers $C1, C2, \ldots Cn$ on a single dataset $D$, which consist of pattern examples. In the first phase, a set of base level classifiers are generated. In second phase, a meta–level classifier is learned that combines the outputs of the base level classifier [5]. In brief, stacking can be visualized as a method which uses a new classifier to correct the errors of previously learned classifier.

The algorithm of stacking [18] is as given in Fig. 2.

| **Algorithm:** ***Stacked generalization*** **(or** ***stacking*****)** | |
|---|---|
| **Step 1.** | Split the training set into two disjoint sets. |
| **Step 2.** | Train several base learners on the first part |
| **Step 3.** | Test the base learners on the second part. |
| **Step 4.** | Using the predictions from 3) as the inputs, and the correct responses as the outputs, train a higher level learner. |
| **Note: the steps 1) to 3) are the same as cross -validation, but instead of using a winner -takes -all approach, here idea is to combine the base learners, possibly nonlinearly.** | |

**Fig. 2** Stacked generalization (or stacking)

## 8 Experimental Setup

In this work we have experimented using the following environment. The main objectives of this work is: (1) to notice the performance of stacking ensemble technique on lung nodule prediction data compared to single classifier model (2) to observe the role of how the performance can be boosted if the dataset is made balanced prior to classification algorithm.

For the first set of experiment we have used REPTree (Reduced Error Pruning Tree), Bagging (Bootstrapped Aggregating) and AdaBoost (Adaptive Boosting) algorithms as base classifiers. We have used the same REPTree as a base learning method also for bagging and AdaBoost. The choice of using REPTree in all the cases is to investigate how stacking method performs in homogenous ensemble condition. We have stacked above said four base models with stacking method using linear regression as a meta-level learner.

In the second set of experiments we have used the following model as base learners. REPTree, Naïve Bayes, PART [19] (rule based classifier), Bagging (here the J48decision tree has been used a base classifier), AdaBoost (here we have used Decision Stump as a base learner), Support Vector Machine (here the sequential minimum optimization is used to train the SVM with polynomial kernel). All the above mentioned six base models are learned and stacked using linear regression meta-level learning algorithm. We have run the experiment twice using above said environments, once on original dataset and once on resampled dataset. We have used m- fold cross validation, where the value of m is set to 5.

## 9 Performance Evaluation

In this work we are evaluating the performance of model using five performances metrics. Accuracy (ACC), Root Mean Squared Error (RMSE), F–Measure, Kappa Statics, Area Under Curve (AUC). Accuracy and F-measure are regarded as thresholded metrics and we have fixed threshold to 0.5 and it means that classifier above the threshold is considered as good performer and classifier which performance below are threshold regarded as under performer. Root Mean squared Error is used as probability metric. Probability metric are minimized when the predicted value for each case is equals to true conditional probability. AUC is used to a rank metric and this metric measures how well the positive cases and negative cases are ordered and viewed. Kappa statics is used as agreement measures, which in turn reflect how well model agrees between the expert prediction and machine prediction. The kappa interpretation scale has been given in Table 5.

**Table 5** Kappa statistics interpretation scale

| K-value | Strength of agreement |
|---|---|
| <0 | Poor |
| 0–0.2 | Slight |
| 0.21–0.4 | Fair |
| 0.41–0.6 | Moderate |
| 0.61–0.8 | Substantial |
| 0.81–1 | Almost perfect |

## 10 Results and Discussion

The experiments have been carried out in two different environments as discussed in previous section. In Tables 6 and 7 we have tabulated the results. Each corresponds to different base model and each column corresponds to the obtained performance metric results. For each performance metric there will be two results, which refer to classifier response to original dataset and resampled dataset. We have used the different base classifier for our experiment. This is because it has been claimed in the literature [20] that construction of ensemble is directly proportional to the choice of base learners, the reason behind this is if the base learner is unstable ensemble will get much diversity and works better, if not, ensemble of classifier faces over fitting problem.

### *10.1 With Homogenous Ensemble Environment*

We have tabulated the results and highlighted the best performing model (best in the column). In the entire performance category stacking has outperformed all other models. Only in the case of AUC it is equals with that of bagging method on both original dataset and resampled dataset. It worth noticing that all the classifiers including stacking have gained improvement in accuracy as well as on other metrics when operated on resampled dataset.

**Table 6** Results from experiment using homogenous ensemble of classifier

| Classifier | Accuracy | | RMSE | | F-measure | | Kappa | | AUC | |
|---|---|---|---|---|---|---|---|---|---|---|
| | OD | RD | OD | RD | OD | RD | OD | RD | OD | RD |
| REPTree | 74.09 | 81.54 | 0.31 | 0.26 | 0.82 | 0.87 | 0.68 | 0.77 | 0.92 | 0.95 |
| Bagging | 80.26 | 87.13 | 0.24 | 0.20 | 0.88 | 0.92 | 0.75 | 0.84 | 0.98 | 0.99 |
| AdaBoost | 74.33 | 83.18 | 0.27 | 0.22 | 0.83 | 0.89 | 0.67 | 0.79 | 0.96 | 0.98 |
| Stacking | 81.66 | 88.18 | 0.23 | 0.19 | 0.89 | 0.94 | 0.76 | 0.85 | 0.98 | 0.99 |

*OD* original dataset, *RD* resampled dataset

**Table 7** Results from experiment using heterogeneous ensemble of classifier

| Classifier | Accuracy | | RMSE | | F-measure | | Kappa | | AUC | |
|---|---|---|---|---|---|---|---|---|---|---|
| | OD | RD | OD | RD | OD | RD | OD | RD | OD | RD |
| REPTree | 74.90 | 81.54 | 0.31 | 0.26 | 0.82 | 0.87 | 0.68 | 0.77 | 0.92 | 0.95 |
| Naïve Bayes | 36.92 | 37.53 | 0.46 | 0.46 | 0.46 | 0.47 | 0.23 | 0.24 | 0.83 | 0.83 |
| PART | 76.98 | 83.56 | 0.29 | 0.25 | 0.84 | 0.88 | 0.71 | 0.79 | 0.92 | 0.95 |
| AdaBoost | 36.51 | 35.48 | 0.38 | 0.38 | 0.70 | 0.65 | 0.13 | 0.13 | 0.80 | 0.78 |
| Bagging | 75.04 | 82.24 | 0.26 | 0.24 | 0.84 | 0.88 | 0.68 | 0.77 | 0.97 | 0.98 |
| SVM | 58.65 | 58.53 | 0.36 | 0.36 | 0.75 | 0.76 | 0.46 | 0.46 | 0.90 | 0.90 |
| Stacking | **81.09** | **86.70** | **0.24** | **0.20** | **0.87** | **0.91** | **0.76** | **0.83** | **0.98** | **0.98** |

*OD* original dataset, *RD* resampled dataset

### *10.2 With Heterogeneous Ensemble Environment*

In the heterogeneous environment we have used different classifier with different base learners. Best example is, we have used REPTree as a base classifier for AdaBoost in previous set up and here we have selected decision stump. When REPTree is used as a base classifier for AdaBoost it has performed very well by obtaining ACC = 74.33 %/83.18 %, RMSE = 0.27/0.22, F- measure = 0.83/0.89, Kappa = 0.67/0.79 and AUC = 0.96/0.98. When we compare the same AdaBoost with decision stump base learner it has given ACC = 36.51 %/35.48 %, RMSE = 0.38/0.38, F- measure = 0.70/0.65, Kappa = 0.13/0.13 and AUC = 0.80/0.78 which shows the choice of base learner is also very important when we deal with ensemble methods. But when we compare the results between the homogenous stacked ensemble and heterogeneous stacked ensemble the results in the all the columns are almost similar. Stacking of classifier can be considered as a information fusion technique. This is because, as we have noticed in our experiments, the meta-learner in the stacking will correct the errors made by the base learners.

## 11 Conclusion

In this work we have presented experiments to observe the role of machine learning techniques on medical data which happens to be imbalanced. We used machine learning techniques both at data level and algorithmic level. At data level we have used resample technique called SMOTE to make data distribution balanced across the classes in the case prior to classification task. At algorithmic level we have used stacking ensemble method which uses stack of classifiers as a base model, gets their scores and in next phase uses meta-learning algorithm which corrects the error which has occurred in previous stage. Results from experiments shows that machine learning methods outperform at all the levels and we also observed the following.

1. Making the dataset balanced before classification task always improves the result significantly; this can be validated with results from [14].
2. With reference to [11] we can claim that performance of ensemble of classifier is always better when compared to performance of single classifier.
3. Stacking method can be considered as information fusing technique since the results from stacking method outperformed that of any other in the group. Stacking method also performed better compared to other available ensemble method such as bagging or AdaBoost.
4. It has been claimed in literature [6], that generally, heterogeneous ensemble of classifier model performs better than homogenous ensemble of classifiers. However, in our experiments when we compare with respect to some performance metric homogenous ensemble of classifier results show marginally better results. This reveals the fact that the choice of base learning algorithm is very much important in creating ensemble. It can easily happen that a particular data maybe best classified by one particular model of classifier and its ensemble may actually improve the result. On the contrary introducing heterogeneous ensemble of classifiers may actually not improve; perhaps degrade the result even if marginal. So the original data, resampling methods, all play subtle but important role in final performance.
5. Use of ensemble method is similar to the process carried out by human expert, since the output labels from ensemble is produced by a combination rule such as voting. In our experiment we can observe that predictions from stacking method is statistically signification and kappa statics interpret the level of agreement between the expert and that of machine prediction is almost perfect.

## References


1. Varutbangkul E, Mitrovic V, Raichu D, Furst J (2008) Combining boundaries abd rating from multiple observers for predicting lung nodule characteristics. In: IEEE international conference on biocomputing, bioinformatics and biomedical technologies, pp 82–87
2. Ebadollahi S, Johnson DE, Diao M (2008) Retrieving clinical cases through a concept space representation of text and images. SPIE Medical Imaging 2008: PACS and Imaging Informatics. 6919(7). ISBN: 9780819471031
3. Nakumura K, Yoshida H, Engelmann R, MacMahon H, Kasturagawa S, Ishida T et al (2000) Computerized analysis of the likelihood of malignancy in solitary pulmonary nodules with use of artificial neural networks. Radiology 214(3):823–830
4. Zinovev D, Raicu D, Furst J, Armato SG (2009) Predicting radiological panel opinions using a panel of machine learning classifiers. Algorithms 2:1473–1502. doi:10.3390/a2041473
5. Oza NC, Tumer K (2008) Classifier ensembles: select real-world applications. Inf Fusion 9(1):4–20
6. Reid S (2007) A review of heterogeneous ensemble methods. Department of Computer Science, University of Colorado at Boulder
7. Kuncheva LI, Rodriguez JJ (2010) Classifier ensemble for fMRI data analysis: an experiment, magnetic resonance imaging, vol 28. Elsevier Publications, pp 583–593

8. Caruana R, Niculescu-Mizil A, Crew G, Ksikes A (2004) Ensemble selection from libraries of models. In: 21st international conference on machine learning, Banff, Canada
9. Datta S, Pihur V, Datta S (2010) An adaptive optimal ensemble classifier via bagging and rank aggregation with application to high dimension data. BioMed Central 1471-2105/11/427, BMC Bioinformatics
10. Dzeroski S, Zenko B (2004) Is combining classifiers with stacking better than selecting the best one? Mach Learn 54:255–273, Kluwer Academic Publishers
11. Vinay K, Rao A, Hemantha Kumar G (2011) Comparative study on performance of single classifier with ensemble of classifiers in predicting radiological experts ratings on lung nodules. In: Indian international conference on artificial intelligence (IICAI). ISBN: 978-0-9727412-8-6, pp 393–403
12. Ting KM, Witten IH (1999) Issues in stacked generalization. J Artificial Intell Res 10:271–289
13. National Center for Biotechnology Information http://www.ncbi.nlm.nih.gov
14. Vinay K, Rao A, Hemantha Kumar G (2012) Sampling driven approaches for lung nodule characteristic rating predication. In: The 3rd international conference on intelligent information systems and management (IISM), ISBN No.: 978-93-90716-96-1
15. Chawla NV, Bowye KW, Hal LO, Kegelmeye WP (2002) SMOTE: synthetic minority over-sampling technique. J Artificial Intell Res 16:321–357
16. Vinay K, Rao A, Hemantha Kumar G (2011) Computerized analysis of classification of lung nodules and comparison between homogeneous and heterogeneous ensemble of classifier model. In: 3rd national conference on computer vision, pattern recognition, image processing and graphics, 978-0-7695-4599-8/11, IEEE doi:10.1109/NCVPRIPG.2011.56, pp 231–234
17. Polikar R (2006) Ensemble based systems in decision making. IEEE Circuits Syst Mag 6(3):21–45. doi:10.1109/MCAS.2006.1688199
18. Wolpert DH (1992) Stacked generalization. Neural Networks 5(2):241–259
19. Frank E, Witten IH (1998) Generating accurate rule sets without global optimization. In: Shavlik J (ed) Machine learning: proceedings of the fifteenth international conference. Morgan Kaufmann Publishers, San Francisco
20. Polikar R (2009) Ensemble learning. Scholarpedia 4(1):2776

# A Comparative Analysis of Intensity Based Rotation Invariant 3D Facial Landmarking System from 3D Meshes

**Parama Bagchi, Debotosh Bhattacharjee, Mita Nasipuri and Dipak Kr. Basu**

**Abstract** In this paper, we propose a novel approach of a rotation invariant 2.5D face landmarking system which is based on generating the nose-tip, which is essential for the registration of 3D face images. Here we compare our system with the manual landmarking system available in the Bosphorus database. Here we have applied the maximum intensity technique as described in Sect. 3 to determine the nose tip and we have applied it in case of both the landmarked model and the face model generated from the raw Bosphorus images. We have experimented on 988 3D face models selected from the Bosphorus database, and our technique displays 100 % of good nose-tip localization and we have also presented the standard deviation between the landmarked and our generated feature localization technique. It has been proved that the standard deviation between the nose-tip localized in case of the 3D mesh image and the 3D landmarked model of a particular individual is less than 1 which supports the fact that there is significantly very less difference between the generated landmark model and the 3D mesh grid, thereby proving that our method gives good performance based on feature detection and that the feature i.e. the nose-tip has been correctly detected.

P. Bagchi (✉) · D. Bhattacharjee · M. Nasipuri · D. Kr. Basu
Department of Computer Science and Engineering, MCKV Institute of Engineering, Kolkata 711204, India
e-mail: paramabagchi@gmail.com

D. Bhattacharjee
e-mail: debotosh@ieee.org

M. Nasipuri
e-mail: mitanasipuri@gmail.com

D. Kr. Basu
e-mail: dipakkbasu@gmail.com

P. Bagchi · D. Bhattacharjee · M. Nasipuri · D. Kr. Basu
Department of Computer Science and Engineering, Jadavpur University, Kolkata 700032, India

P. P. Swamy and D. S. Guru (eds.), *Multimedia Processing, Communication and Computing Applications*, Lecture Notes in Electrical Engineering 213,
DOI: 10.1007/978-81-322-1143-3_12, 

The main aim of localizing the nose-tip is to, make way for a robust 3D registration technique as discussed in Sect. 3.

**Keywords** Landmarked image · Intensity · Registration · Standard deviation

## 1 Introduction

The analysis of 3D faces is important in many applications, especially in the biometric and medical fields. Such applications aim to accurately relate information from different meshes in order to compare them. A common approach to compare 3D meshes is by rigid registration, where two or more meshes are fitted in exact alignment with one another. The localization of specific landmarks and regions on faces often plays important part in these applications. Land marks can help the registration algorithms in achieving rough alignment of meshes. In biometric applications, landmarks are often instrumental in the generation of signatures for faces [1]. In this paper, we propose a robust framework to accurately localize landmark i.e. the nose-tip from three-dimensional faces obtained from the 3D meshes and we have tested our method by applying it on the meshes generated from the landmark points available from the Bosphorus database and finally tested how much deviated are the nose-tips of the generated model from the original landmarked model. This paper is organized as follows. Section 2 discusses some of the related works on 3D land marking. Section 3 describes the proposed algorithm. Section 4 describes the experimental results and discussions. Finally conclusions and future scope are enlisted in Sect. 5. A common approach to compare 3D meshes is by a rigid registration, where two or more meshes are fitted in exact alignment with one another. The localization of specific landmarks and regions on faces often plays a important part in these applications. Land marks can help the registration algorithms in achieving rough alignment of meshes. In biometric applications, landmarks are often instrumental in the generation of signatures for faces [1]. In this paper, we propose a robust framework to accurately localize landmark i.e. the nose-tip from three- dimensional faces obtained from the 3D meshes and we have tested our method by applying it on the meshes generated from the landmark points available from the Bosphorus database and finally tested how much deviated are the nose-tips of the generated model from the original landmarked model. This paper is organized as follows. Section 2 discusses some of the related works on 3D land marking. Section 3 describes the proposed algorithm. Section 4 describes the experimental results and discussions. Finally conclusions and future scope are enlisted in Sect. 5.

## 2 Related Works

In this section, let us discuss some existing approaches on 3D face detection, landmark localization, face registration and statistical models for face analysis. Colombo [2] performed 3D face detection by first identifying candidate eyes and noses in a classifier. However, the authors highlight that their method is highly sensitive to the presence of outliers and holes around the eyes and nose regions. Most of the existing approaches [3, 4] target face localization, rather than detection, where the presence and number of faces is known. In Ref. [4], face localization is performed by finding the nose tip and segmentation is done through a cropping sphere centered at the nose tip. This approach is highly restrictive to the database used, as each input mesh is assumed to contain only one frontal face and no pose variations is taken into consideration. Moreover, the cropping sphere has a fixed radius over the entire database and hence the segmentation is not robust to scale variation. In Ref. [4], 3D point clustering using texture information is performed to localize the face. This method relies on the availability of a texture map and the authors state, reduction in stability with head rotations greater than ±45° from the frontal pose. Once faces are detected and segmented, landmark localization is often used for face analysis. Many existing approaches rely on accurately locating corresponding landmarks or regions to perform a rough alignment of meshes. In Ref. [5], a curvature based "surface signature" image is used to locate salient points for the estimation of the rigid transformation. Shape models used in these works do not involve fitting the model to 3D data, devoid of texture. The dependence on prior knowledge of feature map thresholds, orientation and pose is evident in most existing methods for landmark localization on meshes [6, 7]. In all the above models, pose variance is not taken into consideration very significantly. But in the landmarking model that we have described, our method is invariant to any pose variation.

## 3 Proposed Algorithm

A range image (Fig. 1) is a set of points in 3D each containing the intensity of individual pixels. The technique also holds in case of 2.5D images which may be described as containing at least one depth value for every (x, y) coordinate. The acquisition process is described as follows: Normally, a 3D mesh image is 3D camera such as a Minolta Vivid 700 camera and a range image is generated from the 3D mesh image. The image in our case is generally in the form of z = f(x, y). Next, some pre-processing methods is applied to eliminate unwanted details such as facial hairs, scars etc. And finally using a geometric approach, the nose-tip is being localized.

Now, we shall discuss the proposed algorithm. Figure 2 shows the proposed technique.

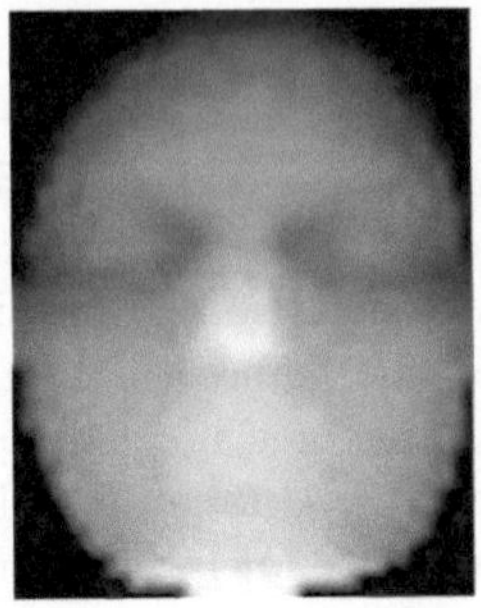

**Fig. 1** Face images: 2.5D range image

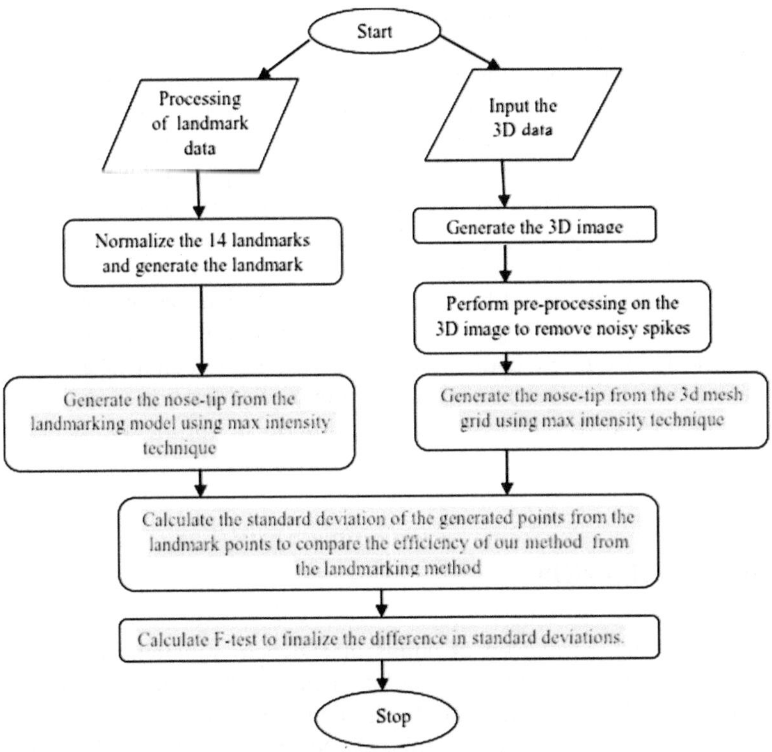


**Fig. 2** An overview of the present proposed method

The present technique makes use of the following steps:

### 3.1 Processing of Landmark Data from Bosphorus Database

We use the 3D facial model from the Bosphorus database. The database composes of a .lm3 file and a .bnt file. With this landmarking model i.e. the .lm3 file, we build a parameterized model, $\Omega = Y(b)$, where $\Omega = \{\omega_1, \omega_2, \ldots, \omega_N\}$, with $\omega i = (x_i, y_i, z_i)$ representing each landmark.

Now as can be seen in Fig. 3, there are 14 landmarks in the Bosphorus database. We have taken the data, discarded the non-face regions and then we have built up the model i.e. the landmarked range image. The landmarked range image, which has been generated by our system is shown in Fig. 4.

In Fig. 4, we demonstrate the landmark models corresponding to frontal pose (a), rotated about y-axis (b), rotated about x-axis (c) and rotated about yz-axis (d). As we can see in Fig. 4, it represents a face because we are building up the entire model with only 14 landmarks. The 14 landmark data of each individual as given in the Bosphorus database have to be normalized and brought within proper coordinate system, otherwise the performance of our comparison based system could not be justified. Here pre-processing of the landmark model is not required because it hardly contains noise.

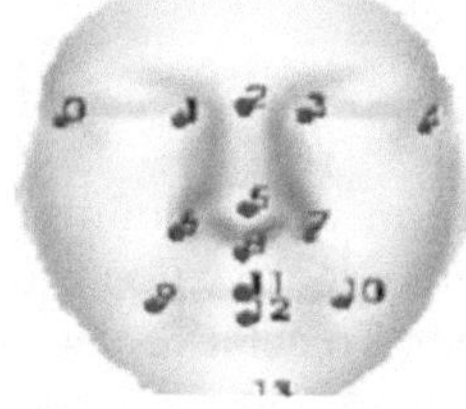

**Fig. 3** Sample face showing the *14* landmarks used to generate the landmark model

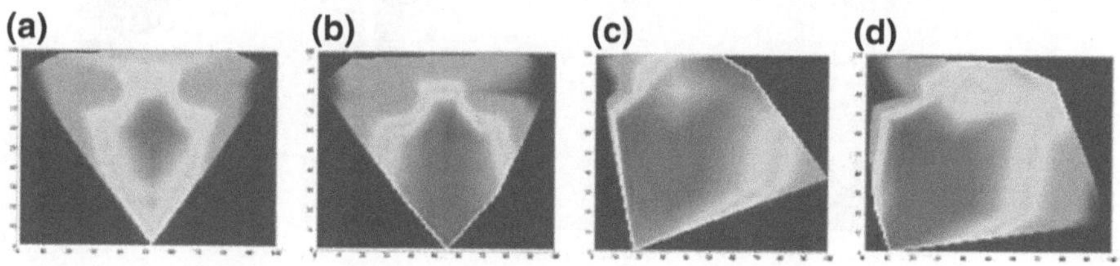


**Fig. 4** Sample face from the Bosphorus database showing the landmark model. **a** Frontal pose. **b** Rotated about x-axis. **c** Rotated about y-axis at 30°. **d** Rotated about yz-axis

### 3.2 3D Range Image Acquisition

A range image is actually an array of numbers where the numbers quantify the distances from the focal plane of the sensor to the surfaces of objects within the field of view along rays emanating from a regularly spaced grid. Different from 3D mesh images, it is easy to utilize the 3D information of range images because the 3D information of each point is explicit on a regularly spaced grid. The Fig. 5 show samples of range images that we have taken for testing from the Bosphorus database.

The mesh-grids corresponding to the range images shown in Fig. 5a–d are shown in Fig. 6.

### 3.3 Perform Pre-Processing on the 3D Image

Sometimes 3D face images are affected by noise and several other factors. So some types of smoothing techniques are to be applied. In our present technique, we have extended the concept of 2D weighted median filtering technique to 3D face images. The present technique performs filtering of 3D dataset using the weighted median implementation [8] of the mesh median filtering. The weighted median filter is a modification of the simple median filter. After smoothing the results corresponding to Fig. 6a–d are obtained in Fig. 7a–d respectively.

### 3.4 Feature Localization

Faces have a common general form with prominent local structures such as eyes, nose, mouth, chin etc. Facial feature localization is one of the most important tasks of any facial classification system. To achieve fast and efficient classification, it is needed to identify features which are mostly needed for classification task.

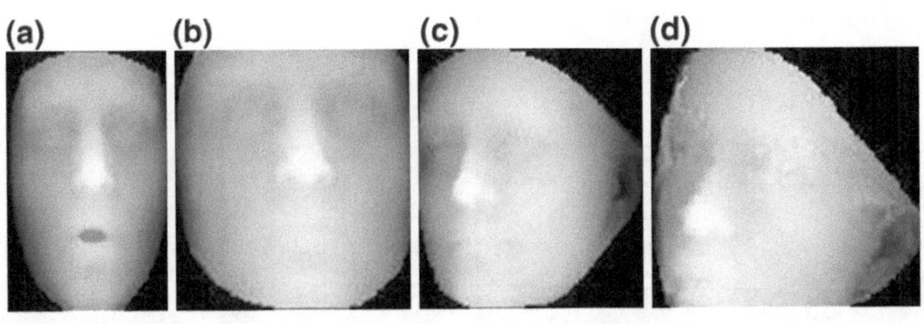


**Fig. 5** Samples from the Bosphorus database corresponding to a single person for frontal pose (**a**), image rotated about y axis (**b**), image rotated about x-axis (30°) (**c**), image rotated about yz-axis (**d**)

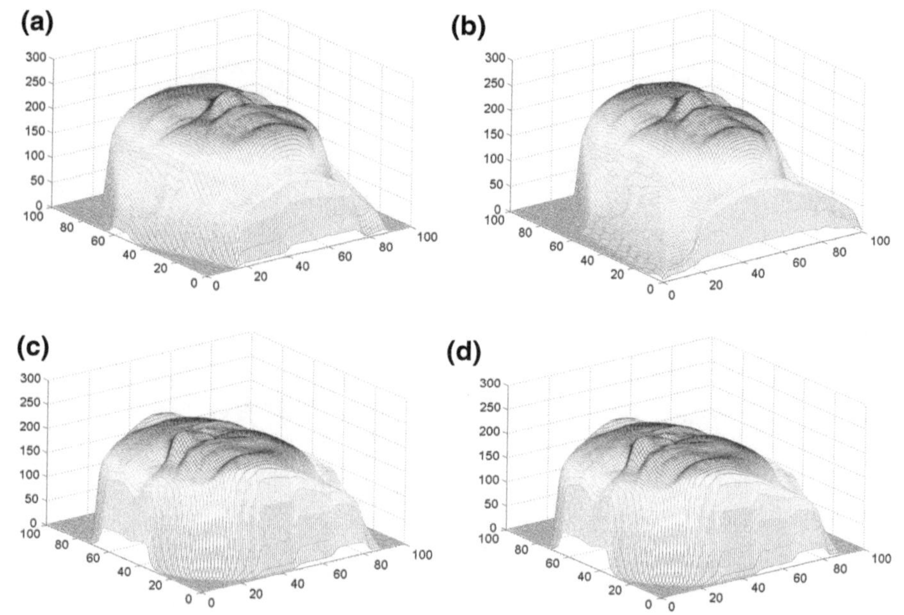


**Fig. 6** Mesh-grids from the Bosphorus database corresponding to a single person (female) for frontal pose (**a**), image rotated about y axis (**b**), image rotated about x-axis (30°) (**c**), image rotated about yz-axis (**d**)

### 3.4.1 Surface Generation

The next part of the present technique concentrates on generating the surface [9] of this 3D mesh image. For the nose tip localization we have used the maximum intensity concept as the tool for the selection process. Each of the landmark models as shown in Fig. 3 were inspected for localizing the nose tip. A set of fiducially points are extracted from both frontal and various poses of face images using a maximum intensity algorithm [9]. As shown in Fig. 8, the nose tips have been labeled on the facial surface, and accordingly, the local regions are constructed based on these points. The maximum intensity algorithm used for our purpose is given below:

```
Algorithm Find_Maximum_Intensity(Image)

  Step 1:- Set max to 0
  Step 2:- Run loop for I from 1 to width_of_Image
  Step 3:- Run loop for J from 1 to height_of_Image
  Step 4:- Set val to sum(image(I−1:I+1,J−1:J+1))
  Step 5:- Check if val is greater than max.
  Step 6:- Set val to max
```

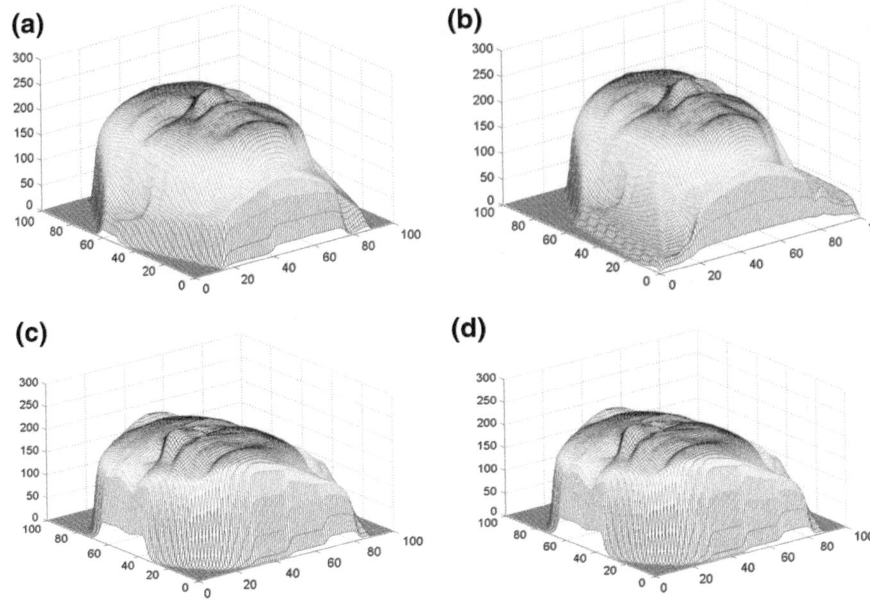

Fig. 7 Smoothed mesh-grids from the Bosphorus database corresponding to a single person (female) for frontal pose (**a**), image rotated about y axis (**b**), image rotated about x-axis (30°) (**c**), image rotated about yz-axis (**d**)

```
    Step 7:- End if
    Step 8:- End loop for J
    Step 9:- End loop for I
End
```

The same approach is hereby applied to the mesh-grids obtained as shown in Fig. 9 which shows the result with the nose-tip localized.

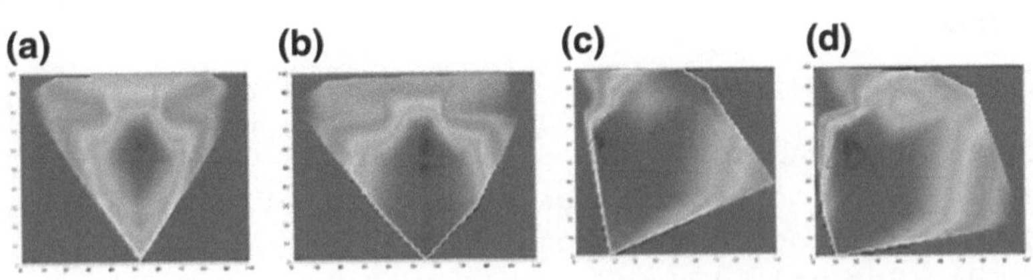

**Fig. 8** Sample face from the Bosphorus database showing the landmark model with nose-tips localized. **a** Frontal pose. **b** Rotated about x-axis. **c** Rotated about y-axis at 30°. **d** Rotated about yz-axis

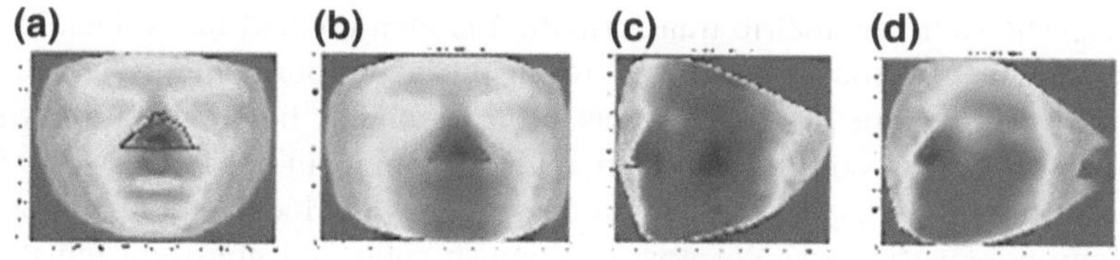

**Fig. 9** Smoothed mesh-grids from the Bosphorus database corresponding to a single person (female) for frontal pose. **a** Image rotated about y axis. **b** Image rotated about x-axis (30°). **c** Image rotated about yz-axis. **d** With nose-tips localized

## *3.5 Calculate the Standard Deviation of the Points Generated from the Landmark Points*

In this step, we have extracted the nose coordinates in the form of (x, y) from the landmark model. The steps that we have performed to check how much our generated nose-tips vary from the coordinates (x, y) extracted from the landmark model are listed in the following Algorithm:

```
Algorithm CompareLandmark
  Step 1:- At first calculate the mean for all the data points.
  Step 2:- The next step is to get the deviations in these
  numbers. Do this by subtracting the mean from each of the
  numbers in the set of data.
  Step 3:- Now, square the deviations calculated in Step 2.
  Step 4:- Next, add all of the squares in Step 3 together.
  Step 5:- Now divide the sum in Step 4 by the total number of
  data in both x and y coordinates.
  Step 6:- Next take the square root of the result in Step 5.
  Step 7:- The result in step 6 is the standard deviation of
  the original dataset from the landmarked model.
  Step 8:- Calculate the co-variance of x and y
  Step 9:- Calculate the standard deviation of x and y
  Step 10:- Finally, apply f test to see the difference of
  standard deviations.
End of Algorithm
```

## *3.6 3D Registration*

Registration basically means transforming shapes in such a way that they can be compared. For 3D face recognition, e.g. it is common to locate a number of landmarks (e.g. eyes, nose, and mouth) in each face and rotate, translate and scale these landmarks in such a way that they are projected to fixed, predefined positions. The same geometric transformation is then applied to the facial image. The facial image is thus transformed to an intrinsic coordinate system. In this paper only the feature localization has been specified, but registration will be performed on the basis of the generated landmark points in our future work. In the present method, we have specified intuitively on the localization of nose-tips across any pose variations. We now distinguish rigid and non-rigid registration. The former only performs rotation and translation (and possibly scaling) of the point clouds. The latter also allows for (small) deformations of the point cloud to realise an optimal registration. Non-rigid registration can be useful in handling facial expressions. Our method of registration would be rigid registration only.

Also there are yet some other classifications of registration technique which are enlisted as follows:

- One-to-all registration (register one face to another).
- Registration to a face model or atlas.
- Registration to an intrinsic coordinate system using geometric properties of the face like landmarks. Our proposed technique for registration would be to some intrinsic coordinate system using geometric properties of the face like landmarks. In the algorithm below, we propose a Algorithm for registration which would be implemented as a part of our future work.

```
Algorithm Registration_3D
  Step 1:- Input the 3D image in frontal pose and the rotated
  image
  Step 2:- Pre-process both the 3D images
  Step 3:- Generate the nose-tips both in case of the 3D image
  and the rotated images.
  Step 4:- while(x coordinate of rotated_image <= x coordi-
  nate of frontal_image)
  Step 5:- if (x coordinate of rotated_image <e)
  Step 6:- Output the registered image and exit
  Step 7:- else
  Step 8:- Rotate the image by 2°.
  Step 9:- End if
  Step 10:-End while loop
End Algorithm
```

# 4 Experimental Results

To evaluate the accuracy o f the landmark localization, we compare the landmarks localized on 988 faces of the Bosphorus database with the provided ground-truth. Since, we are concentrating on the pose-variations, so it is better to say that let us first list on how many 3D range images the max-intensity algorithm is correctly able to detect the nose-tips. We have considered the pose-variations and thus listed the results of nose-tip localization in all frontal poses including expressions as shown in Table 1 (Figs. 10, 11, 12, 13).

In Table 2, we present the result of our nose-tip detection method up to pose-variations of 10° across YZ-axis.

In Table 3, we present the result of our nose-tip detection method up to pose-variations of 5° and 10° across X-axis.

In Table 4 we present the result of our nose-tip detection method up to pose-variations of 5° and 10° across Y-axis.

Here in the following section, we would list the comparison of our technique on both landmarked and generated 3D mesh-grids are listed as below:

As we see in Fig. 14, we have plotted some samples marked in red from the landmarked image and some samples marked in black from the generated 3D mesh. The samples that we have basically taken in Fig. 14 are all expressions faces taken from the Bosphorus database. Figure 15 shows the standard deviation we have obtained for the expressions face i.e. ANGER and DISGUST and we list the standard deviation of the generated range model and the landmarked model.

Now, we have plotted in Fig. 16, the standard deviations that we have obtained in case of expressions, neutral and pose variations of 3D faces.

In the last and final step, we have applied f-test. The result of the test must be one of the two following conclusions:

**Table 1** Results of nose-tip localization in frontal pose with expressions

| No. of nose-tips | No. of nose-tips correctly detected | Percentage of success | Percentage of failures |
|---|---|---|---|
| 798 | 798 | 100 | 0 |

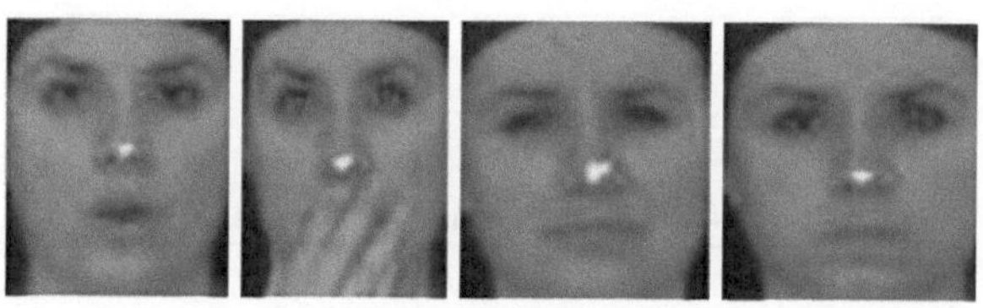

**Fig. 10** Some samples in frontal pose from Bosphorus database

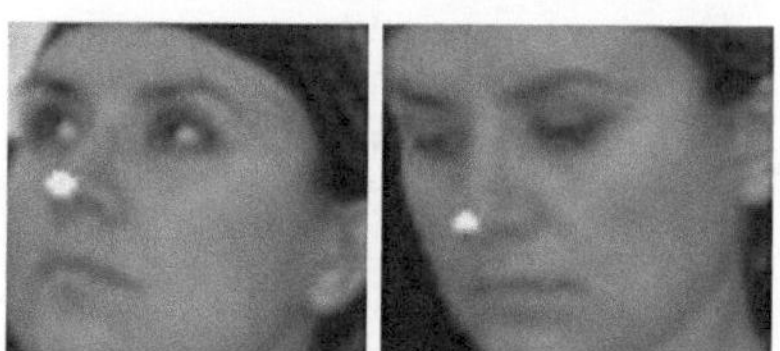

**Fig. 11** Some samples in non-frontal pose from Bosphorus database with images rotated about yz-axis

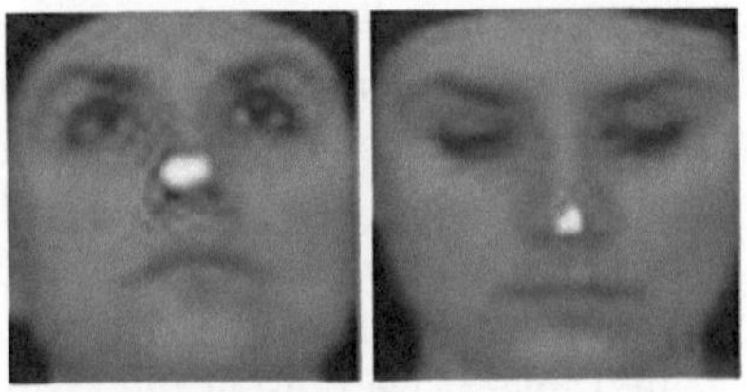

**Fig. 12** Some samples in non-frontal pose from Bosphorus database with images rotated about x-axis

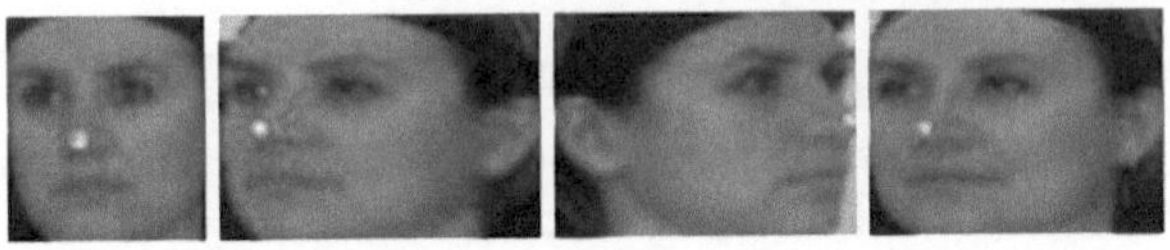

**Fig. 13** Some samples in non-frontal pose from Bosphorus database with images rotated about y-axis

- The population standard deviations are different from each other (if $F > 1$).
- The population standard deviations are not different from each other (if $F < 1$).The test statistic of this curve is in the form $F = s_1^2/s_2^2$ where $s_1$ and $s_2$ are the standard deviations.

The standard deviations calculated falls in the region where $F < 1$ shown in Fig. 17 which means there is significantly very less difference between the above calculated standard deviations. Thus we have proved that the standard deviations between a particular 3D face mesh-grid and its' corresponding landmarked model of a certain individual are same which proves our statistics.

**Table 2** Results of nose-tip localization in rotated pose with respect to yz axes

| Viewpoint around YZ axes | No. of nose-tips | No. of nose-tips correctly detected | Percentage of success | Percentage of failures |
|---|---|---|---|---|
| +10 | 19 | 19 | 100 | 0 |
| −10 | 19 | 19 | 100 | 0 |

**Table 3** Results of nose-tip localization in rotated pose with respect to x axes

| Viewpoint around X axes | No. of nose-tips | No. of nose-tips correctly detected | Percentage of success | Percentage of failures |
|---|---|---|---|---|
| +5 | 19 | 19 | 100 | 0 |
| −5 | 19 | 19 | 100 | 0 |
| +10 | 19 | 19 | 100 | 0 |
| −10 | 19 | 19 | 100 | 0 |

**Table 4** Results of nose-tip localization in rotated pose with respect to y axes

| Viewpoint around Y axes | No. of nose-tips | No. of nose-tips correctly detected | Percentage of success | Percentage of failures |
|---|---|---|---|---|
| +5 | 19 | 19 | 100 | 0 |
| −5 | 19 | 19 | 100 | 0 |
| +10 | 19 | 19 | 100 | 0 |
| −10 | 19 | 19 | 100 | 0 |

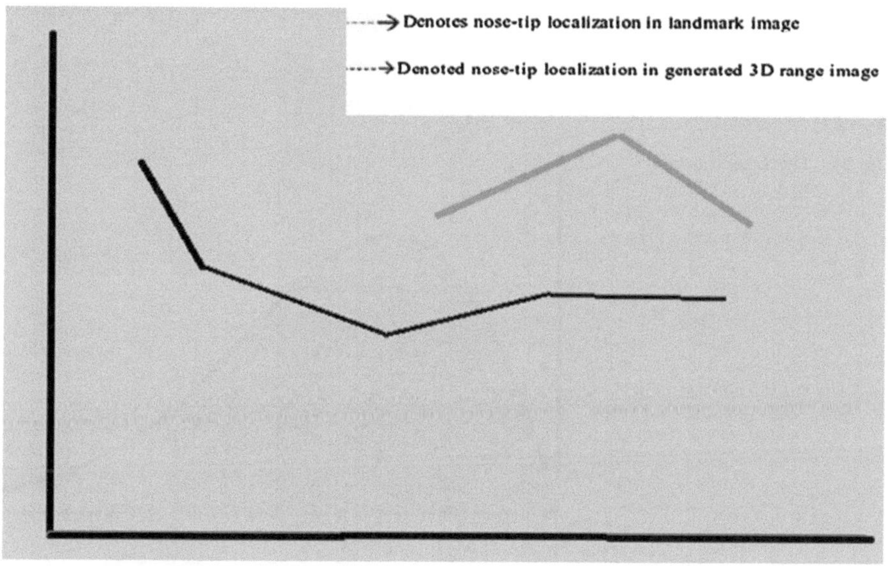


**Fig. 14** Some samples of nose-tips plotted from the 3D landmark model

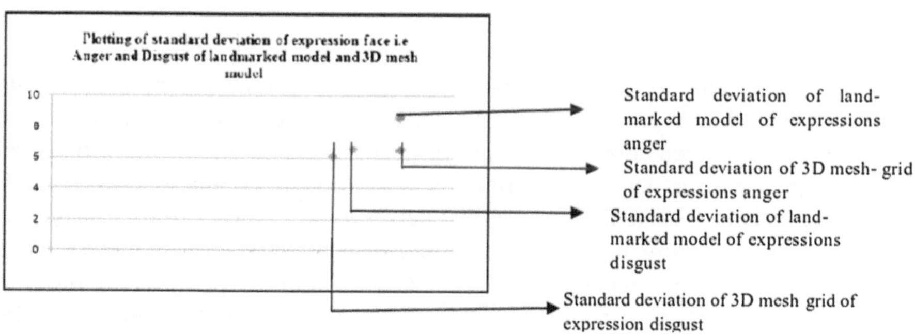


**Fig. 15** Some samples of nose-tips plotted from the 3D landmark model and 3D mesh

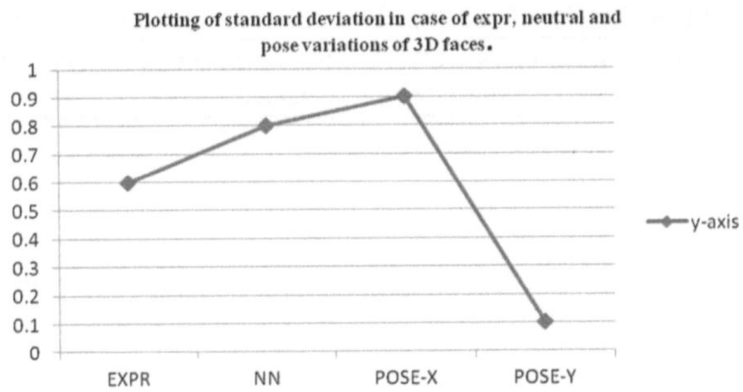


**Fig. 16** Standard deviations of landmarked model from 3D generated model of poses in case of expression faces, neutral faces, poses rotated about x axis and poses rotated about y-axis

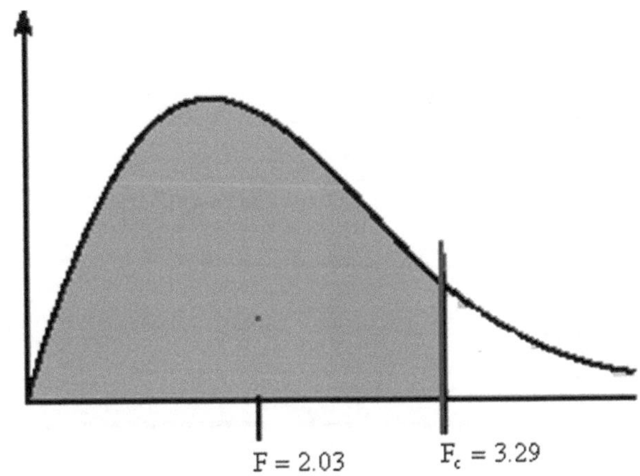


**Fig. 17** The f test curve

# 5 Conclusion and Future Scope

In this paper, we have presented a novel technique for localization of nose-tip in 3D face images and we have compared our method with the landmarked data models available. We have detected the feature i.e. the nose-tip and the entire process of nose-tip detection is invariant to pose variations i.e. we have detected the nose-tip in case of any pose variations across the X, Y and Z axes. Experimental results demonstrate that our method performs well in case of the Bosphorus database. For our future work, the performance evaluation of our method is in progress in case of other databases also. As a part of our future work, we propose to attempt the problem of registration across pose-variations by finding out the translation and rotational parameters. Also, we aim at performing 3D face recognition by using standard classifiers like PCA, LDA etc.

## References


1. Shi J, Samal A, Marx D (2006) How effective are landmarks and their geometry for face recognition. In: International proceedings of computer vision and image understanding, vol 102. pp 117–133
2. Colombo A, Cusano C, Schettini R (2006) 3D face detection using curvature analysis. In: Proceedings of pattern recognition, vol 39. pp 444–455
3. Mian A, Bennamoun M, Owens R (2007) An efficient multimodal 2D–3D hybrid approach to automatic face recognition. Proc IEEE Trans Pattern Anal Mach Intell 29:1927–1943
4. Niese R, Al-Hamadi A, Michaelis B (2007) A novel method for 3D face detection and normalization. J Multimedia 2:1–12
5. Yamany SM, Farag A (2002) Surfacing signatures: an orientation in- dependent free-form surface representation scheme for the purpose of objects registration and matching. Proc IEEE Trans Pattern Anal Mach Intell 24:1105–1120
6. Colbry D, Stockman G, Jain A (2006) Detection of anchor points for 3D face verification. In: Proceedings of IEEE conference on computer vision and pattern recognition, New York, pp 118–125
7. Nair P, Zou L, Cavallaro A (2005) Facial scan change detection. In: Proceedings of European workshop on the integration of knowledge, semantic and digital media technologies, London, pp 77–82
8. Bagchi P, Bhattacharjee D, Nasipuri M, Basu DK (2012) A novel approach for nose tip detection using smoothing by weighted median filtering applied to 3D face images in variant poses. In: Proceedings of the international conference on pattern recognition, informatics and medical engineering, IEEE, Periyar University, pp 272, 21–23 Mar 2012
9. Bagchi P, Bhattacharjee D, Nasipuri M, Basu DK (2012) A novel approach for registration of 3D face images. In: Proceedings of international conference on advances in engineering, science and management (ICAESM), E.G.S. Pillay Engineering College, Nagapattinam, pp 1–7, 30–31 Mar 2012

# Thermal Human Face Recognition Based on Haar Wavelet Transform and Series Matching Technique

**Ayan Seal, Suranjan Ganguly, Debotosh Bhattacharjee, Mita Nasipuri and Dipak Kr. Basu**

**Abstract** Thermal infrared (IR) images represent the heat patterns emitted from hot object and they don't consider the energies reflected from an object. Objects living or non-living emit different amounts of IR energy according to their body temperature and characteristics. Humans are homoeothermic and hence capable of maintaining constant temperature under different surrounding temperature. Face recognition from thermal (IR) images should focus on changes of temperature on facial blood vessels. These temperature changes can be regarded as texture features of images and wavelet transform is a very good tool to analyze multi-scale and multi-directional texture. Wavelet transform is also used for image dimensionality reduction, by removing redundancies and preserving original features of the image. The sizes of the facial images are normally large. So, the wavelet transform is used before image similarity is measured. Therefore, this paper describes an efficient approach of human face recognition based on wavelet transform from thermal IR images. The system consists of three steps. At the very first step, human thermal IR face image is preprocessed and the face region is only cropped from the entire image. Secondly, "Haar" wavelet is used to extract low frequency band from the cropped face region. Lastly, the image classification

A. Seal (✉) · S. Ganguly · D. Bhattacharjee ·
M. Nasipuri · D. Kr. Basu
Department of Computer Science and Engineering, Jadavpur University,
Kolkata 700032, India
e-mail: ayan.seal@gmail.com

S. Ganguly
e-mail: suranjanganguly@gmail.com

D. Bhattacharjee
e-mail: debotosh@indiatimes.com

M. Nasipuri
e-mail: mnasipuri@cse.jdvu.ac.in

D. Kr. Basu
e-mail: dipakkbasu@gmail.com

P. P. Swamy and D. S. Guru (eds.), *Multimedia Processing, Communication and Computing Applications*, Lecture Notes in Electrical Engineering 213,
DOI: 10.1007/978-81-322-1143-3_13, 

between the training images and the test images is done, which is based on low-frequency components. The proposed approach is tested on a number of human thermal infrared face images created at our own laboratory and "Terravic Facial IR Database". Experimental results indicated that the thermal infra red face images can be recognized by the proposed system effectively. The maximum success of 95 % recognition has been achieved.

**Keywords** IR image · Haar wavelet transform · Series matching

## 1 Introduction

Since last three decades there exists many commercially available systems of face recognition technology to identify human faces; however face recognition is still a challenging area in computer vision and pattern recognition. The objectives of the face recognition system is to match the data i.e. faces in the stored database to determine the identity of the possible candidate. Face recognition is a sophisticated problem because of the generally similar shape of faces shared with the numerous variations between images of the same face. Various methods have been used to solve the problem. Every method has its own merits and demerits. Most of the research works in this area have been focused on visible spectrum imaging due to easy availability of low cost visible band optical cameras. But, it requires an external source of illumination. Even though the success of automatic face recognition techniques in many practical applications, the task of face recognition based only on the visible spectrum is still a challenging problem under uncontrolled environments. Thermal IR images [1] have been suggested as a possible alternative in handling situations where there is no control over illumination. Thermal IR images represent the heat patterns emitted from an object and they don't consider the reflected energy. Objects emit different amounts of IR energy according to their body temperature and characteristics. Previously, Thermal IR camera was costly but recently the cost of IR cameras has been considerably reduced with the development of CCD technology [2]; thermal images can be captured under different lighting conditions, even under completely dark environment. Using thermal images, the tasks of face detection, localization, and segmentation are comparatively easier and more reliable than those in visible band images [3]. Humans are homoeothermic and hence capable of maintaining constant temperature under different surrounding temperature and since, blood vessels transport warm blood throughout the body; the thermal patterns of faces are derived primarily from the pattern of blood vessels under the skin. These temperature changes can be regarded as texture features of images and wavelet transform is a very good tool to analyze texture with multi scales and multiple directions. In the recent years, wavelet analysis is being popular to the researchers in the field of both theoretical and applied mathematics, and the wavelet transform in particular has established to be an effective tool for data analysis, numerical analysis, image processing [4] etc. Face recognition is realized by image or feature comparison. The pixels of the whole image are concatenated in row major order or column major order. So, the image can be viewed as a series or vectors or sequences. The problem of image comparison can be transformed into the problem of series comparison. The size of the facial images are generally larger, so the number of pixels of facial images are usually huge, so the wavelet transform is used before image comparison, which can effectively reduce the computational complexity. The paper is arranged as follows. Section 2 presents about the outline proposed system. Section 3 shows the experiment and results. Finally, Sect. 4 concludes and mentions some remarks about different aspects analyzed in this paper.

## 2 Outline of the Proposed System

The proposed Thermal Face Recognition System (TFRS) can be subdivided into four main parts, namely image acquisition, image preprocessing, feature extraction, and classification. In image acquisition stage, a FLIR 7 thermal infrared camera has been used to acquire 24-bits colour thermal face images. The images are saved in JPEG format. A thermal face image depicts interesting thermal information of a facial model. The image pre-processing part involves binarization of the acquired thermal IR image, extraction of largest component as the face region, finding the centroid of the face region and finally cropping of the face region in elliptical shape. In feature extraction part, finds LL band, HL band, LH band and HH band using "Haar" wavelet transform. The LL band contains sufficient information to represent the original image and all other band are to be eliminated here. However, size of the LL band is one-fourth of the original image. So, "Haar" wavelet is used to reduce dimensionality of the original image. Two dimensional LL band image are converted into one-dimensional horizontal vector in row-major order. Then these reduced one-dimensional horizontal feature vectors are fed into a series matching classifier. The block diagram of the proposed system is given in Fig. 1. Different image processing and classification techniques used here are discussed in detail in subsequent subsections.

### 2.1 *Thermal Face Image Acquisition*

In the present work, thermal and visible face images are acquired simultaneously under variable expressions, poses and with/without glasses. Till now 76 individuals have volunteered for this photo shoots and for each individual 39 different templates of RGB color images with Exp1 (happy), Exp2 (angry), Exp3 (sad), Exp4 (disgusted), Exp5 (neutral), Exp6 (fearful) and Exp7 (surprised) are taken.

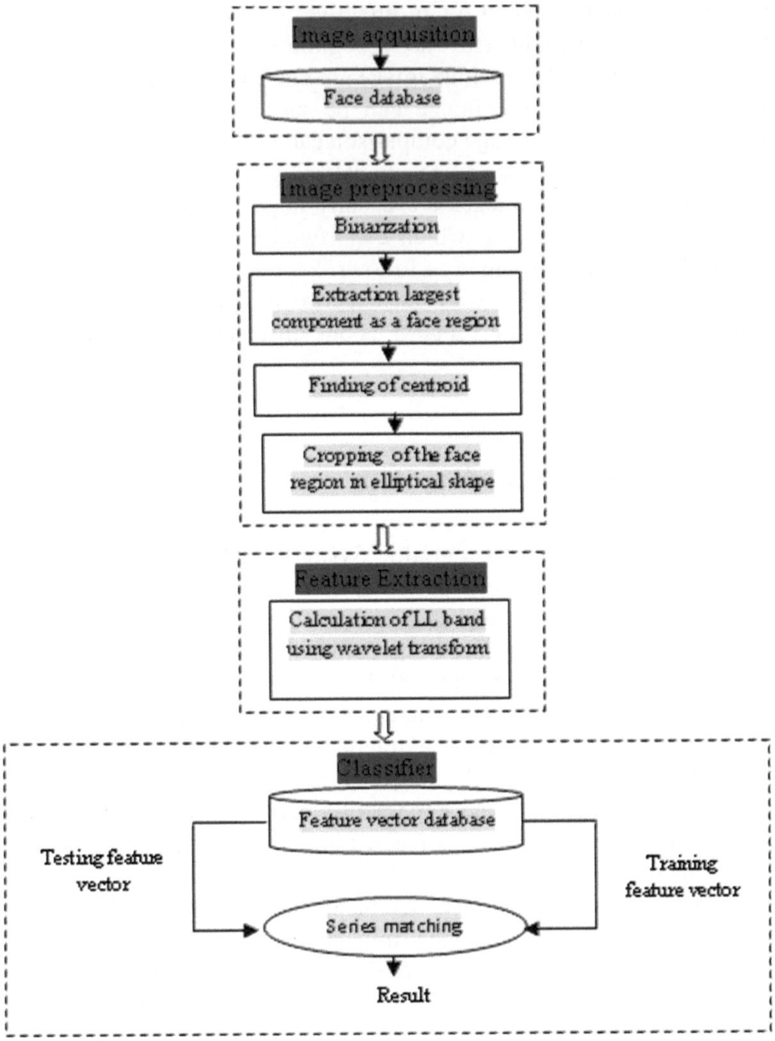


**Fig. 1** Schematic block diagram of the proposed system

Different pose changes about x-axis, y-axis and z-axis are also taken. Resolution of each image is 320 × 240 and the images are saved in JPEG format. Two different cameras are used to capture this database. One is Thermal—FLIR 7 and another is Visible—Sony cyber shot. A typical thermal face image is shown in Fig. 2a. This thermal face image depicts interesting thermal information of a facial model.

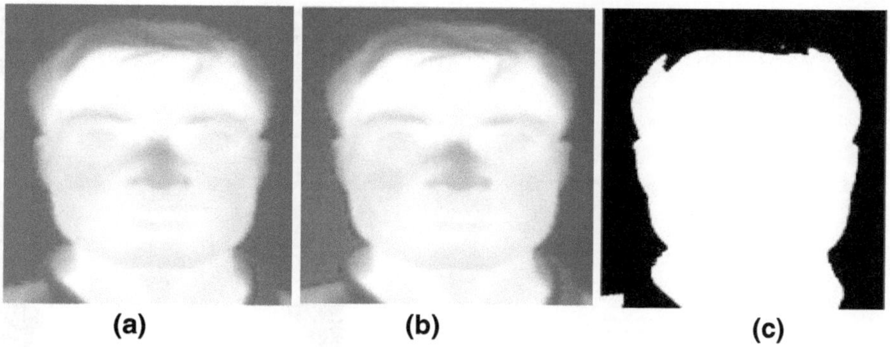


**Fig. 2** Thermal face image and its various preprocessing stages. **a** A thermal face image. **b** Corresponding grayscale image. **c** Binary image

## 2.2 Binarization

Each of the captured 24-bits colour images have been converted into its 8-bit grayscale image. The grayscale image from the previous sample image is shown in Fig. 2b. Then convert those grayscale images into corresponding binary images. The resultant image replaces all pixels in the grayscale image with luminance greater than mean intensity with the value 1 (white) and replaces all other pixels with the value 0 (black). In binary image, black pixels mean background and white pixels mean the face region. The corresponding binary image of the image given in Fig. 2b is shown in Fig. 2c.

## 2.3 Extracted Largest Component

The foreground of a binary image may contain more than one object. Let us consider image, in Fig. 2c, it has three objects or components. The large one represents the face region. The others are at the left bottom corner and small dot on the top. Then largest component has been extracted from binary image using "Connected Component Labeling" algorithm [5]. This algorithm is based either on "4-conneted" neighbours or "8-connected" neighbours method [6]. As an illustrative example, consider a largest component as a face skin region illustrated in Fig. 3 using "Connected component labeling" algorithm. It is a binary image. Here, white means face skin region representing with "1" and black means background representing with "0".

**Fig. 3** A largest component as a face skin region

## 2.4 Finding the Centroid [7]

Centroid has been extracted from the binary image using Eqs. (1) and (2).

$$X = \frac{\sum m_{f(x,y)}x}{\sum m_{f(x,y)}} \tag{1}$$

$$Y = \frac{\sum m_{f(x,y)}y}{\sum m_{f(x,y)}} \tag{2}$$

where x, y is the co-ordinate of the binary image and m is the intensity value that is $mf(x,\ y) = f(x,\ y) = 0$ or 1.

## 2.5 Cropping of the Face Region in Elliptic Shape

Normally human face is of an elliptical shape. Then from the centroid, human face has been cropped in elliptical shape using "Bresenham ellipse drawing" [8] algorithm where, X and Y is x co-ordinate and y co-ordinate respectively for the centroid which is calculated by Eqs. (1) and (2). Distance between the centroid and the right ear is called the minor axis of the ellipse and distance between the centroid and the forehead is called major axis of the ellipse. After finding the centroid of the face, face has been cropped in elliptical shape and mapped to the grayscale image, which is shown in Fig. 4.

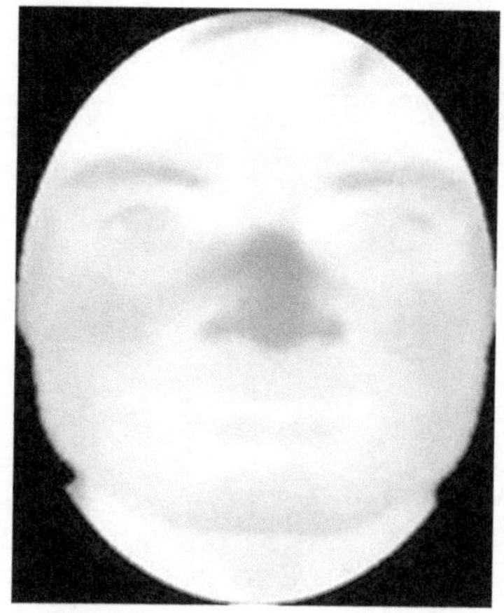

**Fig. 4** Face region in *elliptic shape*

## *2.6 Dimensionality Reduction Using Wavelet Transforms*

The first discrete wavelet transform (DWT) was invented by the Hungarian mathematician Alfréd Haar in 1909. A key advantage of wavelet transform over Fourier transforms is temporal resolution. Wavelet transform captures both frequency and time i.e. location information. The DWT has a huge number of applications in science, engineering, computer science and mathematics. The Haar transformation is used here for dimensionality reduction since it is the simplest wavelet transform of all and can successfully serve our purpose. Wavelet transform has merits of multi-resolution, multi-scale decomposition, and so on. To obtain the standard decomposition [9] of a 2D image, the 1D wavelet transform to each row is applied first. This operation gives an average pixel value along with detail coefficients for each row. These transformed rows are treated as if they were themselves in an image. Now, 1D wavelet transform to each column is applied. The resulting pixel values are all detail coefficients except for a single overall average coefficient. As a result the elliptical shape facial image is decomposed, and then four regions can be gained. These regions are one low-frequency $LL_1$ (approximate component), and three high-frequency region, namely $LH_1$ (horizontal component), $HL_1$ (vertical component), and $HH_1$ (diagonal component), respectively. The low frequency sub-band $LL_1$ can be further decomposed into four sub-bands $LL_2$, $LH_2$, $HL_2$ and $HH_2$ at the next coarse scale. $LL_i$ is a reduced resolution corresponding to the low-frequency part of an image. The sketch map of the quadratic wavelet decomposition is shown in Fig. 5.

As illustrated in Fig. 5, the L denotes low frequency and the H denotes high frequency, and that subscripts named from 1 to 2 denote simple, quadratic wavelet decompositions respectively. The standard decomposition algorithm is given below:

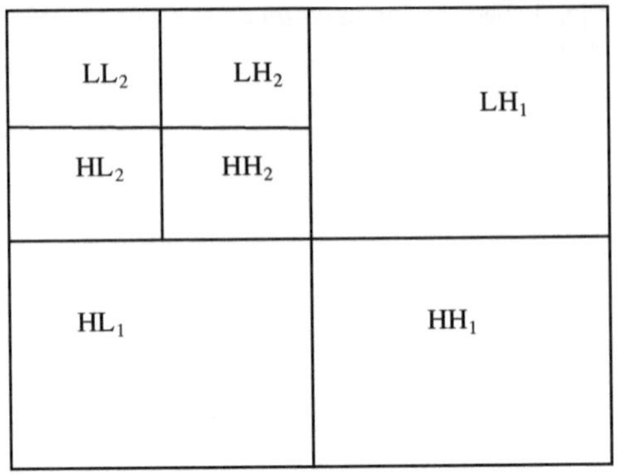


**Fig. 5** Sketch map of the quadratic wavelet decomposition

```
function StandardDecomposition(Im[1 to r,1 to c])
//Im[1 to r,1 to c] is an image realized by 2D array, where r
is the number of rows and c is the number of column.

  for i = 1 to r

     1D wavelet transforms (row-number (i))

  end

  for j = 1 to c

     1D wavelet transforms (column-number (j))

  end

end
```

Let's start with a simple example of 1D wavelet transform [10]. Suppose an image with only one row of four pixels, having intensity values [10 4 9 5]. Now apply the Haar wavelet transform on this image. To do so, first pair up the input intensity values or pixel values, storing the mean in order to get the new lower resolution image with intensity values [7 7]. Obviously, some information might be lost in this averaging process. Some detail coefficients need to store to recover the original four intensity values from the two mean values, which capture the missing information. In this example, 3 is the first detail coefficient, since the computed mean is 3 less than 10 and 3 more than 4. This single number is responsible to recover the first two pixels of original four-pixel image. Similarly, the second detail coefficient is 2. Thus, the original image is decomposed into a lower resolution (two-pixel) version and a pair of detail coefficients. Repeating this process recursively on the averages gives the full decomposition, which is shown in Table 1.

**Table 1** Resolution, mean and the detail coefficients of full decomposition

| Resolution | Mean | Detail coefficients |
|---|---|---|
| 4 | [10 4 9 5] | |
| 2 | [7 7] | [3 2] |
| 1 | [7] | [0] |

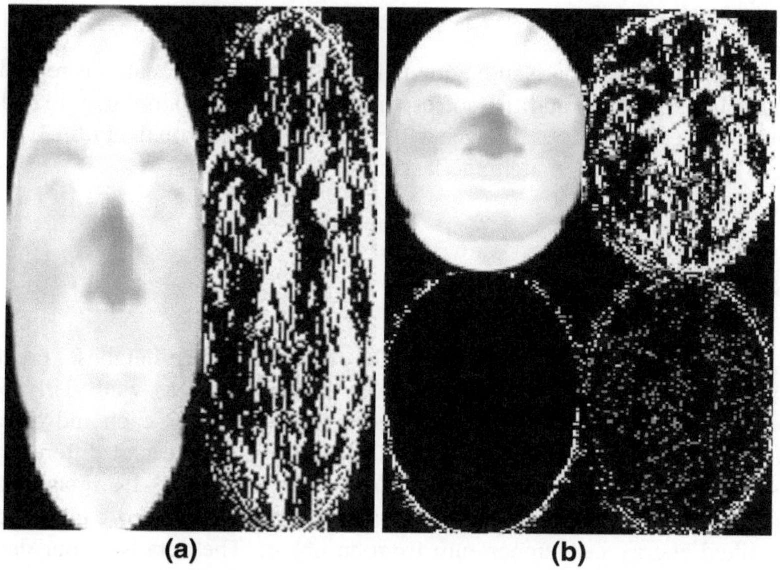


**Fig. 6** Haar wevlet transform. **a** Transform rows. **b** Transform columns

Thus, the one-dimensional Haar wavelet transform of the original four-pixel image is given by [7 0 3 2]. First, 1D wavelet transforms [Standard Decomposed (image) algorithm] is used on Fig. 4 row-wise manner, the resultant figure is shown in Fig. 6a. Secondly, 1D wavelet transforms [Standard Decomposed (image) algorithm] is used on Fig. 6a column-wise manner and the resultant image is shown in Fig. 6b.

The pixels of $LL_2$ image can be concatenated row wise (horizontally) or column wise vertically manner. So the image can be treated as a series. Now the problem of image comparison converts into series comparison.

## *2.7 Series Matching Classifier [11]*

In a gray scale image, each and every pixel is represented by 8-bits, and the range of intensity value is between 0 and 255. Suppose that $l_1 = (l_{11}, l_{12}, l_{13}, \ldots, l_{1k})$ and $l_2 = (l_{21}, l_{22}, l_{23}, \ldots, l_{2k})$ are two numerical series, where $0 \leq l_{ij} \leq 255$, $1 \leq i \leq 2$ and $1 \leq j \leq k$.. Then the similarity between $l_1$ and $l_2$ can be written as

$$\text{sim}(l_1, l_2) = \sum_{j=1}^{k} f(l_1(j), l_2(j)) \tag{3}$$

where f(.) is the matching or comparison function used for similarity measurement. $l_1(j)$ and $l_2(j)$ are the jth elements of each series. The comparison function or matching function is defined as

$$f(l_1(j), l_2(j)) = |l_1(j) - l_2(j)| \tag{4}$$

The matching or comparison having maximal similarity or match is regarded as the optimal comparison, and the corresponding comparison similarity is called as similarity between two series. Images will be classified based on the Eqs. (3) and (4), which is known as series matching classifier.

## 3 Experiment and Results

Experiments have been performed on UGC-JU thermal face database, created at our own laboratory and Terravic Facial IR database [12] as well. Till now 76 individuals have volunteered for this photo shoots and for each individual 39 different templates of RGB color images with different pose and different facial expression changes like happy, surprise, fear, etc. Thermal IR image doesn't depend on surrounding lighting conditions since thermal infrared camera capture only emitted energy i.e. temperature from an object. The details of our database have been discussed in Sect. 2.1. The Terravic facial IR database is composed of 20 persons, each of which has different number of images of 320 × 240 pixels. 10 images of each individual have been selected first. So, 200 images are used for our experiment. If these original images are directly used, the range of value of $\mathrm{sim}(l_1, l_2)$ is too large. For example, according to the Eqs. (3) and (4), when two images are completely different, then their $\mathrm{sim}(l_1, l_2)$ value is $320 \times 240 \times 255 = 19{,}584{,}000$. Whereas, when two images are completely same, their $\mathrm{sim}(l_1, l_2)$ value is regarded as 0. If the image size is large, more computation is required to determine whether two images are same. Therefore, wavelet transform is used to decompose and reduce original images before image comparison. Different orthogonal wavelet filters are available such as Haar, Daubechies, Coiflets, Symlets, etc. Haar wavelet is used here due to simplicity. It is simple and popular as compared to the other wavelet filters. The sizes of the images are reduced significantly after second label wavelet transform. Firstly, the decomposition using Haar wavelet transform is done on the original image and secondly on the low frequency component. LL coefficients are only used here because detail coefficients (LH, HL, HH) may be more useful as features in face recognition to increase the success rate but that will increase processing time enormously. After second label wavelet transform, each image is transformed into horizontal vector, by concatenating pixel in row-wise manner, to form an M × N 2D array (M is the number of images and N is the number of feature vector of images each), where each row corresponds to an image. Clearly, each series has the same length. Each series is composed of only intensity values between 0 and 255 including 0 and 255. Then Eqs. (3) and (4) can be used to compare two images and the corresponding $\mathrm{sim}(l_1, l_2)$. Before image comparison, the whole M × N array is divided into two parts of size (M/2) × N. The odd numbers of rows are taken from the original matrix and put them into the first matrix. Then even

**Table 2** Performance rate for different databases

| Name of the database | Label | Recognition rate (%) |
|---|---|---|
| UGC-JU thermal IR database | Original image | 85 |
| | LL1 | 91 |
| | LL2 | 95 |
| Terravic facial IR database | Original image | 84 |
| | LL1 | 92 |
| | LL2 | 93 |

numbers of rows are taken from the original matrix and put them into the second matrix. This matrix is used for testing purpose. Then average of each rows of the training set is calculated column-wise to form a 1D array or series. Let's say this series or array is named as 'X'. Then this 1D array or series is used to differentiate them with each row of the training set using Eq. (4) and sum of all difference of each row is stored in another 1D array, named as 'Y' using Eq. (3). That means $\mathrm{sim}(l_1, l_2)$ is measured using Eqs. (3) and (4). The above process is also used for testing sets and this series or 1D array is named as 'Z'. Finally, pick one element from 'Z' and find the closest element among 'Y' and then based on this closest element of 'Y', it would be classified. The obtained experimental result is shown in Table 2.

Table 2 presents original image and two different decomposing labels, decomposed by Haar wavelet transform and their recognition rates. Results of face recognition, which are obtained by LL2 component, are better than other components for UGC-JU thermal face database and Terravic Facial IR database, using comparison of each series. The number of pixels of LL2 component is fewer than LL1 component, so that matching time of LL2 components are significantly reduced and the higher recognition rate is also obtained at the same time. A few experimental results based on thermal face images are being given in Table 3, for understanding that our approach is simple and good enough to recognize a person easily.

**Table 3** A comparative study based on performance of different thermal face recognition methods (adapted from [13])

| Method | Recognition rate (%) |
|---|---|
| Segmented infrared images via bessel forms [14] 2004 | 90 |
| PCA for visual indoor probes [15] | 81.54 |
| PCA + LWIR (indoor probs) [15] | 58.89 (Maximum) |
| LDA + LWIR (indoor probs) [15] | 73.92 (Maximum) |
| Equinox + LWIR (indoor probs) [15] | 93.93 (Maximum) |
| PCA + LWIR (outdoor probs) [15] | 44.29 (Maximum) |
| LDA + LWIR (outdoor probs) [15] | 65.30 (Maximum) |
| Equinox + LWIR (outdoor probs) [15] | 83.02 (Maximum) |
| Eigenfaces + LWIR (different illumination but same expression) [16] | 95.0 (Average), 89.4 (Minimum) |
| Eigenfaces + LWIR (different illumination and expression) [16] | 93.3 (Average), 86.8 (Minimum) |

## 4 Conclusion

The proposed human thermal face recognition based on Haar wavelet transform and series matching has been introduced and implemented. The proposed system gave higher recognition rate in the experiments. One of the major advantages of this approach is the ease of implementation. Furthermore, no knowledge of geometry or specific feature of the face is required. However, this system is applicable to front views and constant background only. It may fail in unconstraint environments like natural scenes.

**Acknowledgments** Authors are thankful to a major project entitled "Design and Development of Facial Thermogram Technology for Biometric Security System," funded by University Grants Commission (UGC), India and "DST-PURSE Programme" at Department of Computer Science and Engineering, Jadavpur University, India for providing necessary infrastructure to conduct experiments relating to this work. Ayan Seal is grateful to Department of Science and Technology (DST), India for providing him Junior Research Fellowship-Professional (JRF-Professional) under DST-INSPIRE Fellowship programme [No: IF110591].

## References

1. Socolinsky DA, Selinger A (2002) A comparative analysis of face recognition performance with visible and thermal infrared imagery. In: Proceedings of International Conference on Pattern Recognition, Quebec, 4:217–222
2. Shiqian W, Fang ZJ, Xie ZH, Liang W (2007) Blood perfusion models for infrared face recognition school of information technology. Jiangxi University of Finance and Economics, China
3. Kong SG, Heo J, Abidi BR, Paik J, Abidi MA (2005) Recent advances in visual and infrared face recognition: a review. Comput Vis Image Underst 97:103–135
4. Kunte RS, Samuel RDS (2007) Wavelet descriptors for recognition of basic symbols in printed Kannada text. Int J Wavelets Multiresolut Inf Process 5(2):351–367
5. Bryan SM (1998–2004) Lecture 2: image processing review, neighbors, connected components, and distance
6. Gonzalez RC, Woods RE (2002) Digital image processing, 3rd edn. Prentice Hall, Englewood Cliffs
7. Venkatesan S, Madane SSR (2010) Face recognition system with genetic algorithm and ANT colony optimization. Int J Innov Manag Technol 1(5)
8. Hearn D, Baker MP (1996) Computer graphics C version, 2nd edn. Prentice Hall, Englewood Cliffs
9. Beylkin G, Coifman R, Rokhlin V (1991) Fast wavelet transforms and numerical algorithms I. Commun Pure Appl Math 44(2):141–183
10. Stollnitz EJ, DeRose TD, Salesin DH (1995) Wavelets for computer graphics: a primer, part 1. IEEE Comput Graphics Appl 15(3):76–84
11. Dezhong Z, Fayi C (2008) Face recognition based on wavelet transform and image comparison. International symposium on computational intelligence and design
12. http://www.terravic.com/research/facial.htm
13. Bhowmik MK, Saha K, Majumder S, Majumder G, Saha A, Sarma AN, Bhattacharjee D, Basu DK, Nasipuri M (2011) Thermal infrared face recognition: a biometric identification technique for robust security system. In: Peter M. Corcoran (ed) Reviews, refinements and

new ideas in face recognition. ISBN: 978-953-307-368-2. Tech Open Access Publisher (Open Access publisher of Scientific Books and Journals), Vienaa Office, Zieglergase 14, 1070 Vienna, Austria, Europe
14. Buddharaju P, Pavlidis I, Kakadiaris I (2004) Face recognition in the thermal infrared spectrum. In: Proceeding of the 2004 IEEE computer society conference on computer vision and pattern recognition workshops (CVPRW'04)
15. Socolinsky DA, Selinger A (2004) Thermal face recognition in an operational scenario. In: Proceedings of the IEEE computer society conference on computer vision and pattern recognition (CVPR'04)
16. Socolinsky DA, Wolff LB, Neuheisel JD, Eveland CK (2001) Illumination invariant face recognition using thermal imagery. In: Proceedings of the IEEE computer society conference on computer vision and pattern recognition (CVPR'01), Hawaii

# A Hybrid Approach for Enhancing the Security of Information Content of an Image

**Kumar Jalesh and S. Nirmala**

**Abstract** Recent advancements in communication technologies have resulted in sharing of most multimedia information through internet. Securing such information from eavesdroppers during communication is still challenging task. In this paper, a hybrid approach is proposed for securing the contents of an image. This hybrid approach is a combination of edge detection and encryption process. The proposed approach comprises three stages. In first stage, the input image is divided into two sub images, image with edges and image without edge information. Canny edge detection algorithm is used for detecting edges in the input image. In second stage, sub images obtained from previous stage are encrypted separately using crossover and mutation operations. The encrypted images are fused in the final stage. The performance of the proposed method is measured in terms of correlation coefficient and histogram. The results obtained are compared with method [1]. The comparative study reveals that the image processing technique can be combined with encryption process to provide different levels of security.

**Keywords** Hybrid approach · Edge detection · Selection · Crossover · Mutation · Levels of security

K. Jalesh (✉)
Department of Computer Science and Engineering, J.N.N.C.E, Shimoga, Karnataka 577204, INDIA
e-mail: jalesh_k@yahoo.com

S. Nirmala
Department of Information Science and Engineering, J.N.N.C.E, Shimoga, Karnataka 577204, INDIA
e-mail: nir_shiv_2002@yahoo.co.in

P. P. Swamy and D. S. Guru (eds.), *Multimedia Processing, Communication and Computing Applications*, Lecture Notes in Electrical Engineering 213,
DOI: 10.1007/978-81-322-1143-3_14, © Springer India 2013

# 1 Introduction

From time immemorial, people have been trying to exchange messages without letting others to know about it. In modern times, messages are increasingly being exchanged over computer networks. Secure storage and transmission of digital images are needed in many applications such as medical imaging systems, pay-per-view TV, satellite images, image based document management and confidential video conferencing [2, 3]. Generally, two levels of security for digital image encryption could be considered: low level and high level security encryption [4]. In low level security encryption, the encrypted image has degraded visual quality compared to that of the original one. However, the contents of the image are still visible and understandable to the viewers. In the high level security case, the content is completely scrambled and the image just looks like random noise. Different approaches have been evolved to provide more security for image contents. Security based on edge information is a new approach. Edge information is used in image enhancement, compression, segmentation and recognition [5].

Many encryption techniques are proposed in the past to secure the image contents. But, some of them have been known to be insecure [6]. Evolutionary computation algorithms represent a range of problem solving techniques based on principles of biological evolution, like natural selection and genetic inheritance. Such algorithms can be used to solve a variety of difficult problems, among which are those from the area of cryptography.

In this paper, a hybrid approach is proposed which is a combination of image processing and encryption techniques. In Sect. 2, a related literature survey is carried out. In Sect. 3, the proposed method is described. Experimental results are discussed in Sect. 4. Statistical analysis of the results is presented in Sect. 5. In Sect. 6, comparative study is carried out. Conclusions drawn are summarized in Sect. 7.

# 2 Literature Survey

The security of digital images has become increasingly more important in today's highly computerized and interconnected world. In recent past different techniques have been analyzed on image security. Mitra et al. [7] used a random combination of bit or pixels or permutation of blocks. The permutation of bits decreases the perceptual information, whereas the permutation of pixels and blocks produce high level security. To extract an image, a combinational sequence of permutations and permutation keys using pseudo random index generators should be known. In this investigation the combination of block, bit and pixel permutation are used respectively. Permutation techniques are attractive, but lacks in generated key and security. A block based transformation algorithm based on the combination of image transformation and Blowfish algorithm is discussed in [8]. The original image was divided into uniform blocks which were rearranged into a transformed image using any transformation algorithm. Blowfish algorithm is used for encryption. Transformation process adds additional processing overhead in this technique.

A survey of cryptographic applications that can be developed with the help of evolutionary computation methods are discussed in [9, 10]. Kumar [11] proposed a new approach based on genetic algorithms with pseudorandom sequence to encrypt the data stream. The approach ensures high data security and feasibility for easy integration with commercial multimedia transmission applications. The properties of chaos are used for encryption along with genetic algorithm. Enayatifar [12] described a new method based on a hybrid model composed of a genetic algorithm and a chaotic function for image encryption. Genetic approach is used to get the best encrypted image with the highest entropy and the lowest correlation coefficient among adjacent pixels. Husainy [2] discuss a new image encryption technique using genetic algorithm based on mutation and crossover. Security of this method depends on the use of different vector lengths and number of crossover and mutation operations. In Ref. [3], a new effective method for image encryption which employs magnitude and phase manipulation using differential evolution approach is discussed. Linear feedback shift register is used to select the crossover points. In this approach, discrete fourier transform followed by differential evolution are used for image encryption.

In Ref. [5], a new concept of image encryption which is based on edge information is discussed. The basic idea is to separate the image into the edges and the image without edges, and encrypt them using any existing or new encryption algorithm. The user has the flexibility to encrypt the edges or the image without edges or both of them. Algorithm based on 3D Cat Map is used for encryption process. The type of the edge detection method and its threshold value, the parameters and iteration times of the cat map transform can act as the security keys. The encrypted edges and encrypted image without edges for each 2D component is combines into a format of the complex number to get the encrypted image. Hou [13] have proposed three methods for visual cryptography. Gray-level visual cryptography method first transforms the gray-level image into a halftone image and then generates two transparencies of visual cryptography. We proposed a genetic algorithm for encryption of image contents in [6]. In this work pseudorandom generator, crossover and mutation operation, along with feedback function is used for encryption.

From the literature survey, it is evident that different techniques are evolved for image encryption. Evolution algorithms, visual cryptography and encryption based on image processing techniques provide a new dimension in encryption process. In this paper, a hybrid approach is proposed for securing the contents of an image which is a combination of image processing technique and genetic algorithm. Images are encrypted after dividing into with and without edge information based on genetic process. Two encrypted images are fused using simple exclusive OR function. The encrypted image with or without edge information acts as a key image to retrieve the original image.

## 3 Proposed Method

The proposed approach composes three stages. The block diagram of the proposed method is shown in Fig. 1. In first stage, the input image is divided into two sub images, image with edges and image without edge information. Canny edge detection algorithm is used for detecting edges in the input image. In second stage, sub images obtained from previous stage are encrypted separately. The encryption process used is elaborated in Fig. 2. Encryption process uses key stream generator, crossover and mutation operations as discussed in [1]. In the proposed approach three different cases are considered for securing the contents of source image.

Case 1: Encrypting only the structural information of an image
Case 2: Encrypting the image after removal of structural information
Case 3: Fusion of the information obtained from both the case 1 and case 2

The algorithm of the proposed approach is as given below.

**Algorithm:**
**Stage 1: Division of input image into two sub images**
Input: A color image 'I' of dimension 3 × M × N
Output: Two images with and without edge information

Step 1: Apply canny edge detection algorithm on the input color image

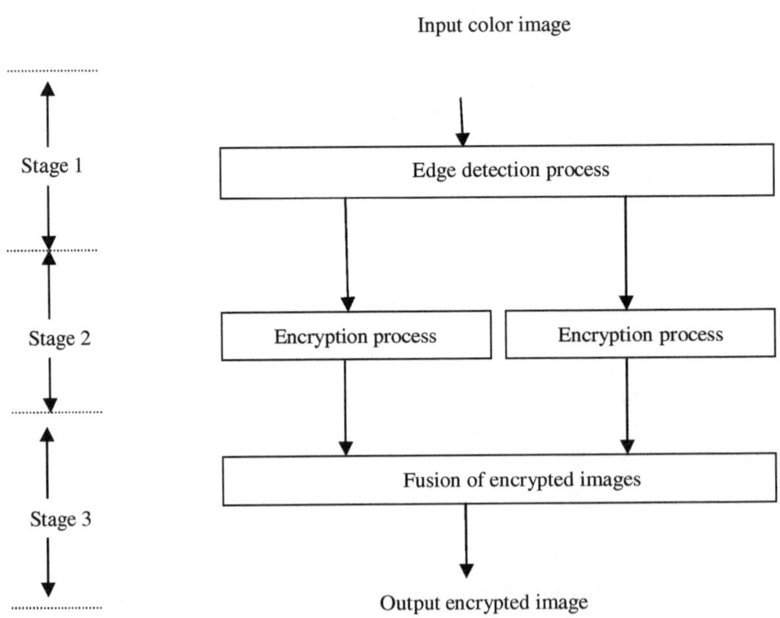


**Fig. 1** Stages in proposed approach

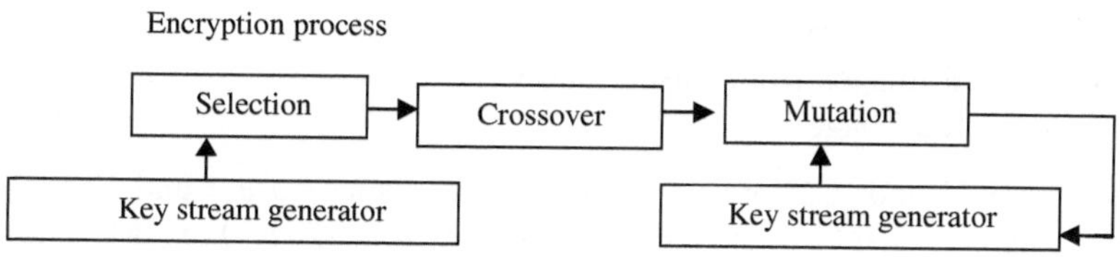


**Fig. 2** Encryption process

Step 2: Obtain two sub images 'A' and 'B' where 'A' contain only edge information and 'B' without edge information.

**Stege 2: Apply the encryption algorithm on both the sub images separately**
Input: Images with and without edge information
Output: Two encrypted images as described in Case 1 and Case 2 of the proposed approach

Step 1: Let each value of pixel in each color channel be in the range 0 to $((M \times N) - 1)$
For $C = 0$ to $((M \times N) - 1)/2$, perform the following operations

1. $v1 = V[i]$, $v2 = V[((M \times N) - 1) - C]$ where 'v1' and 'v2' are variables to store pixel values.
2. Use a linear congruential pseudorandom generator for the selection of the crossover point, say 's'. According to selection point 's', divide 'v1' into 'v11' and 'v12', where $v11 = s$ and $v12 = v1 - s$ divide 'v2' into 'v21' and 'v22', where $v21 = s$ and $v22 = v2 - s$
3. Apply the crossover operation between 'v1' and 'v2', New generation obtained after crossover is stored in position

$$V[i] = v1 \text{ and } V[((M \times N) - 1) - i] = v2$$

Step 2: Generate initial value using a congruential pseudorandom generator, say 'K'. For $C = 0$ to $(M \times N) - 1$

$$V[i] = V[i] \oplus K$$

$$K = V[i]$$

Step 3: Write an encrypted image on the basis of vector 'V' obtained after Step 1 and 2.

**Stage 3: Fusion of encrypted images**
Input: Encrypted images from Case 1 and Case 2
Output: Encrypted image

Step 1: Let 'C1' and 'C2' are encrypted images using Case 1 and Case 2. Each value of pixel in each color channel be in the range 0 to $((M \times N) - 1)$ in both images.

Step 2: For each $i = 0$ to $((M \times N) - 1)$ in 'C1'
For each $j = 0$ to $((M \times N) - 1)$ in 'C2'

$$D[i] = C1[i] \oplus C2[((M \times N) - 1) - 1]$$

Step 3: Write an encrypted image on the basis of vector 'D' obtained after Step 2.

The proposed approach is as shown in Fig. 3.

## 4 Experimental Results

We have created an image corpus for the experimental study. The image corpus consists 60 color images of different sizes. Some images in the corpus are downloaded from web [14]. Image samples in the corpus are of multi colors. Images in the corpus contain text information and non text/graphical information. The results of the proposed approach on sample images in image corpus are shown in Fig. 4.

Three different types of images with uniform color, random textured and complex background are considered. The results obtained after Case 1, Case 2 and Case 3 are shown in Fig. 4. Case 1 shows the sub image contains only edge information for the corresponding input image along with the encrypted image. Further, the results obtained after encrypting the image without edge information are shown in Case 2. As described in Sect. 3, encrypted images obtained in Case 1 and Case 2 are fused together and result is shown in Case 3. From results it is observed that the encrypted images will not reveal any identity of the original image. Encrypted images obtained are completely different from the original images. Image after decryption process is also shown in Fig. 4.

In this work, the decryption is the reverse process of encryption. The result obtained after decryption is shown in Fig. 5. Original input image is retrieved without any loss of information. Two encrypted images are received by the receiver. By performing reverse process of encryption original input image would be obtained. For example, Fig. 5a and b are encrypted images from Case 3 and Case 1 respectively. Figure 5c is obtained after fusion of two encrypted images. Decryption process is applied on the images in Fig. 5b and c to obtain image with and without edge information Fig. 5d and e. From These two images the original input image could be constructed Fig. 5f.

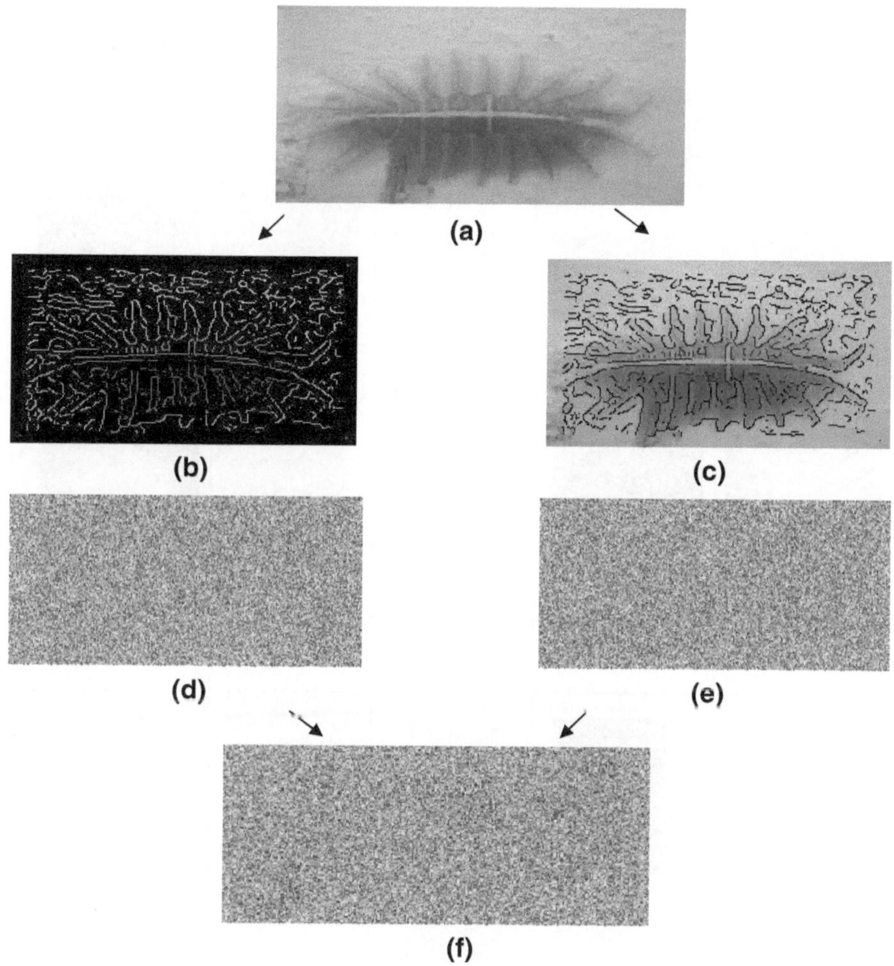


**Fig. 3** The outcomes of all the three stages of the proposed approach for a sample input *color image*. **a** Input color image. **b** Image with only edge information. **c** Image without edge information. **d** Encrypted image with *Case 1*. **e** Encrypted image with *Case 2*. **f** Encrypted image with *Case 3*

## 5 Statistical Analysis

From the literature, it is known that many ciphers have been successfully analyzed with the help of statistical analysis. Several statistical attacks are proposed on images [3]. To prove the robustness of the proposed method, statistical analysis is performed by plotting the histograms of the original and encrypted images. Correlation coefficients between original and encrypted images are also measured.

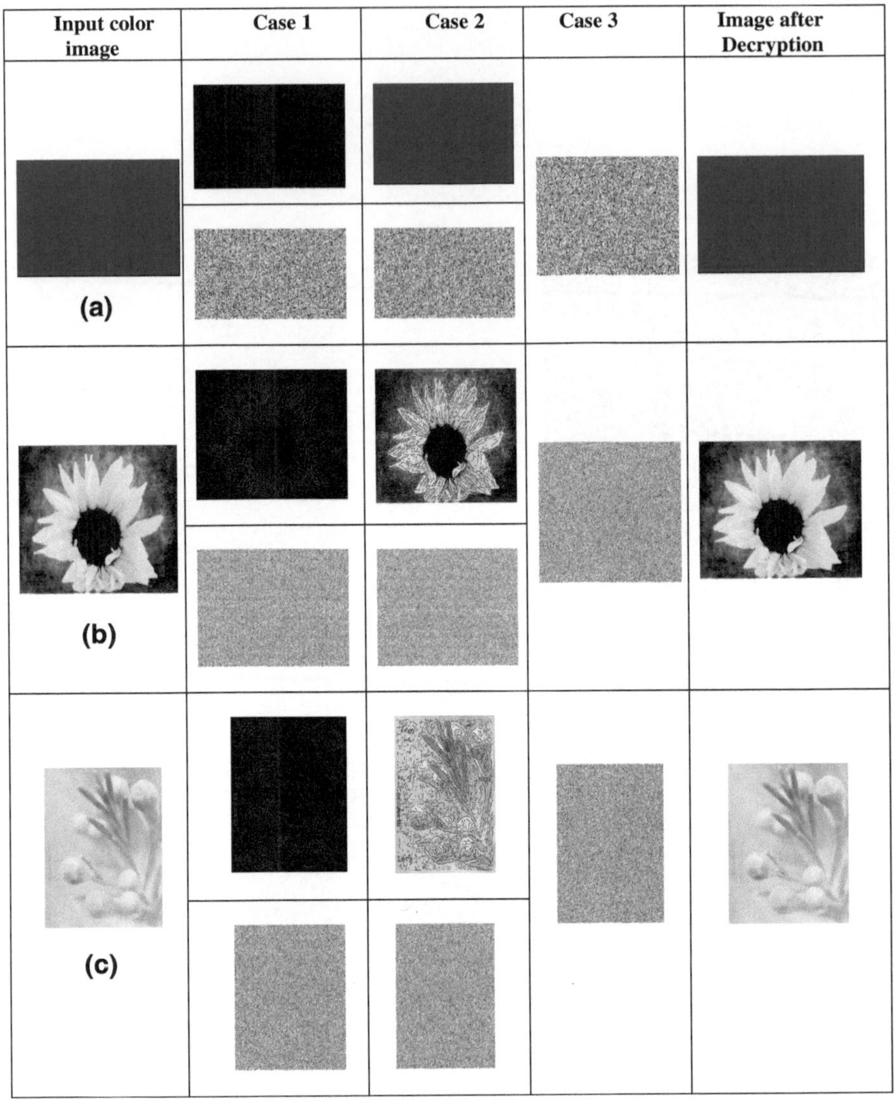


**Fig. 4** Results of the proposed method on sample *color images* in image corpus. **a** Uniform color. **b** Random textured. **c** Complex background

## *5.1 Histogram Analysis*

To prevent the leakage of information from an opponent, it is also advantageous if the cipher image bears little or no statistical similarity to the plain image [3]. The histogram is a graphical representation showing a visual impression of the distribution of data. To prevent an attack, the cipher obtained should not give any clue

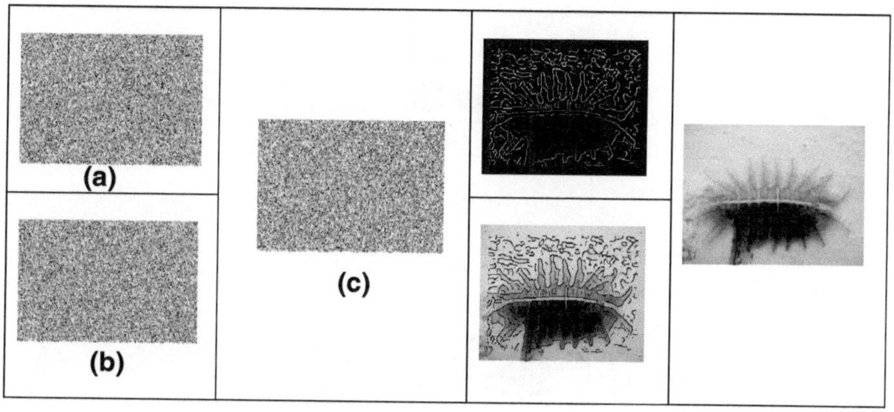


**Fig. 5** Decryption process

about the original image. In the proposed approach, the cipher obtained does not give any clue about the original image which is analyzed through histograms. For sample image and corresponding encrypted images in Fig. 3a–f the histograms are as shown in Fig. 6a–f, respectively. Levels of RGB channels of original image are unequally distributed which is shown in Fig. 6a. Further, Fig. 6b and c represents a histogram for the image with edge and without edge information. Histograms obtained after Case 1 and 2 operations are in Fig. 6d and e, which shows that all pixels are uniformly distributed. Histogram in Fig. 6f represents the encrypted image obtained after fusion operation (Case 3). It is revealed from the histogram that all pixels in R, G and B channels of sample image are distributed uniformly. Hence, the cipher image does not provide any clue to statistical attack in all the cases.

## *5.2 Correlation Coefficient Analysis*

]Correlation coefficient factor is used to measure the relationship between two variables: the image and its encryption. This factor demonstrates to what extent the proposed encryption algorithm strongly resists statistical attacks. Correlation coefficient 'r' between two images is computed using an Eq. (1) [6, 8, 15]. If the correlation coefficient equals one, that means the original image and its encryption is identical. If the correlation coefficient equals zero, that means the encrypted image is completely different from the original. If the correlation coefficient equals minus one that means the encrypted image is the negative of the original image [6].

$$\mathrm{r} = \frac{n\sum xy - (\sum x)(\sum y)}{\sqrt{n(\sum x^2) - (\sum x)^2}\sqrt{n(\sum y^2) - (\sum y)^2}} \tag{1}$$

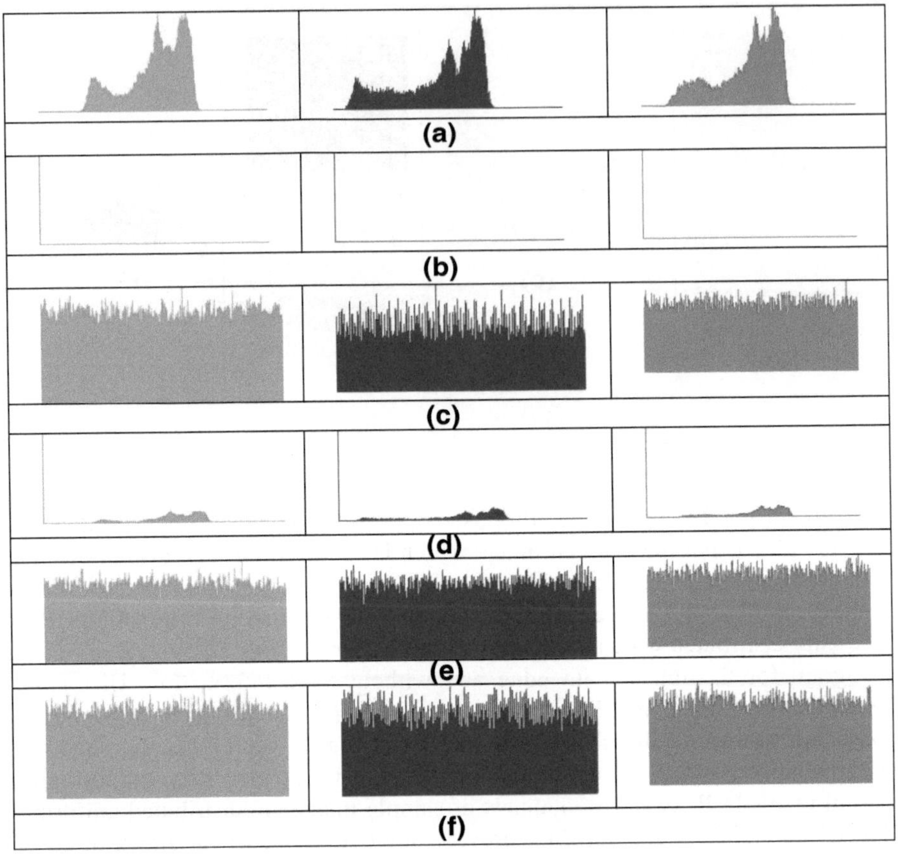


**Fig. 6** *Histogram* of images shown in Fig. 3. **a** For original image. **b** *Case 1* For the image with edges only. **c** *Case 1* For the encrypted image with edges only. **d** *Case 2* For the image without edge information. **e** *Case 2* For the encrypted image without edges information. **f** *Case 3* For the encrypted image after fusion operation

where

| | |
|---|---|
| 'r' | Correlation value |
| 'x' and 'y' | Pixel values of the original and encrypted images |
| 'n' | Number of pairs of data |
| $\sum x\,y$ | Sum of the products of paired data |
| $\sum x$ | Sum of x data |
| $\sum y$ | Sum of y data |
| $\sum x^2$ | Sum of squared x data |
| $\sum y^2$ | Sum of squared y data |

We computed correlation coefficient of input color image and encrypted image obtained from Case 3 and encryption method in Ref. [1]. Images of type uniform

# Recognition of Limited Vocabulary Kannada Words Through Structural Pattern Matching: An Experimentation on Low Resolution Images

S. A. Angadi and M. M. Kodabagi

**Abstract** Text recognition at character/word level is one of the very important steps for development of automated systems for understanding low resolution display board images which facilitate several new applications such as blind assistants, tour guide systems, location aware systems and many more. In this paper, a new approach for recognition of Kannada words in low resolution natural scene images from a limited vocabulary is presented. The proposed method uses structural patterns of vertical and horizontal cuts as features, which are tolerant to font variability, uncertainty, noise and other degradations. These structural representations characterize the shape of the word image. The method works in two phases; In the training phase, several patterns of vertical and horizontal cut features that can occur generally even in the presence of uncertainty are determined from training word images and templates are constructed, one for each word under study. Further, these templates are organized into knowledge bases, one for each set of word images of equal size in terms of number of characters. During testing, a test word image is processed to obtain vertical and horizontal cut features and a newly defined pattern matching procedure that measures the maximum similarity between test sample and pre-constructed templates of word images in the knowledge base is used to recognize the word. The proposed methodology is evaluated for 1,200 Kannada word images and an overall recognition accuracy of 97.67 % is achieved. The proposed method is found to be robust and insensitive to the variations in size and style of font, thickness and spacing between characters, noise, and other degradations.

---

S. A. Angadi (✉) · M. M. Kodabagi
Department of Computer Science and Engineering,
Basaveshwar Engineering College, Bagalkot, Karnataka 587102, India
e-mail: vinay_angadi@yahoo.com

M. M. Kodabagi
e-mail: malik123_mk@rediffmail.com

P. P. Swamy and D. S. Guru (eds.), *Multimedia Processing, Communication and Computing Applications*, Lecture Notes in Electrical Engineering 213,
DOI: 10.1007/978-81-322-1143-3_15, 

**Keywords** Word recognition · Structural pattern matching · Limited vocabulary · Low resolution images · Display boards

## 1 Introduction

In recent years, understanding of text in low resolution images of display boards captured from camera embedded hand held systems such as smart mobile phones, tablets and PDA's has gained significant attention in computer vision. Automatic text understanding can be employed in various applications that are useful in our daily life. One such example/application is an intelligent translation system that recognizes/understands text in display board images and provides translated information in a known language. Such systems are of great help to people who travel across different places in the world for field work and business activities, as they face problem in understanding written text on display boards particularly in foreign environment. This is especially true in countries like India, which are multilingual. These reasons, demand for an automatic recognition and translation system for text written in low resolution natural scene images of display boards.

The written matter on display boards/name boards provides important information for the needs and safety of people, and may be written in unknown languages. The written matter can be street names, restaurant names, building names, company names, traffic directions, warning signs etc. Researchers have focused their attention on development of techniques for understanding written text on such display boards. There is a spurt of activity for development of techniques that are useful in web based intelligent hand held systems for such applications.

In the reported works [1–10] on intelligent systems for hand held devices, not many works pertain to understanding of written text on display boards, and scope exists for exploring such possibilities. The text understanding involves several processing steps; text detection and extraction, preprocessing for line, word and character separation, script identification, text recognition and language translation. Therefore, text recognition at word/character level is one of the very important processing steps for development of such systems prior to further analysis. The recognition of text in low resolution images of display boards is a difficult and challenging problem due to various issues such as font size, style and spacing between characters, skew, perspective distortions and other degradations. The state of art character/word recognition techniques work on clean documents containing well structured/formatted text and are not suitable for text in scene images. Recently, few approaches are explored for recognition of text in natural scene images and are summarized in the next section.

In this paper, a new approach that uses structural patterns of vertical and horizontal cut features and a newly defined pattern matching procedure for word recognition of Kannada text in low resolution natural scene images from a limited vocabulary is presented.

The rest of the paper is organized as follows; the detailed survey related to text recognition in natural scene images is described in Sect. 2. The proposed method is presented in Sect. 3. The experimental results and analysis are given in Sect. 4, Sect. 5 concludes the work and lists future directions.

## 2 Related Works

The recognition of text in low resolution natural scene images of display boards is a necessary step for development of various tasks of text understanding and translation system. Some of the related works are summarized in the following.

A robust approach for recognition of text embedded in natural scenes is given in Zhang et al. [11]. The method extracts features from intensity of an image directly and utilizes a local intensity normalization to effectively handle lighting variations. Then, Gabor transform is employed to obtain local features and linear discriminant analysis (LDA) is used for selection and classification of features. This work is further extended integrating sign detection component with recognition [12]. The extended method embeds multi-resolution and multi-scale edge detection, adaptive searching, color analysis, and affine rectification in a hierarchical framework for sign detection and recognition.

A framework that exploits both bottom-up and top-down cues for scene text recognition at word level is presented in Mishra et al. [13]. The method reports an accuracy of only 73 % and requires further improvement. A Semi-Markov model for recognizing scene text that integrates character and word segmentation with recognition is proposed in Weinman et al. [14]. The probabilistic model for scene text recognition that integrates similarity, language properties, and lexical decision is employed in Weinman et al. [15]. But, the results are shown on a simpler data set. An extension to this work that uses bi-gram model of character widths for recognition of signs is presented in Weinman [16].

Automatic detection and recognition of Korean text in outdoor signboard images is described in Park et al. [17]. The experimental results show that the proposed method has been successfully applied to recognize and translate Korean shop names into English with their outdoor signboard images. The recognition of words in scenes with a head-mounted eye-tracker is described in Kobayashi et al. [18]. However, the method does not report experimental results on scene images having font variability, noise and other degradations, and recommends further investigation for reducing the computational cost. Many other recent works on scene text recognition are also reported in [19–23].

After the thorough study of literature, it is found that few methods [15, 16] work on limited data set, and other cited works [11–14, 17] do not report experimental results dealing with various challenges of processing scene text and recommends further investigation for improving recognition accuracy. It is also not reported at what distance the images are captured from mobile phones/hand held devices, as the perspective distortion will be more, if the images are captured at a larger distance, thus affecting the recognition accuracy. Moreover, most of the lexicon driven text recognition methods are constrained by the lexicon size. Though, the method in Weinman et al. [15] reduces the lexicon words considered with no loss in accuracy, but the service is not available in other techniques. It is also noticed/observed that, the features employed in state of art methods for English, Korean, and Chinese data set are script specific and may not be suitable for Kannada data set due to its differing characters. Therefore, more research is desirable to obtain discriminative features that are invariant to font size, style, noise and other degradations suitable for Kannada Scene text. In the current work, new structural patterns of vertical and horizontal cut features are employed for word recognition of scene text in low resolution images of display boards. The detailed description of the proposed methodology is given in the next section.

# 3 Methodology for Word Recognition

The proposed method uses following processing steps namely Preprocessing, Feature Extraction, Templates Construction, Rearrangement of Templates into Knowledge Bases, and Structural Pattern Matching for Word Recognition. The block schematic diagram of the proposed model is given in Fig. 1. The detailed description of each processing step is presented in the following subsections;

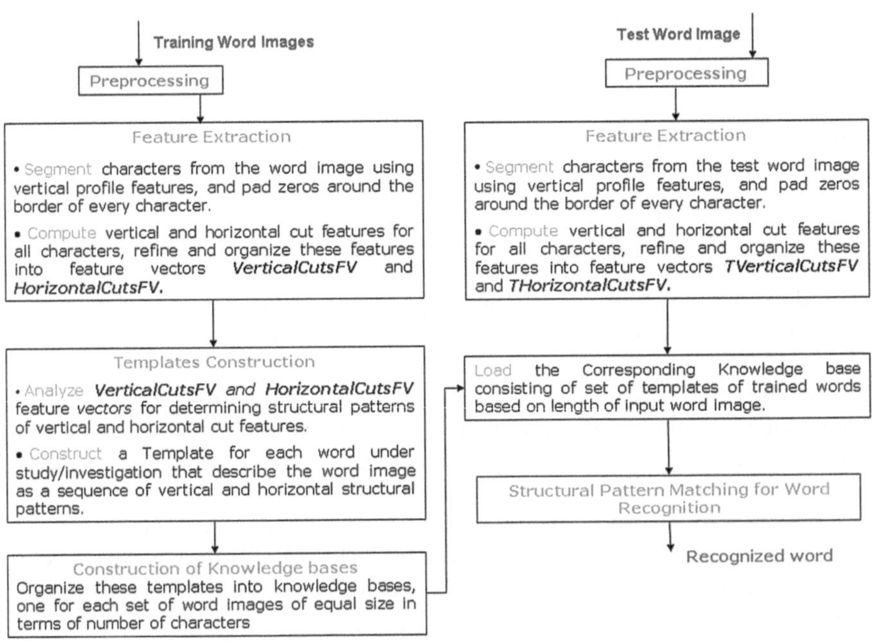


**Fig. 1** Block diagram of proposed method

### 3.1 Preprocessing

In this work, an attempt is made to evaluate performance of new structural patterns of vertical and horizontal cut features extracted directly from variable sized word images without preprocessing them for removal of noise, skew and other degradations. The preprocessing is done for binarization, bounding box generation and padding zeros around the border of the word image. The binarization is done using Otsu's global thresholding method [24]. After binarization, the additional processing is also carried out to ensure that, the foreground objects are represented with white pixels over the black background.

### 3.2 Feature Extraction

In this phase, the word image is processed to obtain vectors of vertical and horizontal cut features, which are further used during template construction to decide structural patterns of primitives/segments that describe the shape of the word image. The vectorization of the word image is detailed below;

Initially, the characters are segmented from the word image using vertical profile features, and every character image is padded with zeros around the border. Then, vertical cut features are computed for all characters of the word image and organized into a feature vector ***VerticalCutsFV***. A vertical cut feature is defined as number of times the vertical scan line enters the component from the background. The vertical cut features for a sample Kannada preprocessed word image (***Panchayat***) is illustrated in Fig. 2. The feature vector ***VerticalCutsFV*** thus obtained is further refined such that the two adjacent feature values are distinct; this is done to eliminate repetitions of values in the successive columns of the word image. The feature vector ***VerticalCutsFV*** is described in Eq. (1)

$$\boldsymbol{VerticalCutsFV} = [\boldsymbol{Vf}_i \; \mathbf{1} \le \boldsymbol{i} \le \boldsymbol{M}] \tag{1}$$

where,

$\boldsymbol{Vf_i}$ represents **i**th vertical cut feature in the vector.
$\boldsymbol{M}$ represents number of vertical cut features in the vector.

The vertical cut features vector for the image in Fig. 2 after refinement is given in Table 1.

**Fig. 2** Vertical cut features

**Table 1** Vertical cut features vector

| Vector Name | Vertical cut Features of word image in Fig.2. |
|---|---|
| *VerticalCutsFV* | [0 2 3 2 4 3 2 1 0 1 2 1 0 1 2<br>3 2 1 3 2 1 2 3 1 2 3 4 3 2 1<br>0 1 2 1 2 4 3 2 1 2 1 0 2 3 4<br>2 3 2 1 2 3 2 1 0] |

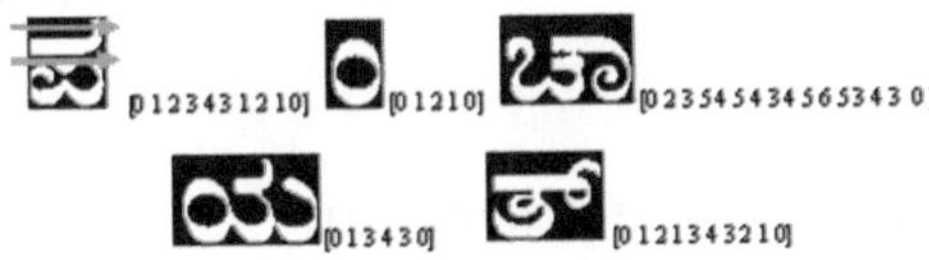


**Fig. 3** Horizontal cut features of characters of word image in Fig. 2

Further, the horizontal cut features of all characters are computed (as shown in Fig. 3) and stored into a feature vector ***HorizontalCutsFV.*** A horizontal cut feature is defined as number of times the horizontal scan line enters the component from the background. The feature vector ***HorizontalCutsFV*** is also refined such that the two adjacent values are distinct. The feature vector ***HorizontalCutsFV*** is depicted in Eq. (2).

$$\boldsymbol{HorizontalCutsFV} = [\boldsymbol{Hf_i}\ \mathbf{1} \leq \boldsymbol{i} \leq \boldsymbol{N}] \tag{2}$$

where,

$\boldsymbol{Hf_i}$ represents **i**th horizontal cut feature in the vector.

$\boldsymbol{N}$ represents number of horizontal cut features in the vector.

The horizontal cut features of all characters of image in Fig. 2 are shown below. And the way these cut features are stored/organized into feature vector ***HorizontalCutsFV*** is given in Table 2.

The vectors thus obtained from training samples in the word database are further used for construction of templates, one for each word under study.

**Table 2** Horizontal cut features vector

| Vector Name | Horizontal cut Features of word image in Fig.2. |
|---|---|
| *HorizontalCutsFV* | [ 0 1 2 3 4 3 1 2 1 0 1 2 1 0 2<br>3 5 4 5 4 3 4 5 6 5 3 4 3 0 1<br>3 4 3 0 1 2 1 3 4 3 2 1 0 ] |

### 3.3 Templates Construction

In this phase, a template consisting of structural patterns of vertical and horizontal cut features is constructed for each word under investigation/study. The templates of all different word images are later organized into knowledge bases as detailed in the next section. The template construction procedure for a word image is as follows.

Initially, the vectors of vertical and horizontal cut features are determined from training samples of the word image under study. Then, these vectors of all training samples are manually analyzed to determine structural patterns which generally occur even in the presence of uncertainty in the input. The structural patterns are subsets of cut features which describe the shape of segments/primitives of the characters in the word image. The manual analysis process/observations revealed that, in most of the training samples certain structural patterns describing segments/primitives generally occur, where as in some other samples some patterns may not occur (cannot be found) due to font variability, uncertain segments/primitives, presence of noise, edge distortions, slant, skew and other degradations. Hence, the structural patterns that generally occur are suitable for describing the shape of the word image and used for construction of the template. However, the patterns representing uncertain segments/primitives may or may not be included in the template. If included, they must exist as subsets, where each subset contains single cut feature value (for example: pattern {4} in Table 3). The inclusion of patterns of uncertain segments will improve the performance of the system, if the uncertain segments appear in the input samples; otherwise performance degrades to small extent. Such intelligent decisions of pattern selection make the method more robust and invariant to the variations in the input. The structural patterns thus decided by analyzing vectors of cut features of training samples of the word image under study are further organized to construct a template. The structure of the template is described in Eqs. (3–7). And the templates of 2 sample word images are also given in Table 3.

$$\boldsymbol{Template} = \{\boldsymbol{VPatternsV}, \boldsymbol{HPatternsV}\} \tag{3}$$

$$\boldsymbol{VPatternsV} = \{\boldsymbol{VCharPatterns}_i \ \mathbf{1} \leq \boldsymbol{i} \leq \boldsymbol{Q}\} \tag{4}$$

$$\boldsymbol{VCharPatterns}_i = \{\boldsymbol{VP}_j \ \mathbf{1} \leq \boldsymbol{j} \leq \boldsymbol{R}\} \tag{5}$$

$$\boldsymbol{HPatternsV} = \{\boldsymbol{HCharPatterns}_i \ \mathbf{1} \leq \boldsymbol{i} \leq \boldsymbol{Q}\} \tag{6}$$

$$\boldsymbol{HCharPatterns}_i = \{\boldsymbol{HP}_j \ \mathbf{1} \leq \boldsymbol{j} \leq \boldsymbol{T}\} \tag{7}$$

where,

***VpatternsV*** is a set of vertical structural patterns of all characters of word image.

***VCharPatterns***$_{\mathbf{i}}$ is a set of vertical structural patterns of **i**th character.

***Q*** represents number of characters in the image.

**Table 3** Templates of two sample word images

| Image | Templates |
|---|---|
| 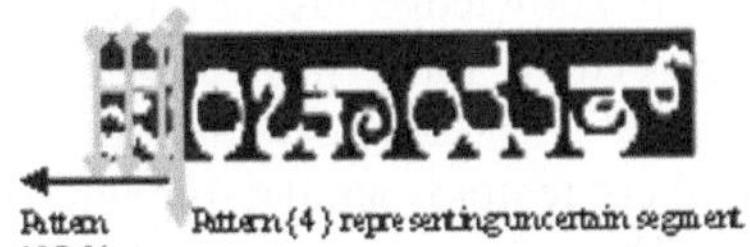<br> | $Template = \{VPatternsV, HPatternsV\}$<br>Where,<br>$VPatternsV = \{VcharPatterns_1, VcharPatterns_2, VcharPatterns_3, VcharPatterns_4, VcharPatterns_5\}$<br>$HPatternsV = \{HcharPatterns_1, HcharPatterns_2, HcharPatterns_3, HcharPatterns_4, HcharPatterns_5\}$<br>$VcharPatterns_1 = \{2\,3\,2\},\{4\},\{3\},\{2\,1\}$<br>$VcharPatterns_2 = \{1\,2\,1\}$<br>$VcharPatterns_3 = \{23\},\{2\,3\},\{2\,1\},\{3\},\{23\},\{4\},\{3\,2\,1\}$<br>$VcharPatterns_4 = \{1\},\{2\,1\},\{2\,3\},\{4\},\{2\,1\},\{2\,1\}$<br>$VcharPatterns_5 = \{2\,3\,4\},\{23\},\{2\,1\},\{3\,2\,1\}$<br>$HcharPatterns_1 = \{1\,2\},\{3\,2\},\{3\,4\},\{3\},\{2\}$<br>$HcharPatterns_2 = \{1\,2\,1\}$<br>$HcharPatterns_3 = \{2\,4\},\{34\},\{5\},\{6\},\{5\},\{4\,3\}$<br>$HcharPatterns_4 = \{1\,2\},\{34\},\{3\}$<br>$HcharPatterns_5 = \{1\,2\,1\},\{3\,2\},\{32\},\{2\,1\}$ |
|  | $Template = \{VPatternsV, HPatternsV\}$<br>Where,<br>$VPatternsV = \{VcharPatterns_1, VcharPatterns_2, VcharPatterns_3, VcharPatterns_4, VcharPatterns_5\}$<br>$HPatternsV = \{HcharPatterns_1, HcharPatterns_2, HcharPatterns_3, HcharPatterns_4, HcharPatterns_5\}$<br>$VcharPatterns_1 = \{1\,2\},\{3\,2\,1\},\{2\,1\}$<br>$VcharPatterns_2 = \{2\,3\},\{2\,3\},\{2\,1\},\{23\},\{2\,1\}$<br>$VcharPatterns_3 = \{2\,1\,2\},\{32\},\{3\,2\},\{3\,4\},\{3\,2\},\{2\,3\},\{2\,1\}$<br>$VcharPatterns_4 = \{23\},\{4\,3\},\{4\,3\},\{4\,3\},\{5]\},\{4\}$<br>$VcharPatterns_5 = \{2\,1\,2\},\{2\,1\},\{2\,1\}$<br>$HcharPatterns_1 = \{23\},\{4\,5\},\{4\,5\},\{4\},\{3\},\{2\}$<br>$HcharPatterns_2 = \{1\},\{2\},\{4\},\{5\,6\},\{7\},\{23\}$<br>$HcharPatterns_3 = \{1\,2\},\{3\},\{4\},\{2\},\{4\,3\},\{5\},\{3\,2\,1\}$<br>$HcharPatterns_4 = \{1\,2\,1\},\{3\,4\}\{1\},\{2\,1\}$<br>$HcharPatterns_5 = \{1\,2\},\{3\,2\},\{1\}$ |

| | |
|---|---|
| ***R*** | represents number of vertical structural patterns in the **i**th character. |
| ***HpatternsV*** | is a set of horizontal structural patterns of all characters of word image. |
| ***HCharPatterns***$_i$ | is a set of horizontal structural patterns of **i**th character. |
| ***T*** | represents number of horizontal structural patterns in the *i*th character. |

For the purpose of template construction, the images are captured from display boards of Government Offices of Bagalkot City in Karnataka state of India. Then, the words are segmented from display board images and a Kannada word image database is constructed. The Kannada word image database consists of 1,200 samples of nearly 102 different words, of which there are 240 samples of 12 different "two" character words, 250 samples of 25 different "three" character words, 260 samples of 23 different "four" character words, 150 samples of 15 different "five" character words, 100 samples of 7 different "six" character words, and 200 samples of 20 words of "seven to ten" character lengths. The images are captured from mobile phone camera at various resolutions 240 × 320, 600 × 800,

900 × 1,200 at a distance of 1–6 m. All these images are used for evaluating the performance of the proposed model. The images captured with camera resolution 240 × 320 at a distance of 1–3 m are found to be clear when the viewing angle is parallel to the text plane, perspective distortions and other degradations occur beyond 3 m with other viewing angles. But the images captured at a distance of 1–6 m with other stated resolutions are clear, perspective distortions still occur when the viewing angle is not parallel. The images in the database are characterized by variable font size and style, uneven thickness, minimal information context, small skew, noise, perspective distortion and other degradations. The word images containing segmentation errors are also available in the database. Then, 50 % of the different samples from each word image are chosen to train the system. And testing is carried out for all word images of database containing 50 % trained and 50 % test samples.

Thus obtained templates of 102 different word images are further organized into Knowledge Bases as detailed in the next section.

## 3.4 Knowledge Bases for Word Recognition

The templates of all different word images are organized into knowledge bases, one for each set of templates of equal length word images. The knowledge bases are described in Eq. (8).

$$\boldsymbol{WordKB_u} = \{\boldsymbol{Template_v}, \mathbf{1} < = \mathbf{v} < = \boldsymbol{W}\} \tag{8}$$

where,

| | |
|---|---|
| $\boldsymbol{WordKB_u}$ | is a knowledge base of set of templates of word images of length $\boldsymbol{u}$. |
| $\boldsymbol{Template_v}$ | is a template of a **v**th word image in the knowledge base. |
| $\boldsymbol{u}$ | represents the length of word images of stored templates in the knowledge base and varies in the range 2–10. |
| $\boldsymbol{W}$ | represents number of templates in the knowledge base. |

These knowledge bases are further used by newly defined structural pattern matching procedure for word recognition of test image.

## 3.5 Structural Pattern Matching for Word Recognition

In this stage, the test image is processed to obtain vectors of vertical and horizontal cut features as described in Eqs. (9) and (10).

$$\boldsymbol{TVerticalCutsFV} = [\boldsymbol{Vf_i}\ \mathbf{1} \leq \boldsymbol{i} \leq \boldsymbol{M}] \tag{9}$$

where,
**$Vf_i$** represents **i**th vertical cut feature in the vector.
**M** represents number of vertical cut features in the vector.

$$\textit{THorizontalCutsFV} = [Hf_i\ 1 \leq i \leq N] \tag{10}$$

where,
**$Hf_i$** represents **i**th horizontal cut feature in the vector.
**N** represents number of horizontal cut features in the vector.

Then, the similarity between test image and every stored template in the knowledge base is determined by matching template structural patterns with the vertical and horizontal cut features of test image using structural pattern matching procedure. The knowledge base is chosen based on length of test word image. The structural pattern matching procedure works as follows; the template structural patterns of every character are matched with the cut features of corresponding character in the test image. And every matched pattern is voted with a value 1. At the end of matching, the procedure returns number of template structural patterns found/matched in the test image. After processing all templates with test image, the input word image is recognized computing similarity values as described in Eqs. (11) and (12).

$$\textit{Recognised Word} = \max_{\forall k}\{\textit{Similarity}_k\}. \tag{11}$$

$$\textit{Similarity}_k = \frac{\textit{Number of matched patterns of Template}_k \textit{ in the test Image}}{\textit{Total Number of Patterns in the Template}_k} \tag{12}$$

where,

***$Similarity_k$*** is a normalized value representing similarity between ***$Template_k$*** and test image. As, the number of stored patterns in the templates vary from one word to another, the similarity value is normalized using the total number of patterns in the corresponding ***$Template_k$*** and lies in the range 0–1.
And ***k*** lies in the range ***$1 \leq k \leq W$***.

The maximum similarity between test image and template in the knowledge base is used to recognize the word image. The proposed methodology performs well for variability in font size, style, and image resolution. The approach also recognizes nonlinear and noisy text words and results are presented in the next section. However, the method requires sufficient training of all variations in font size, style and other degradations.

# 4 Results and Analysis

The proposed methodology for word recognition of Kannada text employing structural patterns of horizontal and vertical cut features has been evaluated for 1,200 low resolution word images. The experimental results of processing several display board images dealing with various issues and computed similarity values with the stored templates in the knowledgebase are described in Table 4. The proposed methodology has produced good results for low resolution word images containing text of different font size, style, resolution and alignment with varying background. The approach also recognizes small skewed text word images (as given in Table 4). It is also invariant to the presence of noise and other degradations (as described in the second row of Table 4).

**Table 4** The performance of the system of processing different images dealing with various issues

| Input Sample Image | Description | Similarity Value between input word image and pre-constructed corresponding Template in the knowledgebase | Similarity Values between input word image and pre-constructed templates of words of same length in the knowledgebase |
|---|---|---|---|
|  | Word Recognition of an image having minimal information content with horizontally stretched characters. The image was captured at a distance of 2-3 meters. | 0.5667 | [illegible] = 0.5667 [illegible] = [illegible] [illegible] = 0.2238 [illegible] = [illegible] [illegible] = [illegible] [illegible] = [illegible] [illegible] = 0.1552 [illegible] = [illegible] [illegible] = 0.3111 [illegible] = [illegible] [illegible] = 0.2421 [illegible] = [illegible] |
|  | The effectiveness of the method in processing degraded Kannada word Image containing characters of uneven thickness, lighting, noise, small skew and other degradations. The image was captured at a distance of 1 meter. | 0.8025 | 0.4229, 0.3417, 0.4479, 0.4565, **0.8025**, 0.2573, 0.6000, 0.1717, 0.5362, 0.3250, 0.4000, 0.2588, 0, 0, 0.3825 0.3217, 0.3844, 0.2510, 0.2941, 0, 0.2830, 0.4468, 0, 0.4197, 0.3356 |
|  | The robustness of the method in recognizing word image containing 5 characters and white noise. The image was captured at a distance of 2-3 meters. | 1.0000 | 0.5138, 0.8093, 0.6094, 0.8576, 0.7083, 0.7184 0.9583, 0.9659, 0.5642, 0.7449, 0.6675, **1.0000**, 0.5192, 0.2856, 0.7453 |
|  | The ability of method in modeling and recognizing word image with text of different font style containing descenders. The image was captured at a distance of 2-3 meters. | 0.8843 | 0.6569, 0.7189, 0.8313, 0.7547, 0.7833, 0.6789 0.5233, 0.5556, 0.5477, 0.6279, 0.8279, 0.4956, 0.7346, 0.2918, **0.8843** |
|  | The method processes a blurred image with small skew and perspective distortion captured at a distance of 5-6 meters. | 0.9368 | 0.6605, 0.5852, 0.6312, 0.8928, 0.7513, 0.8579 0.4553, **0.9368**, 0.7167, 0.5909, 0.8889, 0.7170, 0.5683, 0.8734, 0.7346 |

The experimental values in Table 4 demonstrate that, the similarity value of input image with the corresponding template is higher/larger compared to similarity values with other templates in the knowledgebase. For illustration, the computed similarity values of input image with all *two* character word templates are shown in the first row of Table 4. And the similarity values of images with the corresponding templates in other rows are indicated in the "third" column and also highlighted in the "fourth" column. It can also be seen that, the similarity values of images in the 2, 3 and 4th rows are comparatively larger because of more information content and presence of descender segments, which significantly improve efficacy of the system in most of the cases. And processing images captured at a distance of 5–6 m does not affect the system performance (as shown in 5th row) in the presence of perspective distortion, if the viewing angle is parallel to the text plane.

The method is also tested on word images containing segmentation errors and gives higher degree of accuracy when the information content is more. The experimentations also revealed that, the method works efficiently for all images captured at various resolutions at a distance of 1–6 m. Hence, the proposed method is robust and achieves an average recognition accuracy of 97.67 %, and reports an individual accuracy of 95 % for 2 character words, 95.6 % for 3 character words, 98.04 % for 4 character words, and 100 % for 5–10 character words. A closer examination of results revealed that misclassifications arise due to minimal information content, more noise and larger skew, which affect the structural pattern of primitives/segments of region of text and performance of the approach. It is also found that, if the templates are trained for all variations and degradations, better performance can be obtained. And more information content improves accuracy of the system. The overall performance of the system is reported in Table 5.

The structural patterns and their storage order in the template model the relation between various primitives/segments of the text region/word image. Therefore, the sequences of patterns representing the features made samples separable in the feature space and significantly improved recognition accuracy. During experiments it is also noticed that, modeling the sample with more structural patterns

**Table 5** Overall system performance

| # Samples of word images | Number of different words (word size) | Number of words correctly recognized | Number of misclassified words | Recognition accuracy (%) |
|---|---|---|---|---|
| 240 | 12 (2 Character words) | 228 | 12 | 95 |
| 250 | 25 (3 Character words) | 239 | 11 | 95.6 |
| 260 | 23 (4 Character words) | 255 | 05 | 98.07 |
| 150 | 15 (5 Character words) | 150 | – | 100 |
| 100 | 07 (6 Character words) | 100 | – | 100 |
| 200 | 20 (7–10 Character words) | 200 | – | 100 |

representing more segments/primitives will better characterize the shape of the word image and takes care of more uncertainty in the input. And the character descendants (Ottaksharas) in the word images yield more discriminating features contributing sufficiently in increasing efficiency/accuracy of the system. The structural pattern matching procedure is also found to be effective in using the knowledge of templates for improved classification accuracy.

## 5 Conclusions and Future Work

In this paper, an approach for word recognition of Kannada text in low resolution images of display boards employing structural patterns of horizontal and vertical cut features is proposed. The method recognizes word images without applying techniques for removal of noise and other degradations. This aspect of work makes it more robust and efficient. The proposed set of new features tend to model the shape of a region of text in a better way and thus provide sufficient characterization for improved recognition accuracy specifically for Kannada words. The testing of methodology for 1,200 low resolution word images containing text of different size, font, and alignment with varying background has yielded an average recognition accuracy of 97.67 %. The system is found to be resilient to the presence of small skew and other degradations. This is a significant result, which makes this work suitable for text understanding from low resolution display board images.

The method can be extended for word recognition of images belonging to other scripts. Only modification needed is construction of new templates. And further investigations can focus on automation of structural patterns selection and template construction, which is carried out manually in the current work.

## References

1. Abowd GD, Atkeson CG, Hong J, Long S, Kooper R, Pinkerton M (1997) CyberGuide: a mobile context-aware tour guide. Wireless Netw 3(5):421–433
2. Marmasse N, Schamandt C (2000) Location aware information delivery with *comMotion*. In: Proceedings of conference on human factors in computing systems, pp 157–171
3. Tollmar K, Yeh T, Darrell T (2004) IDeixis—image-based deixis for finding location-based information. In: Proceedings of conference on human factors in computing systems (CHI'04), pp 781–782
4. Leetch G, Mangina E (2005) A multi-agent system to stream multimedia to handheld devices. In: Proceedings of the 6th international conference on computational intelligence and multimedia applications
5. Premchaiswadi W (2009) A mobile image search for tourist information system. In: Proceedings of 9th international conference on signal processing, computational geometry and artificial vision, pp 62–67
6. Ma C-J, Fang J-Y (2008) Location based mobile tour guide services towards digital dunhaung. In: International archives of phtotgrammtery, remote sensing and spatial information sciences, vol XXXVII, Part B4, Beijing

7. Wu S-H, Li M-X, Yanga P-C, Kub T (2010) Ubiquitous wikipedia on handheld device for mobile learning. In: 6th IEEE international conference on wireless, mobile, and ubiquitous technologies in education, pp 228–230
8. Yeh T, Grauman K, Tollmar K (2005) A picture is worth a thousand keywords: image-based object search on a mobile platform. In: Proceedings of conference on human factors in computing systems, pp 2025–2028
9. Fan X, Xie X, Li Z, Li M., Ma WY (2005) Photo-to-search: using multimodal queries to search web from mobile phones. In: Proceedings of 7th ACM SIGMM international workshop on multimedia information retrieval
10. Hwee LJ, Chevallet JP, Merah SN (2005) SnapToTell: Ubiquitous information access from camera. In: Mobile human computer interaction with mobile devices and services, Glasgow, Scotland
11. Zhang J, Chen X, Hanneman A, Yang J, Waibel A (2002) A robust approach for recognition of text embedded in natural scenes. In: Proceedings of 16th international conference on pattern recognition, vol 3, pp 204–207
12. Chen X, Yang J, Zhang J, Waibel A (2004) Automatic detection and recognition of signs from natural scenes. IEEE Trans Image Process 13(1):87–99
13. Mishra A, Alahari K, Jawahar CV (2012) Top-down and bottom-up cues for scene text recognition. In: Proceedings of the IEEE conference on computer vision and pattern recognition
14. Weinman JJ, Learned-Miller E, Hanson A (2008) A discriminative semi-Markov model for robust scene text recognition. In: 19th international conference on pattern recognition ICPR 2008, vol 8–11, pp 1–5
15. Weinman JJ, Learned-Miller E, Hanson A (2009) Scene text recognition using similarity and lexicon with sparse belief propogation. IEEE Trans Pattern Anal Mach Intell 31(10): 1733–1746
16. Weinman JJ (2010) Typographical features for scene text recognition. In: 20th international conference on pattern recognition, vol 23–26, pp 3987–3990
17. Park J et al (2010) Automatic detection and recognition of Korean text in outdoor signboard images. Pattern Recogn Lett 31:1728–1739
18. Kobayashi T, Toyamaya T, Shafait F, Iwamura M, Kise K, Dengel A (2012) Recognizing words in scenes with a head-mounted eye-tracker. In: 10th IAPR international workshop on document analysis systems, vol 27–29, pp 333–338
19. Ghoshal R, Roy A, Parui SK (2011) Recognition of Bangla text from scene images through perspective correction. In: 2011 international conference on image information processing (ICIIP), vol 3–5, pp 1–6
20. Coates A, Carpenter B, Case C, Satheesh S, Suresh B, Wang T, Wu DJ, Ng AY (2011) Text detection and character recognition in scene images with unsupervised feature learning. In: 2011 international conference on document analysis and recognition (ICDAR), vol 18–21, pp 440–445
21. Wang K, Babenko B, Belongie S (2011) End-to-end scene text recognition. In: 2011 IEEE international conference on computer vision, vol 6–13, pp 1457–1464
22. Chung H, Sihn K-H, Hong S, Song HJ, Kim D (2011) Scene text recognition system using multigrain parallelism. Consumer. In: 2011 IEEE communications and networking conference, vol 9–12, pp 865–869
23. Wang, X, Ding X, Liu C (2001) Character extraction and recognition in natural scene images. In: Proceedings of 6th international conference on document analysis and recognition, pp 1084–1088
24. Otsu N (1978) A threshold selection method from gray-level histogram. IEEE Trans Syst Man Cybern 19(1):62–66

# Stained Blood Cell Detection and Clumped Cell Segmentation Useful for Malaria Parasite Diagnosis

**Dhanya Bibin and P. Punitha**

**Abstract** This study presents a new method for splitting clumped blood cells effectively into individual cells and develops a complete framework for automating the detection of malaria parasites in Leishman-stained thin peripheral blood sample images. The images are segmented to extract the foreground information to isolate the RBCs using Chan Vese segmentation algorithm. The noise present in the image is removed using a thresholding method based on the average size of the blood cell. The preprocessed image is subjected to dominant color extraction to detect the stained cells in it. To separate the clumped blood cells, each clumped object is processed by Laplacian of Gaussian edge detection algorithm and then the major axis of the clumped object is computed. The two halves of the segmented object lying above and below the major axis is traversed to find the overlapping points of two clumped cells and the affected cell is separately found out. The robustness and effectiveness of this method has been assessed experimentally with various images collected from Public Health centers.

**Keywords** Malaria parasite diagnosis · Clumped blood cell · Leishman-stained peripheral blood sample · Chan Vese segmentation · Laplacian of Gaussian edge detection

D. Bibin (✉) · P. Punitha
Department of Master of Computer Applications, PES Institute of Technology, Bangalore, Karnataka 560085, India
e-mail: dh.bibin@gmail.com

P. Punitha
e-mail: punithaswamy@pes.edu

P. P. Swamy and D. S. Guru (eds.), *Multimedia Processing, Communication and Computing Applications*, Lecture Notes in Electrical Engineering 213,
DOI: 10.1007/978-81-322-1143-3_16, © Springer India 2013

# 1 Introduction

Malaria is a serious infectious disease caused by a blood parasite named Plasmodium spp. It is considered as most threatening and wide spread parasitic disease among all parasitic diseases [1]. According to the World Health Organization, malaria causes more than 1 million deaths arising from approximately 300–500 million infections every year [2]. Although there exists many techniques to diagnose malaria [3] manual microscopy for the examination of peripheral blood smears [4] is currently "the gold standard". A peripheral blood smear (peripheral blood film) is a glass microscope slide which is coated with a thin layer of venous blood on one side. Thin films and thick films are two types of blood films traditionally used for malaria diagnosis. The thick film is used for quick identification and quantification of parasites whereas the thin film is used for differentiation of parasite species. This paper focus on thin peripheral blood smears analysis.

Diagnosis using a microscope requires manually counting the ratio of infected red blood cells to the number of red blood cells in a slide. A pathologist typically observes around 100 High Power Fields i.e., the area of the blood slide visible under the maximum magnification power of the microscope, to conclude the presence/absence of infection. The time taken to diagnose a patient in this manner demand 15–20 min of a pathologist's time [5]. The quality and accuracy of the final diagnosis ultimately depends on the skill and experience of the pathologist and also the time spent in analyzing each slide. So microscopic examination of the blood smears requires special training and considerable expertise [6]. This process is time-consuming, laborious, and leads to fatigue, especially in peak infection seasons. This may result in wrong diagnosis which leads to wrong treatment and can also result in death of the patient [5]. So manual microscopy is not considered as a reliable screening method and recent study has shown that the agreement rates among various clinical experts for the diagnosis of malaria are surprisingly low.

With this backdrop, the main aim of this research is to design a semi automated malaria diagnosis system by understanding the pathologist's diagnostic expertise and representing it by image processing and pattern recognition algorithms. This research focuses on designing malaria diagnosis system, which replicates the manual microscopy diagnosis procedure to detect the presence of stained blood cell components in thin peripheral blood smear images of malaria infected samples. This is considered as the primary task in malaria diagnosis. Such a diagnostic system can provide a huge assistance to the pathologist in the diagnostic procedure of malaria. With the help of such a semi automated diagnostic system, there will be a significant reduction in the number of slides the pathologist needs to examine especially in regions where malaria is an endemic.

In addition, determining the presence of malaria parasite in images with overlapping clumped blood cells remains an obstacle to be overcome in automated diagnosis of malaria. Hence the main objective of this work is not only to detect the presence of stained blood cell components in malaria affected thin peripheral blood smear images but also this work focuses on identifying the presence of stained blood cell components in clumped/overlapped blood cell clusters. This study detects all the stained blood cell components present in clumped/overlapped blood cell clusters by segmenting the overlapping cells into individual cells and identifies the stained pixels in it. Compared to the existing clump splitting methods, this scheme provides a more simplified approach to the clumped blood cell segmentation problem and the results obtained in this study are comparable to the existing works.

The remaining part of the paper is organized as follows. In Sect. 2 we give a brief introduction to the problem and to the existing manual microscopy diagnosis system. Section 3 explains the work done so far in this area. In Sect. 4 we propose a new method to detect the presence of stained blood cell components in clumped blood sample images. Section 5 illustrates the proposed method and Sect. 6 gives the conclusion.

## 2 Existing Manual Microscopy Diagnosis System

Malaria is transmitted by the infected female Anopheles mosquitoes which carry Plasmodium sporozoites in their salivary glands. Falciparum, vivax, ovale, and malariae are the four species of genus Plasmodium which can cause malaria infections in humans. In peripheral blood of an infected person these different species are observed under four different life-cycle stages such as ring, trophozoite, schizont and gametocyte (peripheral blood is the flowing, circulating blood of the body). The different species of the parasite can be identified by 3 characteristics namely (1) the shape of the infected cell, (2) the presence of some characteristic dots called Schüffner's dots, Maurer's clefts, Ziemann's Stippling and (3) the morphology of the parasite in some of the life-cycle-stages. On the other hand the different life-cycle-stage of the parasite can be identified by its morphology, size, and the presence or absence of malarial pigment [4, 7].

The WHO practical microscopy guide for malaria provides detailed procedures for laboratory practitioners [4]. According to WHO microscopy diagnosis initially requires determining the presence (or absence) of malarial parasites in the examined specimen. Then, if parasites are present two more tasks must be performed: (1) identification of the species and life-cycle stages causing the infection and (2) calculation of the degree of infection, by counting the ratio of parasites versus healthy components (i.e., parasitaemia) [4].

In peripheral blood sample microscopy diagnosis is possible and efficient via a chemical process called staining [8]. Common stains used are Field's stain, Giemsa stain and Leishman stain [9]. The staining process slightly colorizes the red blood cells (RBCs) but highlights Plasmodium parasites as shown in Fig. 2. White blood cells (WBC), platelets and artifacts are also highlighted by the staining process [see Fig. 1]. So the whole diagnosis process requires an ability to differentiate non-parasitic stained components (e.g., red blood cells, white blood

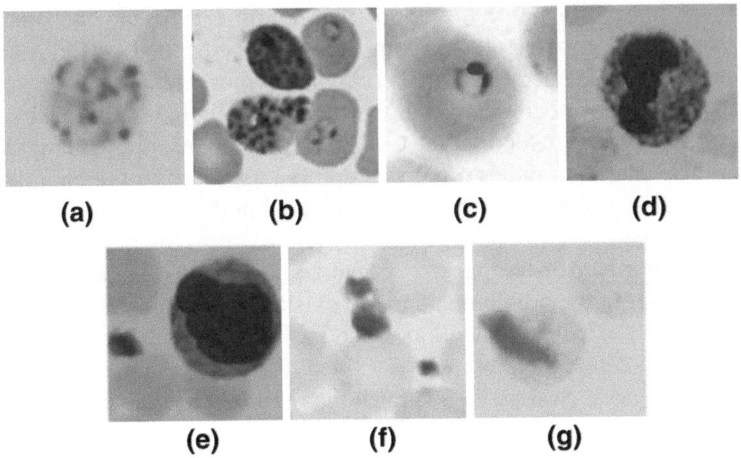


**Fig. 1** Examples of stained objects. **a–c** Parasites, **d–e** WBC, **e** includes a platelet, **f** platelet, **g** artifact [13]

**Fig. 2** Input image

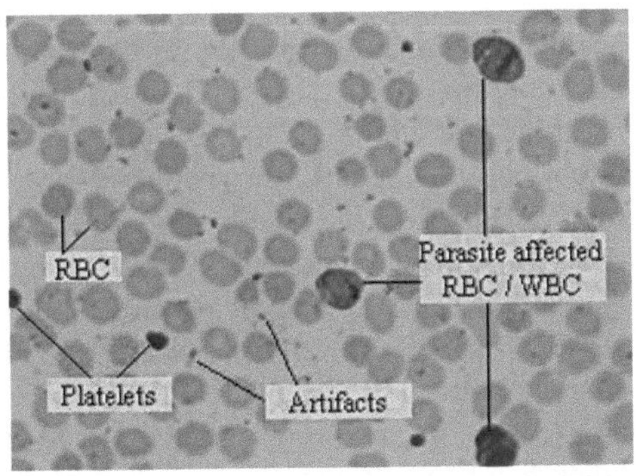


cells, platelets, and artifacts) and the malarial parasites using visual information. The detection of Plasmodium parasites requires the detection of the stained objects. These stained objects need to be analyzed further to determine whether they are parasites or not.

In microscopic diagnosis of malaria, the pathologists often need to examine peripheral blood sample slides with clumped/overlapping blood cell clusters. If the blood sample is not treated well, it is very easy for red blood cells to form clumped clusters [10]. Detection of parasite infected stained blood cell components in clumped blood cell clusters through visual examination is a difficult task for the pathologist. But it is an important task to find the number of uninfected and infected individual blood cells in a clumped cluster to calculate the degree of infection (also called parasitaemia value which is obtained by counting the ratio of parasites versus healthy components). And parasitaemia value is very important in determining the correct medication to the patient.

## 3 Literature Support

Researchers have made a noticeable number of computer vision studies on automatic malaria parasite detection as well as clumped blood cell segmentation [5,8,11–29]. Di Ruberto et al. [12–14] employed morphological regional extrema to detect the stained pixels. They used morphological opening to extract the object regions marked by the stained pixels. They identified the WBCs, platelets, and artifacts and excluded these from further processing. They had done the parasite detection based on morphological operations. Rao and Rao et al. [11, 15] had addressed automatic parasite detection problem by selecting thresholds based on color histograms. They performed life-cycle stage analysis directly on the stained objects of the Plasmodium falciparum in vitro cultured images. In vitro cultured images consist of blood samples grown in a laboratory environment. Hence, they do not contain artifacts, platelets and WBCs. But the diagnosis of malaria must be performed on actual peripheral blood specimens of the patients which should contain other stained bodies such as WBCs, platelets and artifacts. Sio et al. [16] had experimented on Plasmodium falciparum in vitro images. They had developed a "malaria parasite counting" software to count the stained objects in the peripheral blood specimens. But they had not addressed species and life-cycle stage identification problems. Tek et al. [8] had proposed a solution for the parasite detection problem with two consecutive classifications such as stained/non-stained pixel classification and parasite/nonparasite classification. They had employed the Bayesian decision rule to determine whether a pixel is stained or non-stained. The stained pixels are further classified as parasite or nonparasite. A two level global and local thresholding method was used by Ross et al. [17] to locate the stained pixels in the images. They used morphological opening to recover the object binary masks to identify the parasites present on the images. Makkapati and Rao [5], presented an ontology-based classification to detect the presence of parasites and had identified the four species that cause the infection.

Di Ruberto et al. [13, 14] experimented on images with clumped cells for the first time. They used mathematical morphology to separate the overlapping cells. Their methodology had been improved later by Mohana Rao and Dempster [18] and Scotti [18, 19]. Watershed techniques [20, 21], concavity analysis [22, 23], boundary information [24, 25] and model-based approaches [26, 27] are also used extensively by the researchers to study about the clumped blood cell segmentation problem. These approaches still have some drawbacks which need to be improved. Watershed technique is relatively time-consuming because of the partitions being over-split in cluster locations and contour parts being merged together. The concavity analysis methods are too sensitive in using threshold to recognize regions. The object will be over-split at lower threshold, and under-split at higher threshold. Likewise, model-based approach requires initialization of model parameters and considerable computational expense [10].

# 4 Proposed Method

This section describes the proposed methodology to detect the stained blood cell components in Leishman-stained thin peripheral blood smear images of malaria infected samples. It also presents a novel scheme to identify the presence of staining even in clumped or overlapped clusters of blood components. This scheme segments the clumped blood cell components into individual blood cell components and identifies the stained blood cells present in it. The detection procedure is divided into five stages; foreground–background segmentation, artifact and platelet Elimination, feature extraction and stained blood cell detection, clumped/overlapping blood cell segmentation and stained blood cell detection in clumped blood cells.

## *4.1 Foreground–Background Segmentation*

In Leishman-stained thin peripheral blood smear images of malaria infected samples. The foreground of the image contains regular blood cell components such as RBCs, WBCs, platelets and artifacts and the background of the image contains plasma. The input image is segmented into foreground and background regions to eliminate the background information and to extract the foreground region. Chan Vese segmentation algorithm [30] is used to segment the image and extract the foreground information. Chan and Vese [30] proposed energy minimization of the image to detect edges of objects embedded within an image. When Chan Vese algorithm is applied on a stained blood sample image, the entire background information is eliminated and only information about regular blood cell components, i.e.; RBCs, WBCs, platelets and artifacts are retained. These regular blood cell components segmented out in this stage are further processed to detect the stained blood cells in them.

Among regular blood cell components: artifacts represents bacteria, spores, crystallized stain chemicals and particles due to dirt; platelets are small, irregularly shaped bodies; RBCs are the commonest cells in thin peripheral blood films. There are about 5,000,000 RBCs present in each microliter (μl) of blood where as the number of WBCs are much fewer than RBCs. It is normally 6,000–8,000 per microliter of blood. When the peripheral blood sample slides are stained using Leishman protocol, it colors the parasite free red blood cells in a pale-greyish to light-pink color, but colors parasite affected cells, white blood cells (WBC), platelets, and various artifacts in a bright blue tone [4]. Since Leishman staining gives parasite affected cells, artifacts and platelets the same bright blue color, it gives rise to possible misdiagnosis when automated systems are being devised. Hence artifacts and platelets have to be eliminated before further diagnosis of stained components. Empirically it has been observed that the artifacts and platelets are comparatively smaller than the regular blood cells and have irregular shape, which can be used as the significant features to eliminate these irrelevant components.

## 4.2 Artifact and Platelet Elimination

As explained in the previous section, the presence artifacts and platelets in the segmented image may cause error in diagnosis. To eliminate the artifacts and platelets, a thresholding based method which depends on the size of the regular blood cell components is proposed. To find the optimum threshold, the image is subjected to the following operations. Firstly the segmented image is subjected to binarization, then the foreground components are extracted using connected component labeling. The size of each connected component (RBCs, WBCs, Platelets and artifacts) is computed by counting the number of pixels in each connected component. The connected components are sorted in increasing order of their size, the point where there is a drastic change from the size of one component to the other component is selected as the threshold for separating artifacts and platelets from RBC and WBC. Empirically it has been observed that the difference in the size of artifacts and platelets from RBCs and WBCs in terms of number of pixels is in between 115 pixels to 120 pixels. We chosen 117 as the optimum threshold. The proposed approach eliminated majority of the artifacts and platelets. The processed image retains only RBCs and WBCs. To identify the presence of stained blood cell components in the processed image, the image is subjected to feature extraction explains in the next sub section.

## 4.3 Feature Extraction and Stained Blood Cell Detection

The presence of malarial parasites in the image can be detected by the presence of stained blood components which can be identified by the shape (morphology) and color properties of the blood components [4]. This work uses color properties of the blood components for stained object identification. As explained in Sect. 4.1 the Leishman stain colors, red blood cells (RBCs) in a pale-greyish to light-pink color, but gives the parasite affected cells and white blood cells (WBC) a bright blue tone. The processed image obtained from the previous section contains only RBCs and WBCs. This image is subjected to dominant color extraction to identify the stained blood cell components present in it. A color histogram is computed for each connected component (RBCs and WBCs) in the processed image and based on the colors of RBCs and WBCs due to staining (pale-greyish to light pink and bright blue) a threshold is decided to differentiate the connected components as stained or not stained.

After this stage all the stained blood components in the processed image are identified. If the input image contains any clumped/overlapping blood cell cluster, and if there exists a stained blood cell component in this cluster, there is a possibility that the whole overlapping blood cell cluster is identified as stained component. As explained in Sect. 2, clumped cell segmentation is very important to compute the parasitaemia value and hence to prescribe correct medication to the patient. To address this problem, we propose a simple segmentation technique that separates the stained overlapping/clumped blood cell components into individual blood cells. This segmentation technique is explained in the following sub section.

### 4.4 Clumped/Overlapping Blood Cell Segmentation

This section illustrates a new scheme to segment the clumped/overlapping blood cell components into individual cell components. To segment the overlapping cells, in this work we propose a method which detects the meeting points of the overlapping cells and segments the overlapping cells into individual components at the meeting points.

The boundary points are extracted from the clumped blood cell components using Laplacian of Gaussian (LoG) edge detection algorithm. It is observed from the boundary of the clumped component that the overlapping region can be detected by finding the concave points on the boundary. In order to find these concave points the major axis of the clumped blood cell component is computed using the boundary points. Considering this major axis as the reference axis the boundary of the clumped object which lies above and below the major axis can be traversed in clockwise and anticlockwise directions respectively. In order to make this traversal simple the boundary of the clumped blood cell component is rotated by an angle θ such that the major axis is parallel to the horizontal axis. '$\theta$' is the slope of the major axis which is calculated using the boundary points $p1(x_1, y_1)$ and $p2(x_2,y_2)$ which form the end points of the major axis using Eq. (1).

$$\theta = \tan^{-1}\left(\frac{y_2 - y_1}{x_2 - x_1}\right) \tag{1}$$

Then the boundary pixels lying above and below the major axis in the rotated component are traversed in search of the concave points. When there is a significant fall and raise in the boundary curve, a concave point is said to be identified.

### 4.5 Stained Blood Cell Detection in Clumped Component

The overlapping/clumped blood cell components in the input image are segmented into individual blood cell components as explained in the previous section. So in the processed image, all clumped blood cell components are segmented into individual blood cell components. To detect the stained blood cell component from the processed image, the image is subjected to dominant color extraction and an optimal threshold is decided to differentiate the segmented blood cell components as stained or not stained using the methodology explained in Sect. 4.2.

## 5 Illustration

In this section we illustrate the proposed scheme for stained blood cell component detection and clumped blood cell segmentation in malaria parasite affected blood sample images and demonstrate how the proposed identification and segmentation takes place.

Thin peripheral blood smear images obtained from Co-Operative Medical College, Cochin, Kerala are used for the experimental purpose. The samples are stained using Leishman protocol to highlight the parasites and are initially examined by haematopathologists with expertise in malaria diagnosis. Slide images were acquired using a digital microscope (Leica DM500) with maximum magnification. These are RGB color images of 2,048 $\times$ 1,536 pixels resolution. All experiments were performed using Matlab 2010a (Version 7.10.0) and Image Processing Toolbox. Each image represents a section of microscopic field at 1,000X magnification and containing approximately 200 RBCs.

Figure 2 shows Leishman stained thin peripheral blood sample image affected with malaria. Figure 2 contains RBCs, WBCs, parasites, platelets and artifacts. Firstly the input image is subjected to background foreground segmentation using Chan Vese segmentation algorithm as explained in Sect. 4.1. The foreground region contains blood cell components like RBCs, WBCs, platelets, artifacts and parasites. The foreground extracted image obtained after Chan Vese segmentation is shown in Fig. 3. Figure 3 contains artifacts and platelets that are not required for any further processing and may also lead to misdiagnosis. So these artifacts and platelets are eliminated by using the method explained in Sect. 4.2. Figure 4 shows the artifact and platelet eliminated image.

Since majority of the artifacts and platelets which causes error in the diagnosis are eliminated at this stage, it is possible to detect the presence of stained blood cell components in this image. Figure 5 shows the image in which the stained blood cell components are identified as per the methodology explained in Sect. 4.3. In Fig. 5 it is observed that three clumped/overlapping/coinciding blood cells are identified as stained blood cells. When compared to the original input image in Figs. 2 and 5 shows uninfected blood cells merged with the infected blood cells and being identified as stained components. This kind of misdetection will lead to wrong parasitaemia (degree of infection) value and consequently lead to wrong medication to the patient as explained in Sect. 4.3. So we segment the clumped blood cell components into individual components as explained in Sect. 4.4 and then detect the presence of stained blood components in it as explained in Sect. 4.5.

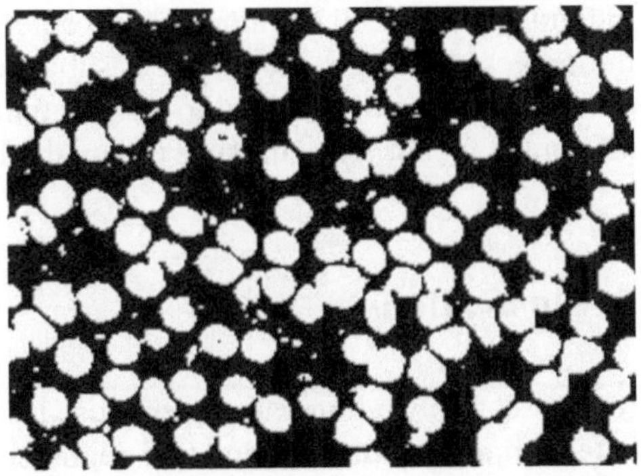

**Fig. 3** Segmented image

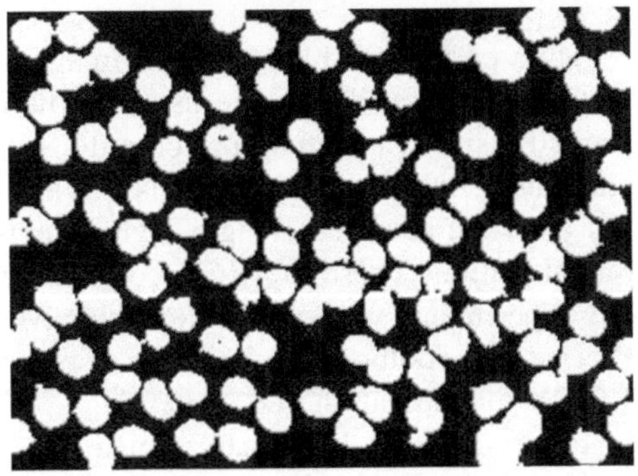

**Fig. 4** Artifacts and platelets eliminated image

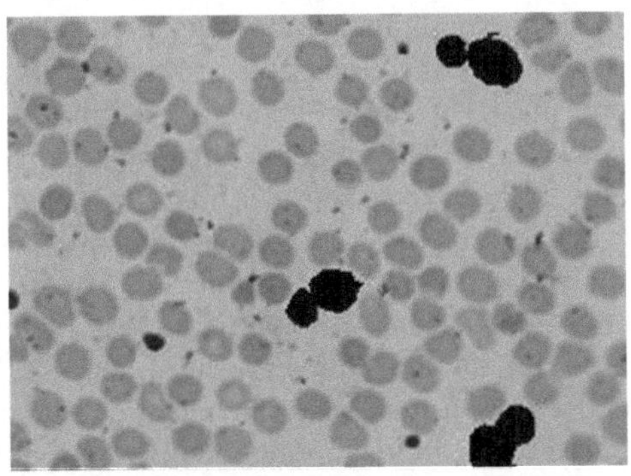

**Fig. 5** Stained blood cells detected image

Figure 6a, d and g shows the coinciding blood cell components those are identified as stained components in Fig. 5. When compared to the original input image in Fig. 2, it is evident that, among these three coinciding blood cell components, only one blood cell is actually stained. So in order to identify the blood cell component that is actually stained, the three coinciding blood cell components are segmented using the methodology explained in Sect. 4.4. The various steps in clumped blood cell segmentation are explained in Fig. 6a–i. Figure 6b, e and h

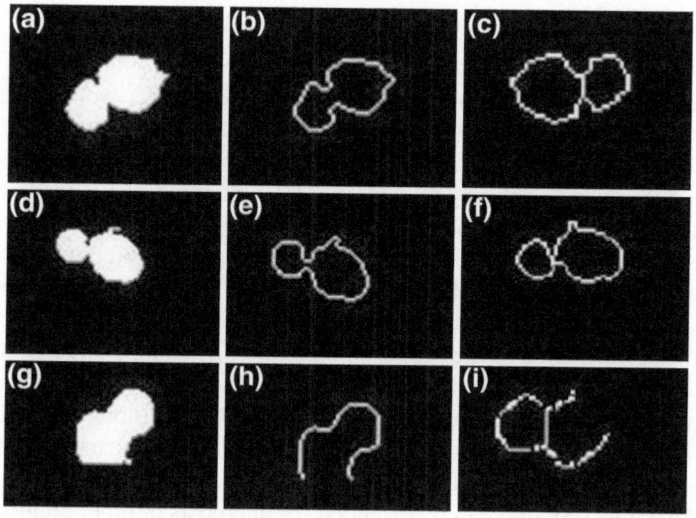


**Fig. 6** Various stages in Clumped blood cell Segmentation for 3 coinciding blood cells. **a**, **d** and **g** Original coinciding blood cells. **b**, **e** and **h** Edge images of original coinciding blood cells. **c**, **f** and **j** Rotated and segmented images of the coinciding clumped blood cells

gives the edge images of Fig. 6a, d and g extracted using LoG edge detector operator. These three edge images are rotated along the horizontal axis and the concave points are determined using the method explained in Sect. 4.4. The rotated and segmented images of the coinciding blood cell components are shown in Fig. 6c, f and i.

After segmenting the coinciding/clumped blood cell components into individual cell components as shown in Fig. 6c, f and i, the blood cell that is actually stained is detected using the methodology explained in Sect. 4.5. The stained blood cell identified image after segmenting clumped/overlapping blood cell components are shown in Fig. 7. Compared to the image shown in Fig. 5, in this image, only the blood cell components those are actually stained in the input image are detected.

**Fig. 7** Stained blood cell identified image after clumped blood cell segmentation

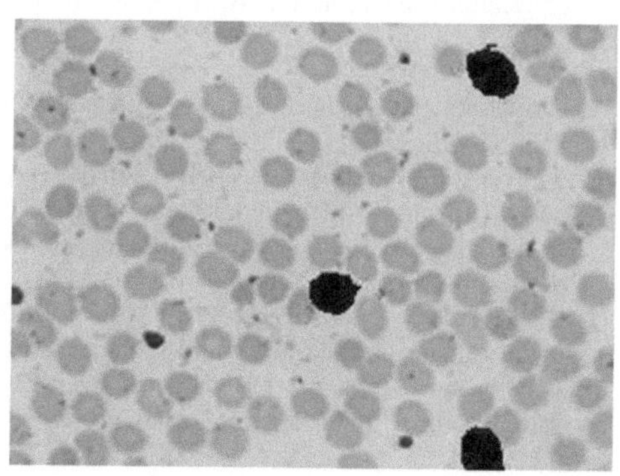

## 6 Conclusion

It has been observed that, the proposed methodology works well for the sample images considered in our experiments. For the experimental purpose we have considered 10 Leishman stained thin peripheral blood sample images of size 2,048 × 1,536, where each image consists of 12 clumped/overlapping blood cell components. Using the methodology explained in this paper, all the clumped/overlapped blood cell components in the images are segmented out and the stained cells are identified correctly. However, the performance of the proposed method has not been evaluated on a large dataset. So the future aim of this research is to evaluate the performance of the proposed methodology using a large dataset which contains images collected from different public health care centers and images acquired using different imaging equipments. In addition, a long term goal of this research is to develop a fully automated malaria diagnosis tool which can be deployed in different diagnostic laboratories in a consistent and cost effective manner. Also, this research can be extended for diagnosis of other abnormalities of peripheral blood.

## References

1. Aregawi M, Cibulskis R, Otten M, Williams R, Dye C (2008) World malaria report 2008. World Health Organization, WHO Press, Geneva
2. Korenromp E, Miller J, Nahlen B, Wardlaw T, Young M (2005) World malaria report. Tech Rep World Health Organization, Geneva
3. Hanscheid T (2003) Current strategies to avoid misdiagnosis of malaria. Clin Microbiol Infect 9:497–504
4. WHO (1991) Basic malaria microscopy Part I. Learner's Guide World Health Organization
5. Makkapati VV, Rao RM (2009) Segmentation of malaria parasites in peripheral blood smear images. ICASSP Acoust Speech Sig Process
6. Kettelhut MM, Chiodini PL, Edwards H, Moody A (2003) External quality assessment schemes raise standards: evidence from the UKNEQAS parasitology subscheme. J Clin Pathol 56:927–932
7. Coatney G, Collins W, Warren M, Contacos P (1971) The primate malarias. U.S. Department of Health, Education and Welfare, Washington, DC
8. Tek FB, Dempster AG, Kale I (2009) Computer vision for microscopy diagnosis of malaria. Malaria J 8:153
9. Guidelines on standard operating procedures for haematology, http://www.searo.who.int/en/Section10/Section17/Section53/Section480_1732.htm
10. Nguyen NT, Duong AD, Vu HQ (2011) Cell splitting with high degree of overlapping in peripheral blood smear. Int J Comp Theory Eng 3(3)
11. Rao KNRM (2004) Application of mathematical morphology to biomedical image processing. PhD thesis. U. Westminster
12. Di Ruberto C, Dempster A, Khan S, Jarra B (2000) Automatic thresholding of infected blood images using granulometry and regional extrema. ICPR, pp 3445–3448
13. Di Ruberto C, Dempster A, Khan S, Jarra B (2002) Analysis of infected blood cell images using morphological operators. IVC 20(2):133–146

14. Di Ruberto C, Dempster A, Khan S, Jarra B (2001) Morphological image processing for evaluating malaria disease. In: Proceedings of the international workshop vision form, Capri, Italy
15. Rao KNRM, Dempster AG, Jarra B, Khan S (2002) Automatic scanning of malaria infected blood slide images using mathematical morphology. In: Proceedings of the IEE seminar on medical applications of signal process, London, UK
16. Sio SWS, Sun W, Kumar S, Bin WZ, Tan SS, Ong SH, Kikuchi H, Oshima Y, Tan KSW (2007) Malariacount: an image analysis-based program for the accurate determination of parasitemia. J Microbiol Methods 68:11–18
17. Ross NE, Pritchard CJ, Rubin DM, Duse AG (2006) Automated image processing method for the diagnosis and classification of malaria on thin blood smears. Med Biol Eng Comput 44:427–436
18. Mohana Rao KNR, Dempster AG (2002) Use of area-closing to improve granulometry performace. International symposium on video/imgae processing and multimedia communications
19. Scotti F (2005) Automatic morphological analysis for acute leukemia identification in peripheral blood microscope images. CIMSA, IEEE international conference on computational intelligence for measurement systems and applications
20. Jiang K, Liao QM, Dai SY (2003) A novel white blood cell segmentation scheme using scale-space filtering and watershed clustering. In: 2nd international conference on machine learning and cybernetics
21. Tek FB, Dempster AG, Kale I (2005) Blood cell segmentation using minimum area watershed and circle radon transformations. Mathematical Morphology: 40 years on, Springer
22. Kumar S, Ong SH, Ranganath S, Ong TC, Chew FT (2002) Automated clump slitting in digital spore images. In: 7th international congress on aerobiology
23. Kumar S, Ong SH, Ranganath S, Ong TC, Chew FT (2006) A rule based approach for robust clump splitting. Pattern Recogn 39
24. Ongun G, Halici U, Leblebicioglu K, Atalay V, Beksac M, Beksac M (2001) Feature extraction and classification of blood cells for an automaticed differencial blood count system. In: Proceeding IJCNN
25. Ritter N, Cooper J (2007) Segmentation and border identification of cells in images of peripheral blood smear slides. In: Proceedings of the 30th Australasian conference on computer science
26. Díaz G, González FA, Romero E (2007) Automatic clump splitting for cell quantification in microscopical images. CIARP
27. Díaz G, González FA, Romero E (2009) A semiautomatic method for quantification and classification of erythrocytes infected with malaria parasites in microscopic images. J Biomed Inform 42:296–307
28. Mitiku K, Mengistu G, Gelaw B (2003) The reliability of blood film examination for malaria at the peripheral health unit. Ethiop J Health Dev 17(3):197–204
29. Gonzalez RC, Woods RE (2008) Digital image processing, 3rd edn. Prentice Hall
30. Chan TF, Vese LA (2007) Active contours without edges. IEEE Trans on Image Process 10:266–277

# A Novel Approach for Shot Boundary Detection in Videos

**D. S. Guru, Mahamad Suhil and P. Lolika**

**Abstract** This paper presents a novel approach for video shot boundary detection. The proposed approach is based on split and merge concept. A fisher linear discriminant criterion is used to guide the process of both splitting and merging. For the purpose of capturing the between class and within class scatter we employ $2D^2$ FLD method which works on texture feature of regions in each frame of a video. Further to reduce the complexity of the process we propose to employ spectral clustering to group related regions together to a single there by achieving reduction in dimension. The proposed method is experimentally also validated on a cricket video. It is revealed that shots obtained by the proposed approach are highly cohesive and loosely coupled.

**Keywords** Shot boundary detection · Split and merge · FLD · Texture · Spectral clustering

## 1 Introduction

Video shot boundary detection is a major step in the automation of content based video indexing and retrieval. A video can be viewed as a hierarchy of frames, shots and scenes as shown in Fig. 1. Shots are sequence of frames captured by a single

D. S. Guru (✉) · M. Suhil · P. Lolika
Department of Studies in Computer Science, University of Mysore, Manasagangothri, Mysore, India
e-mail: dsg@compsci.uni-mysore.ac.in

M. Suhil
e-mail: mahamad45@yahoo.co.in

P. Lolika
e-mail: lolika_18@yahoo.co.in

P. P. Swamy and D. S. Guru (eds.), *Multimedia Processing, Communication and Computing Applications*, Lecture Notes in Electrical Engineering 213, DOI: 10.1007/978-81-322-1143-3_17, © Springer India 2013

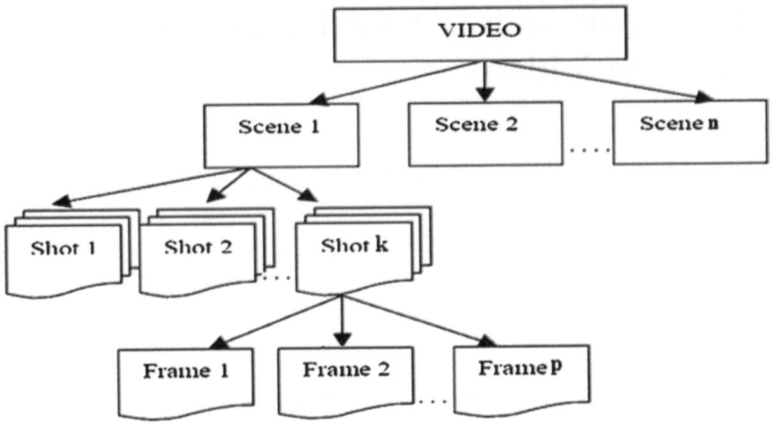


**Fig. 1** Hierarchical structure of a video

camera in a particular time period. Identification of the transitions between two successive shots is called as shot boundary detection.

There are mainly two types of shot boundaries or shot transitions [1] namely, abrupt transitions and gradual transitions. In abrupt transitions a sudden discontinuity in the frame sequence is seen, where as in gradual transitions a slow (change) transition is seen. Fade Out, Fade In, Wipe, and dissolve are the different techniques in gradual transitions.

Various shot boundary detection algorithms are available in the literature. They are mainly classified into temporal based techniques which measure the differences in video frames with respect to time and content based techniques which detect the shot boundary by studying the content present in the frames. Many algorithms have been proposed for temporal based video segmentation including pixel based, Block based and Histogram based comparisons [1–4] for both compressed and uncompressed videos; and for content based video segmentation including color [5–8], texture [1, 9–11] and shape features [1, 9, 10].

Most of the shot boundary detection algorithms use different features like color histogram using local color features (LCF) [1], interest points [12], edge based features [13], sift features [14]. Some transformations like Cosine Transform, Fourier Transform, Wavelet transforms etc., are also used [15, 16]. Block motion techniques are employed to extract the motion vectors for motion features [16–20]. Also, combinations of features from different modalities are used [21–23].

Even though there are lots of methods available in literature for shot boundary detection, most of them have a common drawback: thresholding. So, Koumaras et al. [16] proposed discrete cosine transform (DCT)-based and low-bit-rate encoded clips, which exploit the perceptual blockiness effect detection on each frame without using any threshold parameter. Manjunath et al. (2011) exploited the eigen gap analysis to preserve the variations among the video frames. Damnjanovic et al. in [24] explored scene change detection based on the information from eigenvectors of a similarity matrix. Goyal et al. exploited the use of split and merge mechanism to find story boundaries based on visual features and text transcripts [25].

Keeping all merits and demerits of the existing shot boundary detection algorithms in mind, in this paper, we propose a novel approach for shot boundary detection based on split and merge philosophy. The proposed method uses spectral clustering for grouping of objects in each frame subsequently apply $(2D)^2$ FLD based split and merge approach for the detection of shots.

The rest of the paper is organized as follows: In Sect. 2 the proposed split and merge based shot boundary detection is presented. The dataset and experimental results are discussed in Sect. 3. The conclusion and future work are given in Sect. 4.

## 2 Proposed Method

In this section we propose a novel split and merge based shot boundary detection algorithm. The main motivation is to view the video as a finite set of groups where in which each group consists of a set of adjacent frames with visually or contextually similar content preserving temporal continuity in a video. Figure 2 shows the main steps involved in the proposed method.

### *2.1 Region Identification in Each Video Frames*

First, we identify regions in each video frame and represent them by a set of attributes to achieve dimensionality reduction. We use block division for region

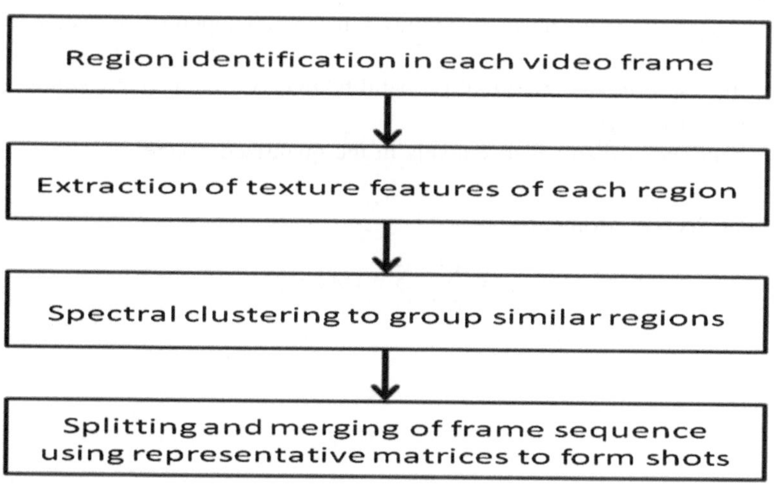


**Fig. 2** Block diagram of the proposed method

**Fig. 3** Block based region identification

identification in video frames. Each video frame is first converted from RGB to a gray space and then is divided into equal sized blocks which themselves are considered as the frame regions. We resize every frame into 256 × 256 and divide it into blocks of size 32 × 32 as shown in Fig. 3.

## *2.2 Textural Features for Video Frame Representation*

Statistical texture features proposed by Haralick et al. in [26], are used in our work for the representation of frame regions. Initially Gray Level Co-occurrence matrix (GLCM) is computed for each frame region using the pairwise occurrences of image intensities. Using the GLCM we calculate 14 different texture features for each frame region. A video frame with r regions will then be represented by r × 14 feature matrix.

Let us assume that P is the gray level co-occurrence matrix obtained from the image region r, the expressions for different texture features which we have used are as follows.

Notation

p(i, j) (i, j) th entry in a normalized gray-tone spatial dependence matrix, = P (I, j)/R

$p_x(i)$ ith entry in the marginal-probability matrix obtained by summing the rows of p(i,j)

$N_g$ Number of distinct gray levels in the quantized image.

Angular Second Moment:

$$f_i = \sum_i \sum_j \{p(i,j)\}^2 \quad (1)$$

Contrast:

$$f_2 = \sum_{n=0}^{N_{g-1}} n^2 \left\{ \sum_{i=1}^{N_g} \sum_{\substack{j=1 \\ |i-j|=n}}^{N_g} P(i,j) \right\} \tag{2}$$

Correlation:

$$f_3 = \frac{\sum_i \sum_j (ij)p(i,j) - \mu_x \mu_y}{\sigma_x \sigma_y} \tag{3}$$

Where $\mu_x$, $\mu_y$, $\sigma_x$ and $\sigma_y$ are the means and standard deviations of $p_x$ and $p_y$.

Sum of Squares: Variance

$$f_4 = \sum_i \sum_j (i - \mu)^2 p(i,j) \tag{4}$$

Inverse Difference Moment:

$$f_5 = \sum_i \sum_j \frac{1}{1 + (i-j)^2} P(i,j) \tag{5}$$

Sum Average:

$$f_6 = \sum_{i=2}^{2N_g} i p_{x+y}(i) \tag{6}$$

Sum Variance:

$$f_7 = \sum_{i=1}^{2N_g} (i - f_8)^2 P_{X+Y}{}^{(j)} \tag{7}$$

Sum Entropy:

$$f_8 = -\sum_{i=2}^{2N_g} p_{x+y}{}^{(i)} \log \left\{ p_{x+y}{}^{(i)} \right\} \tag{8}$$

Entropy:

$$f_9 = -\sum_i \sum_j p(i,j) \log(p(i,j)). \tag{9}$$

Difference Variance:

$$f_{10} = variance\ of\ p_{x-y'} \tag{10}$$

Difference Entropy:

$$f_{11} = -\sum_{i=0}^{N_{g-1}} p_{x-y}(i) \log\{p_{x-y}(i)\} \tag{11}$$

Information Measures of C

$$f_{12} = \frac{HXY - HXY1}{\max\{HX, HY\}} \tag{12}$$

$$f_{13} = (1 - \exp[-2.0(HXY2 - HXY)]^{1/2} \tag{13}$$

$$HXY = -\sum_i \sum_j p(i,j)\log(P(i,j))$$

where HX and HY are entropies of $p_x$ and $p_y$ and,

$$HXY1 = -\sum_i \sum_j p(i,j)\log\left\{p_x(i)p_y(j)\right\}$$

$$HXY2 = -\sum_i \sum_j p_x(i)p_y(j)\log\{p_x(i)p_y(j)\}$$

Maximal Correlation Coefficient:

$$f_{14} = (\textit{second largest eigenvalue of } Q)^{1/2} \tag{14}$$

where,

$$Q(i,j) = \sum_k \frac{p(i,k)P(j.k)}{p_x(i)p_y(k)}$$

## 2.3 Clustering of Frame Regions Using Spectral Clustering

In this step, we group the similar regions in each frame. Since spectral clustering [27–29], has been very efficiently used in the literature of image segmentation we adapt spectral clustering to group the regions in each frame using the feature matrices of each region obtained in the previous step.

The main problem in using spectral clustering is the curse of dimensionality. Because, when we try to apply spectral clustering to an image of size m × n the size of the affinity matrix will be mn × mn and when the size of an image is large it becomes a tedious job to handle such a huge affinity matrix. Our purpose is to identify different regions existing in each video frame so that we can represent the frame with some features of those regions and to make the matching of frames easier. So, instead of going for pixel based spectral clustering we go for region based spectral clustering. That is, we apply region identification to get the first impression of the regions existing in the frame and then we represent each extracted region by a set of features and spectral clustering is applied using the representative feature vectors of each region to form clusters of regions.

We are performing two levels of dimensionality reduction. First level of dimensionality reduction is achieved by identifying the regions in video frames using block division or segmentation and representing each region with a feature vector consisting of 14 texture features and the second level of dimensionality reduction is by applying spectral clustering to group the regions identified in first level. Irrespective of the size of the video frames we get a representative matrix of size $k \times 14$ where k is the number of clusters obtained after applying spectral clustering. Hence our initial goal of classifying a sequence of frames into group of shots has become classifying the similar feature matrices into a group.

Suppose, FM ($r \times 14$) is the feature matrix obtained after representing a video frame f with r regions where each region is described using 14 statistical texture features. We can now say that, we have r points $x_1,\ldots,x_r$ in a $R^{14}$ dimensional space and the similarity $s_{ij} \geq 0$ between all pairs of data points $x_i$ and $x_j$ are calculated using Euclidean distance measure. We apply spectral clustering procedure proposed by Jordan et al. (2002) by treating the data as a graph $G = (V, E)$ where data points are considered to be the set of vertices V and E is the set of edges connecting the vertices.

As a result of spectral clustering of the r points we obtain $k < r$ clusters. We then form the representative feature matrix $RM \in R^{k\times 14}$ of the frame f, whose ith row is the mean of the vectors belonging to ith cluster.

The value of the k, the number of clusters, which is smaller than the value of r is decided based on the type of the dataset being used. It is fixed up for the entire set of frames so that the variation in the size of the representative feature matrix is maintained. We check the performances of the clustering with different values of k and an optimal k value is selected empirically.

## 2.4 Shot Boundary Detection

The proposed method detects shots in a given video using the concept of split and merge which is driven by the criterion function of Fisher's Linear Discriminant analysis. The recursive splitting and merging of the sequence of frames is done using the representative matrices obtained in the previous section. We first introduce the concept of split and merge in general and its usage in the perspective of shot boundary detection. Subsequently we move onto the concept of $(2D)^2$ FLD and how it is adapted in the proposed method for shot boundary detection.

**The Concept of Split and Merge.** Split and merge is a well known concept in image segmentation where the image is subdivided into a fixed number of regions repetitively until the predicate becomes true for all the regions. Consequently merging is done for any two adjacent regions when the predicate becomes true when they are considered as a single region. In the proposed method the same analogy is used to arrive at the shots from a given video. The larger sequence of video frames is split into 2 smaller subsequences repeatedly using the representative matrices obtained in the previous stage until the predicate becomes true. This process results in a number of smaller subsequences of the original video.

Then any two adjacent subsequences are merged if the predicate is continued to be true when they are considered as a single subsequence.

**$(2d)^2$ FLD for Splitting and Merging of Video Frame Sequences.** The predicate we have considered to split the sequence of frames is the criterion function (J) of Fisher's Linear Discriminant analysis [30]. The two subsequences of frames are treated as two classes of data points and given to the FLD function; the job of FLD is to find an optimal projection axis to project the data points of two classes. If the data points are well separated in the projected space then the value of the criterion function will be maximum. But, we cannot use this criterion to split the sequence since we do not have any initial assumption about the maximum value of J such that we can fix up a threshold for it and if the value of J obtained is greater than the threshold we can allow splitting otherwise not. Actually our task is to somehow find a position for split which gives a maximum value for J. The solution for this problem is to compare the J value obtained for the sequence under consideration and its left neighboring sequence with the corresponding J value for the left sub sequence of the sequence under consideration and the same left neighboring sequence. Similarly, compare the J value obtained for the sequence under consideration and its right neighboring sequence with the corresponding J value for the right sub sequence of the sequence under consideration and the same right neighboring sequence. If both the J values in the current iteration are greater than that of previous then we allow the original sequence to get split into exactly two subsequences and update the values of J for the corresponding sequences. That is, the sequence to be checked for the possibility for splitting is passed to the FLD as two subsequences along with the two adjacent sequences, if the values of J obtained for those two subsequences with their neighboring sequences yields a greater value than the corresponding J values obtained before splitting then only that sequence will be allowed to get split.

The same procedure is considered in reverse for merging of the two adjacent subsequences. That is, any two adjacent subsequences are merged if and only if the value of J between the left subsequence and its neighboring subsequence, and the value of J between the right subsequence and its neighboring subsequence are less than or equal to the values of J at corresponding locations after merging.

The process of splitting and merging is started with a single long sequence of video frame representatives and continued until the J value gets maximized for every pair of adjacent subsequences. Each and every resultant subsequence is declared to be the shot present in the video.

Since we have the 2 dimensional representatives of each video frame, we recommend using the 2D FLD [31, 32] instead of conventional 1D FLD. Hence, We have used the most recent and efficient two dimensional Fisher's Linear Discriminant algorithm the $(2D)^2$ FLD proposed by Nagabhushan et al. [31].

## 3 Results and Analysis

In this section, we conduct experiments on various video samples. Subsequently, quantitative evaluation of the proposed shot boundary detection method in terms of Precision, Recall and F-measures is given.

### 3.1 Dataset and Ground-Truth

To test the performance of the proposed method we have considered two different types of videos, they are Cricket and News. We have considered two from each type downloaded from the internet. Manually identified shots present in each of the testing video sequence are considered as ground truth. The details of the test videos and ground truth for the evaluation of the proposed method are given in the Table 1.

### 3.2 Experimental Results

The experimental results of the proposed algorithm for the cricket and news videos are given in the Table 2. We make use of the following most popular evaluation measures to evaluate the results of the proposed method,

**Table 1** Dataset and Ground-truth

| Video type | Caption of the video sequence | Frame interval | Total number of transitions |
|---|---|---|---|
| 1. Cricket | 1. India vs. Australia 2nd match at Kolkata | 5,400–5,900 | 5 |
| | 2. DLF-IPL-2010 CSK vs KXIP | 1,5000–15,500 | 3 |
| 2. News | 3. NDTV–God Science and the Universe | 1,000–1,500 | 6 |
| | 4. NDTV-God Science and the Universe | 2,000–2,500 | 6 |

**Table 2** Experimental results of the proposed method

| Test video sequence | Number of shots | D | MD | FA | Precision (P)% | Recall (R)% | F-Measure F (%) |
|---|---|---|---|---|---|---|---|
| Cricket-1 | 5 | 3 | 2 | 0 | 100 | 60 | 75 |
| Cricket-2 | 3 | 3 | 0 | 0 | 100 | 100 | 100 |
| News-1 | 6 | 5 | 1 | 2 | 71.42 | 83.33 | 76.91 |
| News-2 | 6 | 4 | 2 | 3 | 57.14 | 66.66 | 61.53 |
| | | | | Average | 82.14 | 77.4975 | 78.36 |

$$\text{Precision} = \frac{D}{(D + FA)} \tag{15}$$

$$\text{Recall} = \frac{D}{(D + MD)} \tag{16}$$

$$\text{F-measure} = \frac{2(Precision \times Recall)}{(Precision + Recall)} \tag{17}$$

where D is the number of shots correctly detected, MD is number of shots missed (MD), and FA is the number of false alarms.

We performed the experimentation on both cricket and news videos with different values of k, the number of clusters ranging from 2 to 64. Since, we divide each video frame into exactly 64 blocks and upon clustering at least we can expect two objects and at most there can be 64 objects. So we performed the experimentation with various values of k and we have empirically set the value of k as 6 for both types of videos. The results obtained for cricket sequences are really appreciable when compared to the results of news video. Though there are few missed detections and false alarms, the overall performance of the proposed algorithm is considerably good. Few of the missed detections and false alarms are due to the gradual transitions present in the video. The main drawback is the parameter fixation that is, fixing up of the value of k for a particular video sequence is necessary. In future we will think of an adaptive approach which can effectively solve the problem of fixing k value.

# 4 Conclusion

Shot boundary detection is the very first step in the automation of content based video indexing and retrieval. It searches for and recognizes the visual discontinuities caused by the transitions between frames, to segment a video stream into elementary uninterrupted content units for subsequent high-level semantic analysis. Despite the long research history and numerous proposed techniques, shot boundary detection is still a challenging and active area of research. In this paper we have proposed a novel approach for the shot boundary detection problem. The novelty of the proposed approach lies in the exploitation of split and merge philosophy which has been proved to be an efficient method to solve many complex problems along with two major concepts namely, fisher's linear discriminant analysis and spectral clustering. In the proposed method we have exploited the beauty of split and merge concept to split the given video sequence into smaller subsequences with temporally identical content based on another well known concept fisher's linear discriminant analysis. We have used spectral clustering to group the regions in video frames based on the textural features. Experiments are conducted on various datasets such as cricket and news. The proposed method is evaluated using Precision, Recall and F-measures and we have achieved 82.14 % of Precision, 77.5 % of Recall and 78.36 % of F-measure for the test data set considered.

## References


1. Idris F, Panchanathan S (1997) Review of image and video indexing techniques. J Vis Commun Image Represent 8(2):146–166
2. Zhang D, Lu G (2001) Segmentation of moving objects in image sequence: a review. Circuits Syst Signal Process 2(2):143–183
3. Koprinska I, Carrato S (2001) Temporal video segmentation: a survey signal processing. IEEE Trans 16(5):413–506
4. Manjunath S, Guru DS, Suraj MG, Harish BS (2011) A non parametric shot boundary detection: an Eigen gap based approach. Proceedings of fourth annual ACM Bangalore conference, vol 1. pp 1030–1036
5. Yasira Beevi CP, Natarajan Dr S (2009) An efficient video segmentation algorithm with real time adaptive threshold technique. Int J Sig Process Image Process Pattern Recognition 2(4):13–28
6. Reddy PVN, Satya Prasad K (2011) Color and texture features for content based image retrieval. Int J Comp Tech Appl 2(4):1016–1020
7. Quynh NH, Ha NTT, Tao NQ (2012) An efficient content based image retrieval method for retrieving images. Int J Innovative Comput Inf Control 8(4):2823–2836
8. Yi T, William I (1997) Object-based image retrieval using point feature maps. Proceedings of the International Conference on Database Semantics (DS-8), Rotorua, pp 59–73
9. Wang H, Divakaran A, Vetro A, Chang SF, Sun H (2003) Survey of compressed-domain features used in audio-visual indexing and analysis. J Visual Commun Image Represent 14:150–183
10. Patel BV, Meshram Shah BB (2012) Anchor Kutchhi Polytechnic, Content based video retrieval systems. Int J Ubi Comp (IJU) 3(2):13–30
11. Liua Y, Zhanga D, Lua G, Mab WY (2007) A survey of content-based image retrieval with high-level semantics. Pattern Recogn 40:262–282
12. Fu X, Xian Zeng J (2009) An effective video shot boundary detection method based on the local color features of interest points. In Proceedings of the 2009 second international symposium on electronic commerce and security, pp 25–28
13. Don Adjeroh MC, Lee NB, Uma K (2009) Adaptive edge-oriented shot boundary detection. EURASIP J Image Video Process, Hindawi Publishing Corporation. doi:10.1155/2009/859371
14. Yuchou C, Lee DJ, Yi H, James A (2008) Unsupervised video shot detection using clustering ensemble with a color global scale-invariant feature transform descriptor. J Image Video Proc 1:1–10
15. Brunelli R, Mich O, Modena CM (1999) A survey on the automatic indexing of video data. J Vis Commun Image Represent 10:78–112
16. Koumaras H, Gardikis G, Xilouris G, Pallis E, Kourtisa A (2005) Shot boundary detection without threshold parameters. Paper 05210LRR 36(2):133–144
17. Boreczky JS, Rowe LA (1996) Comparison of video shot boundary detection techniques. J Electron Imaging 5(2):122–128
18. Alan FS, Palu O, Aiden RD (2010) Video shot boundary detection: Seven years of TRECVid activity. Comput Vis Image Und 114(4):411–418
19. Abdelati MA, Ben AA, Mtibaa A (2010) Video shot boundary detection using motion activity descriptor. J Telecommun 2(1):54–59

20. Yao N (2002) Student member, adaptive rood pattern search for fast block-matching motion estimation. IEEE Trans Image Process 11(12):1442–1449
21. Chen W, Zhang Y-J (2008) Parametric model for video content analysis. Pattern Recogn 29:181–191
22. Jacobs A, Miene A, Ioannidis GT, Herzog O (2004) Automatic shot boundary detection combining color, edge, and motion features of adjacent frames, TRECVID 2004 Workshop Notebook Papers, National Institute of Standards and Technology, Gaithersburg, pp 197–206
23. Lef Evre S Holler J, Vincent N (2003) A review of real-time segmentation of uncompressed video sequences for content-based search and retrieval. Real-Time Imaging 9(1):73–98
24. Uros Damnjanovic (2007) Ebroul Izquierdo and Marcin Grzegorzek, Shot boundary detection using spectral clustering 15th European signal processing conference
25. Anuj G, Punitha P, Frank H, Joemon MJ (2009) Split and merge based story segmentation in news videos. Adv Inf Retrieval 5478:766–770
26. Robuet MH (1973) Shanmugam and its hak dinstein, texture features for image classification. IEEE Trans Man Cybern 3(6):610–621
27. Marco Barreno (2004) Spectral Methods for Image Clustering, CS 281b: Advanced topics in learning and decision-making http://marcobarreno.com/classes/projects/cs281b/
28. Ulrike von Luxburg (2007) A tutorial on spectral clustering. Stat Comput 17(4):395–416
29. Denis Hamad and Philippe Biela, Introduction to spectral clustering
30. MaxWelling, Fisher Linear Discriminant Analysis. Max welling's classnotes in machine learning 16(7):817–830 http://www.ics.uci.edu/~welling/classnotes/classnotes.html
31. Nagabhushana P, Guru DS, Shekara BH (2006) (2D)2 FLD: An efficient approach for appearance based object recognition. Neurocomputing 69:934–940
32. Huilin X, Swamy MNS, Ahmad MO (2005) Two-dimensional FLD for face recognition. Pattern Recogn 38:1121–1124
33. Jordan MI, Andrew Y Ng, Weiss Y (2002) On spectral clustering: analysis and an algorithm. Advances in neural information processing systems, vol 14. MIT Press, Cambridge, pp 849–856

# Enhancing COCOA Framework for Tracking Multiple Objects in the Presence of Occlusion in Aerial Image Sequences

**Vindhya P. Malagi, Vinuta V. Gayatri, Krishnan Rangarajan and D. R. Ramesh Babu**

**Abstract** Multi object tracking in aerial image sequences is a topic of utmost importance in the field of computer vision for both military and civilian applications. In order to extract valid information about moving targets, it is required to detect and track these targets precisely in the input image sequences. Occlusion is one of the prominent problem areas that hinder efficient object tracking. Spatial reasoning literature fails to distinguish various analyses that are prominent to computer vision. However, recognizing valid occlusion states and mining their transition sequences help in analyzing the pose and motion of multiple interacting objects in the scene. In this paper, we propose an enhancement incorporating occlusion in the existing COCOA framework for tracking in aerial image sequence. The contribution of the paper is the novel idea of extracting occlusion cues as a pre-processing step to aid the tracker. We describe approaches to extract occlusion information in the scene and use it as a cue for efficient tracking.

**Keywords** Multi object tracking · COCOA framework · Occlusion sequence mining

V. P. Malagi (✉) · V. V. Gayatri · K. Rangarajan · D. R. Ramesh Babu
Computer Vision Lab, Dayananda Sagar College of Engineering, Bangalore, India
e-mail: vindhyapm@gmail.com

V. V. Gayatri
e-mail: vinuta06.gayatri@gmail.com

K. Rangarajan
e-mail: krishnanr1234@gmail.com

D. R. Ramesh Babu
e-mail: bobrammysore@gmail.com

P. P. Swamy and D. S. Guru (eds.), *Multimedia Processing, Communication and Computing Applications*, Lecture Notes in Electrical Engineering 213,
DOI: 10.1007/978-81-322-1143-3_18, © Springer India 2013

# 1 Introduction

Unmanned Air Vehicles (UAVs) play a critical role in surveillance, target tracking and reconnaissance in both urban and battlefield settings. One of the major challenges in tracking is handling occlusion of the objects being tracked. This paper gives a framework for mining various occlusion scenarios in aerial images from UAV and using them as cue to the tracker. Object tracking, in essence, deals with a set of image sequences that change over time. While the existing algorithms are able to track objects well in controlled environments, they usually fail in the presence of significant variation of the object's appearance or surrounding illumination. Occlusion has also been seen as one of the prominent problem areas in efficient object tracking. Here, the COCOA framework [1] is considered as the underlying framework for tracking moving objects from aerial images.

Tracking objects for long durations in aerial imagery is a challenging task as the objects are small in size, similar in appearance and tend to occlude one another while in motion. Also they may disappear at arbitrary points and then reappear after sometime. Such images invariably contain noise due to the relative movement of the camera with respect to the vehicle often called the 'jitter' and noise induced due to environmental conditions like brightness or glare, noise due to haze etc. Even shadows appearing in the image need to be removed as a preprocessing step. Though a lot of research has gone into addressing this problem of occlusion, a complete solution is still far from being achieved. In this paper we propose extensions to COCOA framework for tracking in the presence of occlusion.

# 2 Tracking Framework

We propose an extension that feeds various occlusion state cues to the tracking module in the existing COCOA framework. Figure 1 shows the proposed extension.

Here COCOA model forms the basis of the framework. We propose to enhance the working of this model by incorporating the occlusion cue module which takes in the appearance models (blobs) from the motion detector, compare them with the mined occlusion states and formulate appropriate occlusion related cues. The details of this processing are discussed in Sects. 3 and 4.

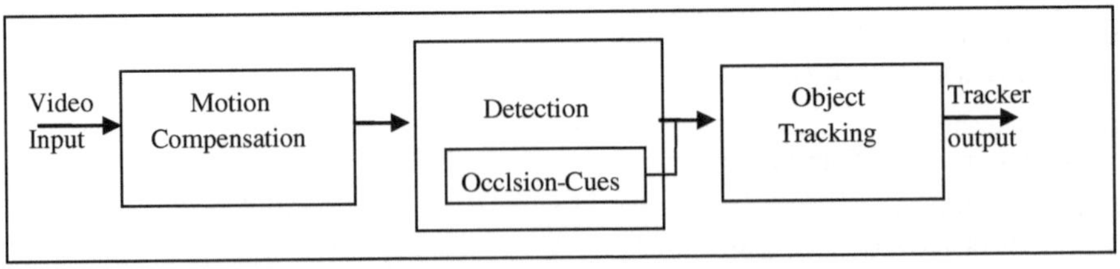


**Fig. 1** Proposed framework with occlusion states as cue to the tracker

## 2.1 Motion Compensation

COCOA framework is an established framework for object tracking in aerial imagery. In this framework, the motion compensation model is used to compensate for the motion of the moving camera. By registering the whole video with respect to one global reference we are able to get a meaningful representation of the object trajectories which can be used to describe the input. Here, the image registration is performed by considering a small block of image set, extracting the features and matching them using SIFT [2]. The tentative matches are then filtered using RANSAC to find the correspondences that best fit a homography.

## 2.2 Motion Detection

After removing global camera motion, we detect local motion of objects moving in the scene. Motion detection provides a classification of the pixels in the video sequence into either foreground (moving objects) or background. A common approach used to achieve such classification is background removal, sometimes referred to as background subtraction [3], where each video frame is compared against a reference or background model, pixels that deviate significantly from the background are considered. Probabilistic modeling of the background has been popular for surveillance videos. However, these methods are found to be inapplicable to aerial image sequences, as they fail by either giving raise to large ghost effects or failing to distinguish between foreground and background pixel accurately. Therefore, we avoid probabilistic models in favor of simple median image filtering [4], which considers a background model with fewer artifacts using fewer frames. We use a 15 frame median image for the purpose with minimum ghost effect.

In surveillance applications, determining occlusion primitives is based on foreground blob tracking, and requires no prior knowledge of the domain or camera calibration. New foreground blobs are identified as putative objects which may undergo occlusions, split into multiple agents, merge back again, etc. Using temporal sequence mining, significant cues on the various occlusion states that can occur due to the above mentioned interaction can be generated that can serve as inputs to the tracker as explained in Sect. 4.

## 2.3 Tracking

In COCOA framework, tracking is critical for obtaining meaningful tracks that capture the motion characteristics over longer durations of time. Here, we have used appearance based tracking approach [5] to perform multi-target tracking. Regions of interest, or blobs, in successive frames are given by the motion detection module. Each blob is represented by a distinct appearance and shape model.

In the proposed framework, the tracker makes use of the occlusion cues which are generated from the temporal sequence mining of occlusion states to refrain from loosing track of the moving objects in cases of partial or complete occlusion scenarios.

## 3 Occlusion Handling

In spatial reasoning, occlusion states are informative on the relative pose and motion of multiple objects. In the application of target tracking, occlusion seems to be inevitable. Former approaches of spatial reasoning are not adequate for such computer vision applications. The work here closely follows the work on occlusion by Guha et al. [6, 7].

Computer vision literature mostly talks about occlusion in terms of split and merge cases. However, OCS-14 paper of Guha [6] well defines 14 different occlusion states that cover almost all computer vision scenarios. But we observe that in particular, the occlusion states like ocSGP (static occlusion with grouping and partial visibility), ocSGF (static occlusion with grouping and fragmentation) and ocSG0 (static occlusion with grouping and no visibility) as stated in the paper cannot be justified as grouping which is interaction between two or more moving objects, brings in dynamic occlusion. Therefore it seems to be appropriate to consider occlusion states based on the practicality of the problem in hand.

Hence in this paper we first look into the simplified occlusion states in object tracking and then look into a way to mine these states. On extensive survey we found that very few researchers have worked explicitly on the problem of occlusion [6, 8, 9]. In the application of tracking, all the occlusions span over two important types of occlusion states—partial occlusion state and complete occlusion state. The rest of the occlusion states can be sub-spanned under these two states, based on visibility of the occluded object and static and dynamic nature of the occluder as follows:

1. Partial Occlusion
    (a) From a static occluder
    (b) From a dynamic occluder
    (c) Due to Split and Merge Scenarios
    (d) Entry/exit scenarios
2. Complete Occlusion
    (a) From a static occluder
    (b) From a dynamic occluder
    (c) Due to Split and Merge Scenarios
    (d) Entry/exit scenarios

Typical example of one of the above scenarios could be the case in which an object approaches a static occluder, undergoes a series of partial occlusions before completely getting occluded (considering their apparent sizes). Similarly at entry/exit points, an object may be partially visible in a sequence of images before it enters/exits the scene completely. Split, merge scenarios are most common in tracking problem and can occur from either static and/or dynamic occluder. A typical merge and split case gives rise to transitions from fully visible to partial or complete occlusion state and vice versa if the split of the occluded object from occluder happens after the merge. When an object is completely occluded over a long period of time before emerging back at a later time, it may be considered as a case of temporal occlusion.

In general, sequence of occlusion state transitions can be considered as important visual signatures of the interaction between objects. As an example, in a scenario where an object is entering a scene, it emerges with partial visibility before being completely visible. This means that the scene undergoes a sequence of transitions from one occlusion state (partial) to another before transitioning to complete visibility and can thus be termed as occlusion state transition. It is possible to use data mining techniques on the occlusion transitions as they emerge in a scene, and one can gain useful abstractions about the scene and the object behaviors.

## 4 Occlusion Mining

For learning the event signatures that can be fed as crucial cues to the tracker, we consider mining the spatio-temporal sequence patterns of the occlusion primitives and model these using substring trees [10]. A spatio-temporal sequence S is a list of locations, $(x_1, y_1, t_1), (x_2, y_2, t_2),\ldots, (x_n, y_n, t_n)$, where $t_i$ represents the time-stamp of location $(x_i, y_i)$ $(1 \leq i \leq n)$. The substring tree is a rooted directed tree whose root links to multiple substring sub-trees. Each node in a sub-tree consists of pattern element (occlusion state) and a counter, which counts the number of substrings (i.e., subsequences of elements) that contribute to the pattern formed by the path from the root to this node.

Querying the substring trees with the time ordered sequences of the occlusion states detects similarities with other events where this type of sequences arose, thereby detecting episodes of the transit-across-a-large-occlusion event.

The novelty of this paper is in the idea that these event signatures can be fed as occlusion cues to the tracker to better the overall tracking performance. For instance, the regions of frequent occlusion clearly reveal important information about the scene depth and also about object behaviors in the given scene context as shown in Fig. 2. Eventually such inferences can be picked up by the tracker to conclude on the upcoming visual scenario thus improving the precision and accuracy of tracking.

**Fig. 2** **a, b** Entry with partial occlusion state (A car entering the scene from the right is partially visible at the image boundary)

## 5 Results

Preliminary implementation on this enhanced framework on aerial images shows promising results. All occlusion states mentioned in Sect. 3 could be identified and a few examples (Figs. 3, 4, 5, 6) are shown below. The video sequences considered for experimentation are of about 2,000 frames length. Frames with moving objects with different occlusion states are shown in figures (a) and their corresponding segmented output in (b).

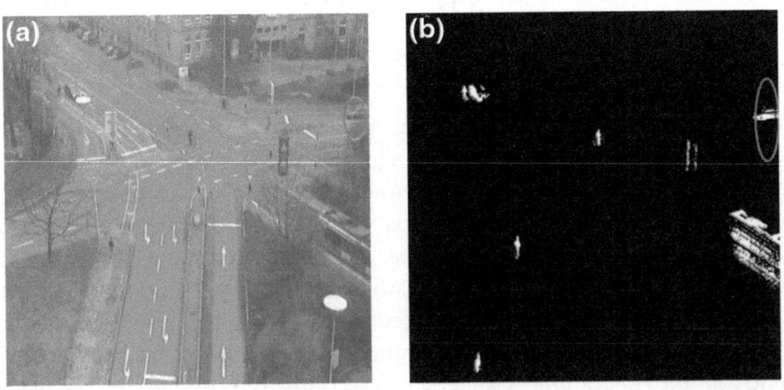


**Fig. 3** **a, b** Entry with partial occlusion state (A car entering the scene from right is partially visible at the image boundary)

**Fig. 4** **a**, **b** Exit with partial occlusion state (A car exiting the scene from right is partially visible at the image boundary)

**Fig. 5** **a**, **b** Partial occlusion with static occluder

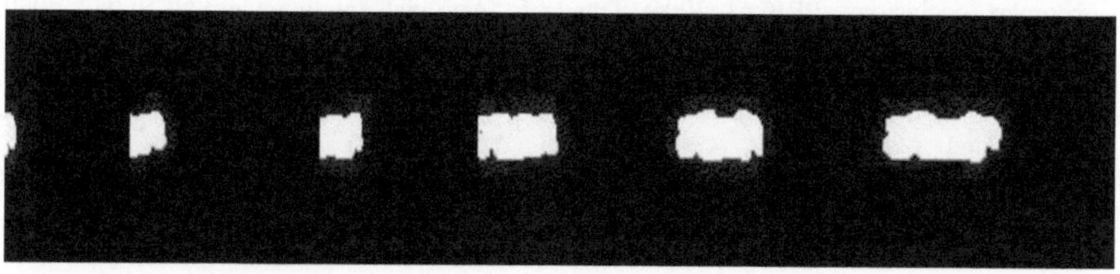

**Fig. 6** Blobs from motion detection in a sequence of images showing the case of entry of a new object and revealing a series of occlusion state transitions of partial occlusion states before complete visibility

## 6 Conclusion

In this paper, a novel approach for efficient object tracking through occlusions has been proposed. This framework is based on the popular COCOA framework for object tracking in aerial images. The work has identified the simplified and valid occlusion states relevant to the tracking application in computer vision. It considers various occlusion state transitions, relevant to spatial–temporal reasoning that can be mined and given as cue to the tracker. The contribution of the paper is the idea of treating occlusion cues as the pre-processing step to the tracker. This proposed idea is likely to better the performance of the tracker. Work on building a strong mathematical model for the proposal and prototype implementations is in progress.

**Acknowledgments** This work is funded by ER & IPR, DRDO ref no: ERIP/ER/1104561/M/01/1353, India.

## References

1. Ali S, Shah M (2006) COCOA—tracking in aerial imagery. SPIE Airborne Intelligence, Surveillance, Reconnaissance (ISR) Systems and Applications, Orlando
2. Lowe DG (2004) Distinctive image features from scale-invariant keypoints. Int J Computer Vision 60(2):91–110
3. Stauffer C, Grimson WEL (1999) Adaptive background mixture models for real-time tracking. In: The proceedings of IEEE Conference on Computer Vision and Pattern Recognition, vol. 2
4. Reilly V, Idrees H, Shah M (2010) Detection and tracking of large number of targets in wide area surveillance. In: The proceedings of 11th European Conference on Computer Vision, LNCS, vol 6313. pp 186–199, Springer
5. Senior A, Hampapur A, Tian Y, Brown L, Pankanti S, Bolle R (2006) Appearance models for occlusion handling. In: 2nd International Workshop on Performance Evaluation of Tracking and Surveillance Systems, Science Direct, Elsevier, Image and Vision Computing, vol 24. pp 1233–1243
6. Guha P, Mukerjee A, Venkatesh KS (2011) OCS-14: you can get occluded in fourteen ways. In: Proceedings of 22nd International Joint Conference on Artificial Intelligence, pp 1665–1675
7. Guha P, Mukerjee A, Venkatesh KS (2006) Appearance based multiple agent tracking under complex occlusions. PRICAI 2006: Trends in Artificial Intelligence LNCS, vol 4099. Springer, Heidelberg, pp 593–602
8. Galton A (1998) Modes of overlap. J Vis Lang Comput 9(1):61–79
9. Kohler C (2002) The occlusion calculus. In: Workshop on Cognitive Vision, Zurich, Switzerland
10. Cao H, Mamoulis N, Cheung DW (2005) Mining frequent spatio temporal sequence patterns. In: 5th IEEE International Conference on Data Mining, pp 82–89

# User Dependent Features in Online Signature Verification

**D. S. Guru, K. S. Manjunatha and S. Manjunath**

**Abstract** In this paper, we propose a novel approach for verification of on-line signatures based on user dependent feature selection and symbolic representation. Unlike other signature verification methods, which work with same features for all users, the proposed approach introduces the concept of user dependent features. It exploits the typicality of each and every user to select different features for different users. Initially all possible features are extracted for all users and a method of feature selection is employed for selecting user dependent features. The selected features are clustered using Fuzzy C means algorithm. In order to preserve the intra-class variation within each user, we recommend to represent each cluster in the form of an interval valued symbolic feature vector. A method of signature verification based on the proposed cluster based symbolic representation is also presented. Extensive experimentations are conducted on MCYT-100 User (DB1) and MCYT-330 User (DB2) online signature data sets to demonstrate the effectiveness of the proposed novel approach.

**Keywords** Feature selection · User dependent features · Fuzzy C means · Symbolic representation · Signature verification

D. S. Guru (✉) · K. S. Manjunatha
Department of Studies in Computer Science, Manasagangothri, University of Mysore, Mysore 570006, Karnataka, India
e-mail: dsg@compsci.uni-mysore.ac.in

K. S. Manjunatha
e-mail: kowshik.manjunath@gmail.com

S. Manjunath
Postgraduate Department of Studies in Computer Science, JSS College of Arts, Commerce and Science, Mysore 570025, Karnataka, India
e-mail: manju_uom@yahoo.co.in

P. P. Swamy and D. S. Guru (eds.), *Multimedia Processing, Communication and Computing Applications*, Lecture Notes in Electrical Engineering 213,
DOI: 10.1007/978-81-322-1143-3_19, 

## 1 Introduction

Signature is the most widely accepted behavioral biometric trait for personal and document authentication in many day-to-day applications. Signature verification has been an active area of research due to its wide acceptance for authentication purpose in many financial and legal transactions like validation of bank cheques, credit card transactions, contracts, and bonds. Depending on the acquisition method, signatures are of two categories offline and online [1]. An offline signature is the static image of the signature available in documents and captured using devices like camera and scanner to have its digital representation. It is easy to forge an off-line signature. In an off-line signature only X and Y co-ordinates (static features) of the signature image are available for verification. On the other hand, online signatures are captured using special devices like digitizing tablet and smart pens which captures both dynamic properties and shape information. Due to the availability of additional dynamic features such as pressure, azimuth, speed, total signature time etc., it is difficult for the forger to imitate both the image shape and the way it has been originally written by the authentic signer and hence it is more reliable compared to offline signature. In an online mode the signature is represented as a time function of various dynamic properties like pressure v/s time [2].

The two stages in any biometric system are (a) Identification (b) Verification [3]. In identification, the test signature is compared with the reference signatures of all the N users in the knowledgebase to establish the identity of the signer. It is a 1: N matching process and hence takes a longer time. In verification, the test signature of the claimed user is compared with other signatures of the claimed user available in the knowledgebase to determine if the given signature is genuine.

Online signature verification is broadly classified into two categories—Parametric approach and Function based approach [4]. In the parametric approach, set of parameters obtained from the signatures are used as the representative of the particular signature. Position, speed, acceleration, number of pen ups and pen-downs, pen down time ratios are the parameters proposed in the literature for online signature verification [5, 6]. In function-based approach, signature is represented as time function of various dynamic properties. In the first category, during verification, the parameters of query and reference signatures are compared to determine whether the query signature is genuine or not. In the function-based approach, query and reference signatures are compared either point-to-point or segment-to-segment basis [7]. A function based approach takes more time as it involves comparing every point in the signature trajectory but gives better performance. In the work of [8], parametric approach shows equally competitive result compared to any function based approach.

Features for on online signatures are categorized into two types (a) local features, which are extracted from specific point in the signature (b) global features describe the whole signature or major part of the signature. Some of the local features for online signatures are curvature change, pressure, speed etc., while the global features are signature writing time, number of strokes, average speed etc., [1].

To establish the authenticity of a test signature, during matching a test signature is compared with the reference signatures stored in the knowledgebase. Different matching techniques proposed in the literature for online signatures are Hidden Markvov Model [9], Support Vector Machine [10], Neural Network [11] and Dynamic Time warping [1] and symbolic classifier [12, 13].

Almost all the signature verification methods proposed in the literature have utilized same features either local or global features for all users. However, signature is a complex biometric trait where each user has his/her own style of signing and hence the same features may not be effective in capturing the typicality of individual user. To the best of our knowledge and from literature survey the concept of user dependent features is not utilized for signature verification. In addition, signature samples of a class have large intra-class variation and there is a need to capture this intra-class variation using suitable representation scheme.

Few works are reported in literature for effective capturing of intra-class variation in signatures. Guru et al. [13] used the concept of cluster based symbolic representation for signature representation which effectively captures intra-class variation. But they have used all the 100 features for all the users which is computationally expensive. In this paper, we propose user dependent features for online signatures. Initially all possible features are extracted for all users and a method of feature selection is employed for selecting user dependent features followed by clustering of signatures based on the features selected. Clustering provides an effective representation in the form of multiple reference signatures for each class. A method of signature verification based on the proposed representation is presented. Instead of storing every signature sample of every user in the database, training signatures are clustered into a number of clusters and each cluster is stored in the knowledgebase by means of symbolic feature vector. The major contribution of this work relies on proposal of user dependent features for signatures which vary from a user to a user instead of a set of common features for all users.

The paper is organized as follows: The proposed model is explained in Sect. 2. In Sect. 3, we summarize the details of experimentation along with the result obtained. A comparative study of our work with other similar work is presented in Sect. 4. Finally in Sect. 5 some conclusions are drawn.

## 2 Proposed Model

The proposed model has three stages, user dependent feature selection, cluster based symbolic representation and signature verification based on the proposed symbolic representation.

### *2.1 User Dependent Feature Selection*

Traditional feature selection methods are either supervised or unsupervised. In supervised mode, features are selected such that the importance of each feature is evaluated by the correlation between class labels and features. Some of the supervised feature selection methods include Pearson correlation coefficient, Fisher score, and information gain. In an unsupervised feature selection method, top ranked features are selected based on a certain score computed for each feature. Here correlation between feature and class labels is neglected and hence the feature selected may not be optimal. In this section we exploit an unsupervised feature selection method suitable for multi-cluster data [14]. Features selected preserve the multi-cluster structure. Tradition feature selection problem selects the features based on certain evaluation criteria, which is computationally expensive as it is a combinatorial optimization problem. The feature selection method we adapted is computationally efficient as it involves a sparse Eigen-problem and L1-regularized least square problem. It uses spectral analysis technique to measure the correlation between different features without class label information.

In spectral clustering, data points are clustered using top eigenvectors of graph laplacian, which is defined on the affinity matrix of the data points. From the perspective of graph partitioning it finds the best cut of the graph so that the criterion function can be optimized. Spectral clustering basically consists of two steps. The first step is "unfolding" the data manifold using the manifold learning algorithm and the second step performing traditional clustering on the "flat" embedding for the data points. The different steps in the feature selection algorithm that we adapted in our work are:

1. Initially graph with one vertex for each data point is created. For each data point $x_i$, p nearest neighbors are identified and an edge is drawn between each data point and all its neighbors. A weight matrix (W) based on the weight of each edge is created. In the weight graph, one of the three weighting scheme can be used: 0–1 weighting, Heat-kernel weighting, dot product weighting [14].
2. From the weight matrix ($W$), a diagonal matrix $D$ is computed whose entries are column or row sums of $W$. $D_{ii} = \sum W_{ij}$.
3. Corresponding graph Laplacian is obtained as $L = D - W$.
4. The "flat"embedding for the data points which "unfold" the data manifold can be found by solving the following generalized eigen-problem $Ly = \lambda Dy$.

Let $Y = [y_1, \ldots y_k]$, $y_i$'s are the eigenvectors of the above generalized eigen-problem with respect to the smallest eigenvalue. Solving the corresponding eigen-problem of step 4 results in a set of eigen vectors corresponding to smallest eigen values. $K$ indicates the dimensionality of the data and each $y_i$ reflects the distribution of data on the corresponding cluster.

After obtaining flat embedding $Y$ for data points, the importance of each feature for differentiating each cluster is measured. Given $y_i$, a column of $Y$, we can find a relevant subset of features by minimizing the fitting error as follows:

$$\min_{a_i} \left\| y_i - X^T a_i \right\|^2 + \beta |a_i| \tag{1}$$

Each $a_i$ contains the combination coefficients for different features in approximating $y_i \cdot |a_i|$ is the $L-1$ norm of $a_i$ and $X$ is the set of data points.

The advantage of using a $L-1$ regularized regression model is to find the subset of features instead of evaluating the contribution of each feature. Each $a_i$ essentially contains the combination of coefficients for different features. It helps in approximating a subset containing the most relevant features corresponding to the non-zero coefficients in $a_i$ with respect to $y_i$. Equation (1) is essentially a regression problem and can be solved using Least Angle regression (LAR) algorithm which results in $K$ sparse coefficient vector $a_i$. Each entry in $a_i$ is a feature and the cardinality of $a_i$ is $d$ which denotes the number of features to be selected.

From the sparse coefficient vector, $d$ features are selected by computing MCFS score with respect to each feature and top $d$ features are selected in the decreasing order of MCFS score.

In our work, we exploited the feature selection method described above for selecting different features for different users. In our proposed method, signature data set of dimension $N \times M \times K$ where $N$ is the number of users, $M$ is the number of samples and $K$ is the number of features is decomposed into $N$ feature vectors of size $M \times K$. Feature selection method discussed above is applied on each of these feature vectors representing an individual user separately. It results in the reduction of dimension of feature vector of a user to a size $M \times d$ where $d$ is the number of features selected $(d < k)$. The $d$ number of features selected varies from a user to user. The indices of the corresponding features selected is also stored in the knowledgebase which is used during verification. The computational complexity of user dependent feature selection is $\mathrm{O(N^2M + Kd^3 + NKd^2 + M\log M)}$ where N is the number of samples, M is the original number of features, K is the number of clusters, d is the number of features selected. For more details on multi-cluster feature selection, readers are referred to Cai et al. [14]. Once the user dependent features are selected, we effectively capture the intra-class variation through the concept of symbolic representation [13] which is described in next section.

## 2.2 Cluster Based Symbolic Representation

Once the user dependent features are selected, training signatures of each user are clustered using the selected features instead of all the original features. Clustering is effective as it provides multiple reference signatures for each user. We have adapted Fuzzy C means for clustering [15]. After the signatures are clustered, each cluster is represented in the form of interval-valued feature vector [13]. This representation is very effective in capturing intra-class variation which is common in signatures.

Let $\{S_1, S_2, \ldots, S_n\}$ be $n$ signature samples of a cluster $C_j$, $j = 1, 2, \ldots, C$ where $C$ is the number of clusters in each class. Let $\{f_{j1}, f_{j2}, \ldots, f_{jd}\}$ be the feature vector representing the cluster $C_j$ where $d$ is the number of features. Let $M_{jk}$, $k = 1, 2, \ldots, d$ and $\sigma_{jk}$, $k = 1, 2, \ldots d$ be the mean and standard deviation of $k$th feature of the cluster $C_j$ i.e.

$$M_{jk} = \frac{1}{n}\sum_{i=1}^{n} f_{ik} \text{ and } \sigma_{jk} = \left[\frac{1}{n}\sum_{1}^{n}\left(f_{ik} - \mu_{jk}\right)^2\right]^{\frac{1}{2}} \quad (2)$$

In order to capture intra-class variation, each feature of the cluster $C_j$ is represented in the form of interval-valued feature. For example $k$th feature of the cluster $C_j$ is represented as $\left[f_{jk}^{-}, f_{jk}^{+}\right]$ where $f_{jk}^{-} = M_{jk} - \alpha\sigma_{jk}$ and $f_{jk}^{+} = M_{jk} + \alpha\sigma_{jk}$ for some scalar $\alpha$ which is used to constrain the upper and lower limits for $k$th feature of cluster $C_j$. Thus, the interval $\left[f_{jk}^{-}, f_{jk}^{+}\right]$ depends on the mean and the standard deviation of the respective individual feature of a cluster. The interval $\left[f_{jk}^{-}, f_{jk}^{+}\right]$ represents the lower and upper limits of the $k$th feature value of a signature cluster in the knowledgebase. In general each of the $d$ features selected is represented in the form of an interval-valued feature. The reference signature for the cluster $C_j$ is thus formed as $RFC_J = \left\{\left[f_{j1}^{-}, f_{j1}^{+}\right], \left[f_{j2}^{-}, f_{j2}^{+}\right] \cdots \left[f_{jd}^{-}, f_{jd}^{+}\right]\right\}$, $j = 1, 2, \ldots C$ where $C$ is the number of clusters in each signature class.

This symbolic feature vector is stored in database as the representative of the entire cluster. Instead of storing every signature of every cluster, it is sufficient to store one feature vector for each of the cluster. If there are $C$ clusters formed for each individual user and $N$ is the number of users then we have totally $NC$ number of reference signatures in the knowledgebase instead of $Nn(>NC)$ number of signatures.

## 2.3 Signature Verification

During verification, we consider a test signature, which is represented in the form of k features of crisp type as $F_t = \{f_{t1}, f_{t2}, \ldots, f_{tk}\}$. Each feature of the test signature is of type crisp in contrast with a reference signature where the corresponding feature is of type interval valued. For authenticating the test signature, we compare the only $d$ features of a test signature with the corresponding d interval valued features of the reference signature stored in the knowledgebase. The indices of the $d$ features of test signatures to be compared with corresponding features reference signature is available in the knowledgebase. Reemploying of feature selection is not required for a test signature as the features selected for the claimed user is known at the time of training.

The total number of features of the test signature which lie within the corresponding interval valued features of a reference signature is called degree of authenticity [13]. Degree of authenticity is expressed by means of an acceptance count, which is a measure of authenticity for the test signature to qualify as genuine or forgery. If a feature of the test signature lies within corresponding interval-valued feature of reference signature, the acceptance count is incremented by one. If the total acceptance count is greater than the predefined threshold, then the test signature is accepted as genuine else, it is considered as forgery.

The acceptance count is defined to be

$$A_c = \sum_{i=1}^{d} C\left(f_{ti}, \left[f_{ji}^{-}, f_{ji}^{+}\right]\right) \quad (3)$$

where

$$C\left(f_{ti}, \left[f_{ji}^{-}, f_{ji}^{+}\right]\right) = \begin{cases} 1 & if \left(f_{ti} \geq f_{ji}^{-} \ and \ f_{ti} \leq f_{ji}^{+}\right) \\ 0 & otherwise \end{cases} \quad (4)$$

Here $\{f_{t1}, f_{t2}, \ldots f_{td}\}$ defines the feature vector of the test signature consisting of values corresponding to the selected features.

Each feature of the test signature which lies within the corresponding interval valued feature of the reference signature contributes a value of 1 towards acceptance count.

## 3 Experimentation and Result

In this section, we discuss on dataset used and on the details of experiments conducted in our work along with the result obtained.

### 3.1 Experimentation

**Dataset**: We have conducted experiments on two data sets MCYT-100 (DB1) consisting of signatures of first 100 users and MCYT-330 (DB2) consisting of signatures of all the 330 users [16]. Both MCYT-100 and MCYT-330 online signature databases consisting of 25 genuine and 25 skilled forgeries for each user. We have used a set of 100 global features of online signatures for our experimentation. The details of these 100 global features of online signature can be found in the work of [17]. The purpose of using the DB1 is to set up the system with a small set of users. Once the different parameters like similarity threshold, number of features to be selected for each user keeping EER minimum are decided with the small database DB1, the experimentations are performed on the whole database. This results in reduction of computation time and avoids the risk of over training. Feature selection experiments are repeated for each user in the training phase which selects best features for the particular user. Experiments are conducted under varying number of features.

**Experimental setup**: Initially we conducted feature selection experiments on DB1 by varying the number of features from 5 to 75 in step of 5 and noted the EER in each case. Further, experiments are conducted by varying the feature numbers in step of 1 to identify the best value of the number of features for achieving a minimum EER. Similarity threshold values are also varied from 0.1 to 0.9. We conducted 20 trials and for every trial, training signatures are randomly selected and EER is noted in each trial. Finally, we considered the average EER of all the twenty trials as the final EER value. The number of features to be selected is empirically fixed so that EER is minimum. We have used DB1 as a validation dataset for fixing up the value the number of features to be selected. Once the value of the number of features to be selected is decided, we conducted experiments on DB2 using the same value of the number of features and noted down the EER. In DB2 also, we randomly selected the training signatures in each of the 20 trials and conducted verification experiments. Details of training and testing signatures in our experimentation for both DB1 and DB2 are tabulated in Tables 1 and 2 respectively. Figure 1a–d shows the variation of FAR and FRR in all the four categories with respect to DB1.

We trained the system with a small training set consisting of 5 genuine signatures (Skilled_05 and Random_05) and with a big training set consisting of 20

**Table 1** Details of the training and testing signatures with DB1

| Training/testing samples | Skilled_05 | Skilled_20 | Random_05 | Random_20 |
|---|---|---|---|---|
| Number of training signatures | 100 users × 5 genuine signatures per user = 500 | 100 users × 20 genuine signatures per user = 2,000 | 100 Users × 5 genuine signatures per user = 500 | 100 Users × 20 genuine signatures per user = 2,000 |
| Number of testing signature | 100 users × 20 genuine signatures per user = 2,000 + 100 users × 25 skilled forgery per user = 2,500 | 100 users × 5 genuine signatures per user = 500 + 100 users × 25 skilled forgery = 2,500 | 100 users × 20 Genuine signatures per user = 2,000 + 100 users × 99 random forgeries = 9,900 | 100 users × 5 Genuine signatures per user = 500 + 100 users × 99 random forgeries = 9,900 |

**Table 2** Details of the training and testing signatures with DB2

| Training/testing samples | Skilled_05 | Skilled_20 | Random_05 | Random_20 |
|---|---|---|---|---|
| Number of training signatures | 330 users × 5 genuine signatures per user = 1,650 | 330 users × 20 genuine signatures per user = 6,600 | 330 Users × 5 genuine signatures per user = 1,650 | 330 User × 20 genuine signatures per user = 6,600 |
| Number of testing signature | 330 users × 20 genuine signatures per user = 6,600 + 330 users × 25 skilled forgery per user = 8,250 | 330 users × 5 genuine signatures per user = 1,650 + 330 users × 25 skilled forgeries per user = 8,250 | 330 users × 20 genuine signatures per user = 6,600 + 330 user × 329 random forgeries = 108,570 | 330 users × 5 genuine signatures per user = 1,650 + 330 user × 329 = 108,570 |

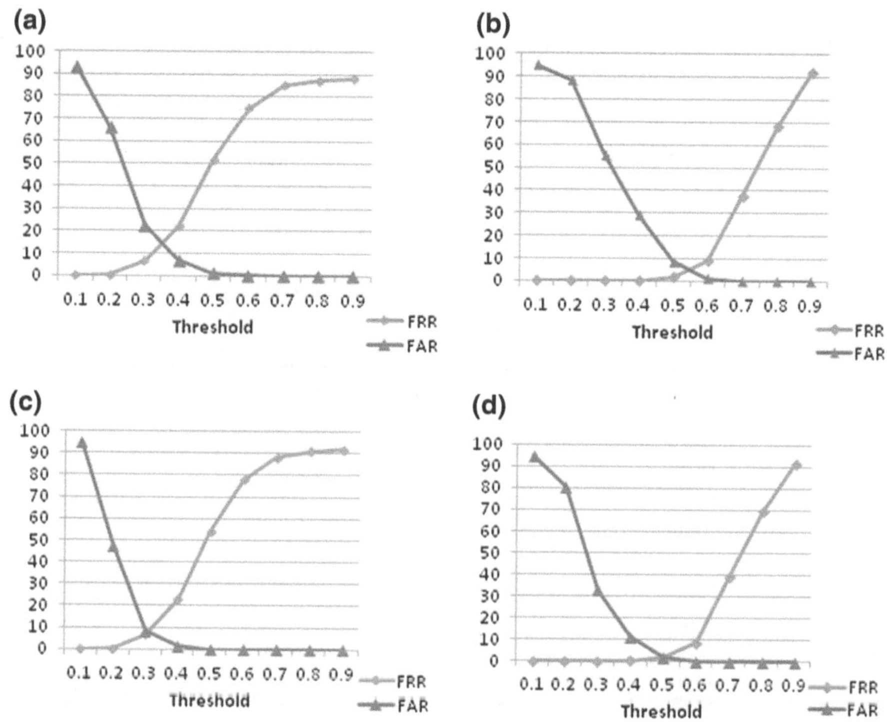


**Fig. 1** Variation of FAR and FRR under varying thresholds for DB1: **a** For Skilled_05 with 60 features. **b** For Skilled_20 with 50 features. **c** For Random_05 with 60 features. **d** For Random_20 with 50 features

genuine signatures (Skilled_20 and Random_20). The test set consists of the remaining genuine signatures and all the forgery signatures. In case of random forgery, genuine signature of every other user is considered as random forgery.

## 3.2 Experimental Results

In our experimentation, training signatures are clustered using Fuzzy C-means as it is distribution free when compared to other well known statistical techniques like KNN classifier, maximum likelihood estimate etc., and also its ability to discover cluster among data. Verification performance of our method for DB1 and DB2 are shown in Table 3. Data in Table 3 shows the minimum EER achieved. The EER value for varying number of features selected in DB1 is shown in Table 4. In case of skilled-05 and Random-05, we achieved a minimum EER for 60 features. Even if the number of features selected is increased in step of 1 up to 75, there was only marginal decrease in the EER and hence we considered only 60 features and similarly with respect to Skilled_20 and Random_20 we achieved a minimum EER for 50 features.

**Table 3** Minimum EER with Skilled and Random forgery

| Dataset | Skilled_05 (60 features) | Skilled_20 (50 features) | Random_05 (60 features) | Random_20 (50 features) |
|---|---|---|---|---|
| MCYT-100(DB1) | 14.90 | 5.06 | 7.98 | 2.02 |
| MCYT-330(DB2) | 15.90 | 6.10 | 1.90 | 1.80 |

**Table 4** Verification performance (EER) with skilled and random forgeries under varying features of DB1

| Features | Skilled_05 | Skilled_20 | Random_05 | Random_20 |
|---|---|---|---|---|
| 5 | 26.05 | 17.77 | 23.28 | 13.93 |
| 10 | 21.20 | 12.86 | 19.52 | 7.63 |
| 15 | 19.19 | 8.92 | 15.73 | 5.82 |
| 20 | 17.66 | 8.08 | 14.06 | 5.77 |
| 25 | 17.55 | 7.42 | 13.48 | 4.58 |
| 30 | 16.75 | 6.98 | 12.37 | 4.22 |
| 35 | 16.68 | 6.14 | 12.07 | 3.08 |
| 40 | 16.00 | 6.00 | 10.56 | 2.77 |
| 45 | 16.20 | 5.45 | 10.45 | 2.33 |
| 50 | 15.47 | 5.06 | 9.09 | 2.02 |
| 55 | 15.47 | 5.07 | 9.015 | 2.15 |
| 60 | 14.90 | 5.22 | 7.98 | 2.00 |
| 65 | 15.05 | 4.76 | 8.01 | 1.96 |
| 70 | 14.52 | 4.91 | 7.25 | 1.86 |
| 75 | 14.42 | 4.56 | 7.305 | 1.87 |

## 4 Comparative Analysis

In this section, we compare the verification performance of our method with that of other existing methods. Since most of the researchers have reported their work on DB1, we have considered DB1 for comparative analysis. Table 5 shows the performance of different verification systems which work on the same dataset as ours. Details of some of these classifiers is found in the work of [2, 17, 18]. From Table 5 it is clear that our method is superior when compared to methods like NND, Base Classifier, MOGD_3 and MOGD_2 in Skilled_20, Random_5, Random_20 categories and when compared to methods like Symbolic classifier, LPD, PCAD and SVD, even though the EER we obtained is slightly high we have used only 50 features (Skilled_20 and Random_20) and 60 features(Skilled_05 and Random_05) while others have used all the 100 global features.

**Table 5** Equal error rates of various online signature verification approaches on DB1

| Method | Skilled_05 | Skilled_20 | Random_05 | Random_20 |
|---|---|---|---|---|
| 1. Proposed method | 14.9 | 5.0 | 7.9 | 2.0 |
| 2. Symbolic classifier [13] | 15.4 | 4.2 | 3.6 | 1.2 |
| 3. Linear programming description (LPD) | 9.4 | 5.6 | 3.6 | 2.5 |
| 4. Principal component analysis description (PCAD) | 7.9 | 4.2 | 3.8 | 1.4 |
| 5. Support vector description (SVD) | 8.9 | 5.4 | 2.8 | 1.6 |
| 6. Nearest neighbour method description (NND) | 12.2 | 6.3 | 6.9 | 2.1 |
| 7. Random ensemble of base (RS) | 9.0 | – | 5.3 | – |
| 8. Random subspace ensemble with resampling of base (RSB) | 9.0 | – | 5.0 | – |
| 9. Base classifier (BASE) | 17.0 | – | 8.3 | – |
| 10. Parzen window classifier (PWC) | 9.7 | 5.2 | 3.4 | 1.4 |
| 11. Mixture of Gaussian description_3 (MOGD_3) | 8.9 | 7.3 | 5.4 | 4.3 |
| 12. Mixture of Gaussian description_2 (MOGD_2) | 8.1 | 7.0 | 5.4 | 4.3 |
| 13. Gaussian model description S | 7.7 | 4.4 | 5.1 | 1.5 |
| 14. Kholmatov model (KHA) | 11.3 | – | 5.8 | – |
| 15. Fusion methods | 7.6 | – | 2.3 | – |
| 16. Regularized Parzen window classifier RPWC [8] | 9.7 | – | 3.4 | – |

## 5 Conclusion

In this work, we introduced a novel concept of user dependent features for online signature verification. The proposed method is very effective in capturing the typicality of individual user. In addition, our method is computationally efficient as it works in lower dimension. We conducted extensive experiments on MCYT database and the result demonstrates the effectiveness of the proposed method. Overall, the idea of adaption of user dependent features for online signature verification is our contribution which is first of its kind in the literature.

**Acknowledgments** We thank Dr. Julian Fierrez Aguillar, Biometric Research Lab-AVTS, Spain for providing MCYT Online signature dataset. We also thank Deng Cai, Associate Professor, Zhejiang University, China for sharing his work on unsupervised feature selection for multi-cluster data.

## References

1. Jain AK, Griess FD, Connel SD (2002) On-line signature verification. Pattern Recogn 35:2963–2972
2. Nanni L, Lumini A (2006) Advanced methods for two-class problem formulation for on-line signature verification. Neurocomputing 69:854–857

3. Ismail MA, Gad S (2002) Offline Arabic signature verification. Pattern Recogn 33:1727–1740
4. Plamondon R, Lorette G (1989) Automatic signature verification and writer identification: the state of the art. Pattern Recogn 2(2):63–94
5. Lee LL, Berger T, Aviczer E (1996) Reliable on-line signature verification systems. IEEE Trans Pattern Anal Mach Intell 18:643–649
6. Kashi R, Hu J, Nelson WL, Turin W (1998) A hidden Markov model approach to on-line handwritten signature verification. IJDAR 1:102–109
7. Zhang K, Prathikakis I, Cornelis J, Nyssen E (2003) Using landmarks to establish point-to-point correspondence between signatures. Pattern Anal Appl 3:69–73
8. Fierrez AJ, Krawczyk S, Ortega-Garcia J, Jain AK (2005) Fusion of local and regional approaches for on-line signature verification. International workshop on biometric recognition system (IWBRS), LNCS, vol 3781. pp 188–196
9. Fierrez AJ, Ortega-Garcia J, Ramos D, Gonzalez-Rodriguez J (2007) HMM-based on-line signature verification: feature extraction and signature modeling. Pattern Recogn Lett 28(16):2325–2334
10. Kholmatov A, Yanikoglu B (2005) Identity authentication using improved online signature verification method. Pattern Recogn Lett 26:2400–2408
11. Baltzakis H, Papamarkos N (2001) A new signature verification technique based on a two stage neural classifier. Eng Appl Artif Intell 14:95–103
12. Guru DS, Prakash HN (2009) Online signature verification and recognition: an approach based on symbolic representation. IEEE Trans Pattern Anal Mach Intell 31(6):1059–1073
13. Guru DS, Prakash HN, Manjunath S (2009) Online signature verification: an approach based on cluster representation of global features. In: Seventh international conference on advances in pattern recognition, pp 209–212
14. Cai D, Zhang C, He X (2010) unsupervised feature selection for multi-cluster data. In: 16th ACM SIGKDD conference on knowledge discovery and data mining (KDD'10), pp 333–342
15. Bezdez JC (1981) Pattern recognition with fuzzy objective algorithms. Plenum, New York
16. Garcia OJ, Fierrez AJ, Simon D (2003) MCYT baseline corpus: a bimodal database. In: IEEE proceedings on vision, image and signal processing, vol 150. pp 395–401
17. Fierrez AJ, Nanni L, Penalba JL, Garcia JO, Maltoni D (2005) An on-line signature verification system based on fusion of local and global information. AVBPA, LNCS 3546:523–532
18. Nanni L (2006) Experimental comparison of one-class classifier for on-line signature verification. Neurocomputing 69:869–873

# An Integrated Filter Based Approach for Image Abstraction and Stylization

**H. S. Nagendra Swamy and M. P. Pavan Kumar**

**Abstract** In this paper, we present a non-photo-realistic image rendering (NPR) technique based on integrated filtering approach. The proposed method integrates 2D anisotropic filter, 2D difference of Gaussian filter, modified coherence shock filter and mean curvature flow (MCF). Coherence shock filter is applied iteratively to enhance the edge information in an image. Dithering with a fixed deviation value is also applied to produce a rendering effect in the abstracted image. The proposed method can be applied to color as well as gray scale images to produce stylized and cartoon like images. The method does not require any kind of post processing for image abstraction. Implementation of the proposed work is carried out in Mat Lab environment using local library functions. Efficacy of the proposed work has been corroborated by conducting experiments on various types of images and the results have also been compared with the other contemporary work. The approach is found to be computationally efficient in producing effective cartoon like images being simple in terms of its implementation.

**Keywords** Non-photo-realistic rendering · Anisotropic filter · Shock filter · Dithering · Mean curvature flow

H. S. N. Swamy (✉)
Department of studies in Computer Science Editorial, University Of Mysore, Mysore, Karnataka, India
e-mail: swamy_hsn@yahoo.com

M. P. P. Kumar
Department of Information Science and Engineering, J.N.N College Of Engineering, Shimoga, Karnataka, India
e-mail: pavankumarjnnce@gmail.com

P. P. Swamy and D. S. Guru (eds.), *Multimedia Processing, Communication and Computing Applications*, Lecture Notes in Electrical Engineering 213,
DOI: 10.1007/978-81-322-1143-3_20, 

## 1 Introduction

Real time captured picture recognition often encloses more information than required to communicate proposed information. So image abstraction can be used to reduce unnecessary information in an image and preserving relevant information for interpretation. More formally, image abstraction refers to the process of simplifying complex scène by removing irrelevant information that is not mandatory for particular event [1, 2]. Image abstraction has been advanced technology under NPR and has been an effective visual tool for many applications. Image abstraction has talented to suggest the assured aspects and real time informative scene more effectively.

In this paper, an integrated filter based approach for image abstraction is proposed. We made an attempt to effectively integrate anisotropic filter, Difference of Gaussian filter and coherence shock filter followed by Mean Curvature Flow (MCF) and dithering to produce more effective abstracted image. We have exploited the features of these filters through integration for better image abstraction useful for many applications. The abstraction process is not confined to any shape or color. The proposed abstraction technique is capable to convey colors, regions, shapes as well as a particular event in an effective manner. Any image abstraction technique is essentially suppress the random and real time noise, preserving image structure, shape and handle poor lighting conditions in an image and hence it is a challenging task in the research area of image processing.

Image abstraction is most useful for numerous applications such as smoothing isophote curves, stipple drawing, mosaics engraving, simplifying visual cues, cubist rendering, cartoon rendering of 3D object, animated movies, noise suppression, pen and pencil sketch illustration to name a few. It is also useful to solve the optimization problem in engineering and scientific applications.

## 2 Related Works

Several researchers have made an attempt to propose effective techniques for image abstraction. Image segmentation, color quantization, or feature preserving smoothing techniques are used for image abstraction. Edge preserving filters [3] are also used to automatically create stylized abstractions from images or videos.

A familiar approach to create non photo realistic representation of an image is to transform an image into abstracted image using an interactive or automatic technique [4]. Another interactive tool approach, where the brush strokes are placed automatically was introduced by Hays and Essa [5] in the year 2004.

DeCarlo et al. [6] proposed stylization and abstraction of photographs based on mean shift color image segmentation. The technique transforms images into a line-drawing style using bold edges and large regions of constant color. But the method may not produce satisfactory results for images with complex structures.

Comaniciu et al. [7] and Collomosse et al. [8] extended the mean shift segmentation to video to produce a temporally coherent cartoon like image sequence. Wang et al. [9] used anisotropic mean shift filter for handling elongated structures often found along temporal axis of a video. Wen et al. [10] also used mean shift segmentation algorithm for generating a colored sketch from a photograph. Lecot and Levy [11] developed a triangle based image segmentation algorithm for abstract and stylistic bitmap-to-vector image conversion. Though the image segmentation is a natural choice of tool for the task of image abstraction, a crude segmentation often results with incomplete abstraction, which further requires some post processing like curve fitting, editing, smoothing and stylizing.

Filter based image abstraction methods have also been proposed in the literature. Winnemoller et al. [12] showed that bilateral filter can be used to abstract color images as well as video. Orzan et al. [13] developed a multi-scale image abstraction system based on gradient reconstruction. Gooch et al. [14] proposed Artistic vision: Painterly Rendering Using Computer Vision Techniques based on Difference-of-Gaussians (DOG) filter. In their work, they have considered raster image as input and obtained a painting—like image composed of strokes rather than pixels. Kang and Lee [15] proposed shape simplifying image abstraction method for producing stylistic abstraction of a photograph. The method used mean curvature flow in conjunction with shock filter to simplify both shape and color simultaneously. But an obvious limitation of this method is that the curvature flow contracts small circular shapes very quickly. If the circular shape is of high importance in certain cases, it needs to be masked before running this algorithm. Kang et al. [16] proposed image and video abstraction using anisotropic Kuwahara filter. This filter effectively removes the details in high contrast edges by preserving shape boundaries in low contrast region. But the Kuwahara filter is unstable in the presence of high noise in the source image and suffers from block of artifacts. Kang et al. [17] proposed Flow-Based Image Abstraction technique based on line and region extraction filters guided by Edge Tangent Flow (ETF) that describes the flow of salient features in the image. The method may not perform satisfactorily when there are large number of irregularities and random noise in a poor intensity images.

From the literature survey, it is found that the methods proposed for image abstraction possess certain limitations in terms of quality output or in handling type of input image. Also the methods have been implemented using GPU devices and CUDA languages, which demands high computation and sophisticated environment for implementation. Since the implementation of the proposed work is carried out in MatLab environment using local library functions, it is simple and does not require any sophisticated high computing environment.

Rest of the paper is organized as follows: Sect. 3 describes the proposed methodology, Sect. 4 explains the experimental analysis and Sect. 5 presents the conclusion.

# 3 Proposed Method

The proposed method of image abstraction involves five major steps. First, anisotropic filter is applied to suppress the irregularities and to preserve the prominent edge boundaries in an image. However, some edge information may not be clear and need to be enhanced. In the second step, a Difference of Gaussian filter is applied to make the uncleared edges more clear. In the third step, we apply coherence shock filter iteratively to provide sharpening effect to the image. The image obtained at this stage may contain surface irregularities due to sharpening and need to be suppressed. Hence in the fourth step, we apply MCF to remove the irregularities in the image surface. Finally, we apply dithering to create the mis-apprehension or non realistic of color depth in an image with a limited color palette, which produces artistic effect without any upsetting effect to human eye. The following sections provide a brief description about the filters used in the proposed methodology.

## *3.1 2D Anisotropic Filter*

Anisotropic filter is used to smooth and enhance edges and texture in an image. This filter effectively suppresses the unwanted noises from input images and produces best rendering and noise free images. Implementation of anisotropic filter involves various steps as follows:

The first step is to calculate the local gradient information of a color image I where I = (x, y) denotes pixel information.

$$\mathrm{I(f)} : \mathrm{R}^2\,\mathrm{grad} \rightarrow \mathrm{R}^3\,\mathrm{grad}$$

Subsequently, the local adoptive smoothing based on the local gradient information of an image is performed. Further, the smoothing kernel is locally calculated choosing Gaussian forms given in [17–20]

$$G(dir) = \exp\left\{\frac{1}{2}\left(\frac{dir.a^{+}}{sigma_1^2}\right) + \left(\frac{dir.a^{-}}{sigma_2^2}\right)\right\} \tag{1}$$

Here, dir is the location vector, where $a^+$ and $a^-$ are the vectors perpendicular to the local gradient. Values $\sigma 1$ and $\sigma 2$ are appropriately chosen. The Color image orientation and region direction derivatives information is calculated using the formula.

$$\frac{\partial f}{\partial x} = \left(\frac{\partial R}{\partial x} \cdot \frac{\partial G}{\partial x} \cdot \frac{\partial B}{\partial x}\right) \quad \frac{\partial f}{\partial y} = \left(\frac{\partial R}{\partial y} \cdot \frac{\partial G}{\partial y} \cdot \frac{\partial B}{\partial y}\right) \tag{2}$$

The local gradient is calculated to estimate important portions in an image by means of Eigen vector and Eigen values. It can be mathematically represented as follows.

$$(dI)^2 = \left(\frac{dx}{dy}\right)^T = \begin{pmatrix} \left(\frac{\partial f}{\partial x}\right)^2 & \frac{\partial f}{\partial x}\frac{\partial f}{\partial y} \\ \frac{\partial f}{\partial x}\frac{\partial f}{\partial y}\cdot & \left(\frac{\partial f}{\partial x}\right)^2 \end{pmatrix} = \left(\frac{dx}{dy}\right) \tag{3}$$

With the help of sequential gradient magnitude, we calculate the largest Eigen value in a given matrix. This can be mathematically represented as

$$\alpha = \left(\frac{\lambda 1 - \lambda 2}{\lambda 1 + \lambda 2}\right) \tag{4}$$

The gradient magnitude is computed using finite difference approximation for partial differential equation as follows.

$$||\nabla f|| = \left\{\left(\frac{\partial f}{\partial x}\right)^2 + \left(\frac{\partial f}{\partial y}\right)^2\right\}^{\frac{1}{2}} \tag{5}$$

Anisotropic filter not only smooth the image but it also gives more importance to preserve the edges gradient strength. In this work, we estimate the corner and edge strength for the purpose of preserving corners and also to smooth edges by using the equation,

$$C = (1 - \alpha)||\nabla I||^2 \tag{6}$$

During smoothing and local orientation protection process, corners should be preserved. A corner is identified as an isotropic with large local gradient strength. We can estimate the corner strength by:

$$C = (1 - \alpha)||I^2|| \tag{7}$$

How To preserve corners, we divide the standard deviations by another 1 + C. The final variances in (1) are then given by:

$$\sigma_1 = \frac{(1-\alpha).\mathrm{a}}{(1+c)} \quad \sigma_2 = \frac{\mathrm{a}}{(1+c)} \tag{8}$$

Where a is an image noise factor. The variance of the image noise $\sigma$ can be estimated globally by calculating the images local variances. Based on local orientation estimation, we can measure the noise factor. Anisotropic filter preserves large homogeneous and heterogeneous noise free regions.

## *3.2 2D Difference of Gaussian Filter*

Difference of Gaussians filter [12] is used to smooth an image as well as useful for line extraction. The filter is also used to show significant high discontinuities. Two Gaussian filters with different blurring radius are formed and the respective images after Gaussian filter are obtained. The resulting images are subjected to subtraction operation to obtain the result. Image smoothing can be performed by convolution using appropriate spatial mask. Applying the spatial mask to an input image suppresses the high frequency spatial information. A general DoG is used to remove unwanted noise and for line extraction. The most important parameters for DoG filter are the smoothing radii ($\sigma$) for the two Gaussian blurs [21, 22]. It is observed that a small increasing in the radius tends to give thicker appearing edges and a small decreasing tends to increase the threshold for recognizing something as an edge. In most cases, a best result is obtained when the value for (radius-2) is smaller than the (radius-1). DoG equation can be mathematically represented as follows.

$$f(I, \mu, \sigma_1, \sigma_2) = \frac{1}{\sigma_{1,}\sqrt{2\prod}}\exp\left(-\frac{(I-\mu)^2}{2\sigma_1^2}\right) - \frac{1}{\sigma_2\sqrt{2\prod}}\exp\left(-\frac{(I-\mu)^2}{2\sigma_2^2}\right) \tag{9}$$

## *3.3 Modified 2D Coherence Shock Filter*

2D Coherence Shock Filter [23–25] is an edge preserving and smoothing filter, which gives more importance to direction of the edges in an image. It involves either a dilation or erosion process depending on whether the pixel is present in the maximum or minimum influence zone.

$$It = -sign(\Delta u)|\nabla u| \tag{10}$$

The filter creates shocks between maximum and minimum influence zone and it represents that the shock filter is within the range of original image. The modified version of the coherence shock filter makes the edges more sharp and helps more accurate segmentation of region of interest. A slight modification to coherence shock filter is accomplished as follows:

$$I_t = -sign(\Delta u^* I)|\nabla u|^* I_{\text{Smooth}} \tag{11}$$

Here, I is the gradient image and $I_{\text{smooth}}$ is an anisotropic filtered image. The modified coherence shock filter combines the gradient image, anisotropic filtered image and −sign (Δu). The modified coherence shock filter gives a shining effect to the image by preserving hidden edges.

### *3.4 Mean Curvature flow*

Mean Curvature flow (MCF) is used to remove the irregularities from complex background images. It effectively protects and conveys directional characteristics of shapes, features and textures. MCF is capable to identify luminance contour on the image for suppressing unwanted irregular curves, removing noise and irregular peaks. It suppresses and elaborates the irregular curves in an input image. MCF can be mathematically expressed as [15],

$$I_{MCF} = \kappa|\nabla I| \tag{12}$$

### *3.5 Dithering*

It is a technique used to create the misapprehension/non-realistic of color depth in images with a limited color palette. The aim of dithering is to decrease the number of colors and to provide artistic effect to an image without any upsetting effect to human eye. It contributes more to convert original image into cartoon, half tone like image. The idea of dithering is mainly based on quantization and color approximation technique [26]. In this work, we used the dithering function supported by Mat Lab.

The proposed method of integrated filter-based approach to image abstraction can be algorithmically expressed as follows:

**Algorithm:** An integrated filter based approach for image abstraction

**Input:** Raw image
**Output:** Abstracted image
**Method:**

**Step 1:** Apply anisotropic filter for preserving low contrast regions in an image.
**Step 2:** Apply Difference of Gaussian filter for extracting dominant edges.
**Step 3:** Apply shock filter for recovering hidden edges.
**Step 4:** Apply mean curvature flow to regularize the irregular isotope curves in an image.
**Step 5:** Apply dithering for reducing the color space.

**Algorithm ends.**

## 4 Experimentation

In order to study the efficacy of the proposed technique for image abstraction, we conducted experiments on various types of natural images. In this section, we are presenting few of the results obtained from the proposed method. In our experimentation, the kernel value $\sigma$ for anisotropic filter is set to 3.0 and anisotropic filtering half-width to 5.0. In Gaussian filter, the $\sigma 1$ and $\sigma 2$ values are set to 3.0 to 2.9 and the deviation value $\mu$ is set to 0.5. For shock filter, the deviation value is set to 3.02 and the number of iterations is set to 40. The MCF is designed using local isophote curvature and an 8 bit dithering function is applied to reduce the color space. Figure 1 shows the results of varies stages of the proposed method of image abstraction.

We have also conducted an experiment on the image used in [15]. Figure 2a is the input image and Fig. 2b, c shows the result of the method in [15] and the proposed method respectively. It can be observed that the proposed method produced the better output when compared to the method used in [15].

We have also conducted an experiment on natural image and the proposed method has produced the encouraging result. Figures 3a and 4a is an input image and Figs. 3b–f and 4b–h show the result of various stages of the proposed method of image abstraction and stylization.

**Fig. 1** **a** Input image. **b** Anisotropic filter output. **c** Difference of Gaussian filter (DoG). **d** Shock filter (90 iterations). **e** Mean curvature flow (*MCF*). **f** 8 bit dithering

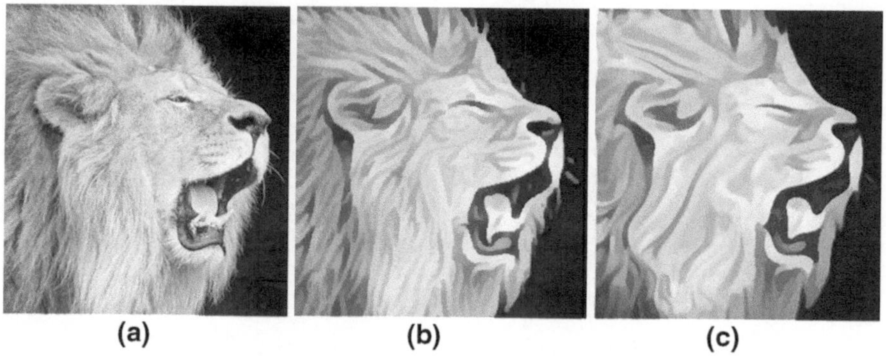


**Fig. 2** **a** Input image. **b** Shape simplifying image abstraction output. **c** Our method

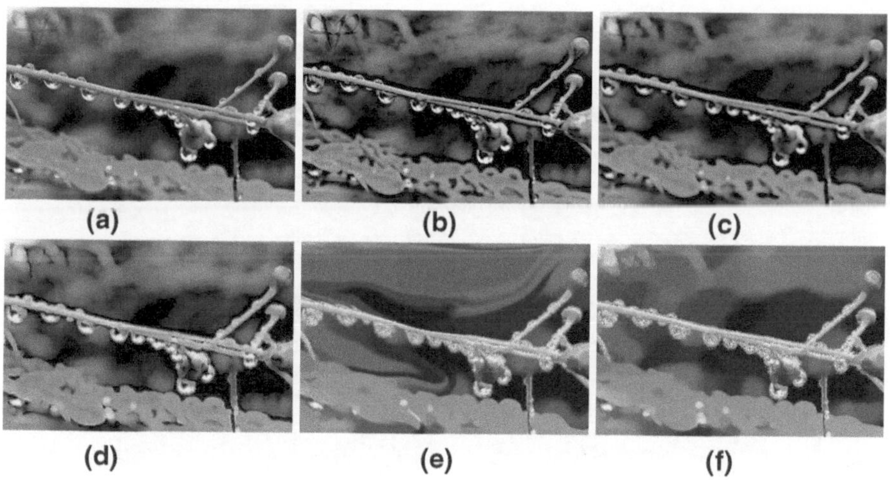


**Fig. 3** **a** Natural image. **b** Anisotropic filter output. **c** DoG filter. **d** Shock filter (30 iterations). **e** Mean curvature flow (MCF). **f** 16 bit Dithering output

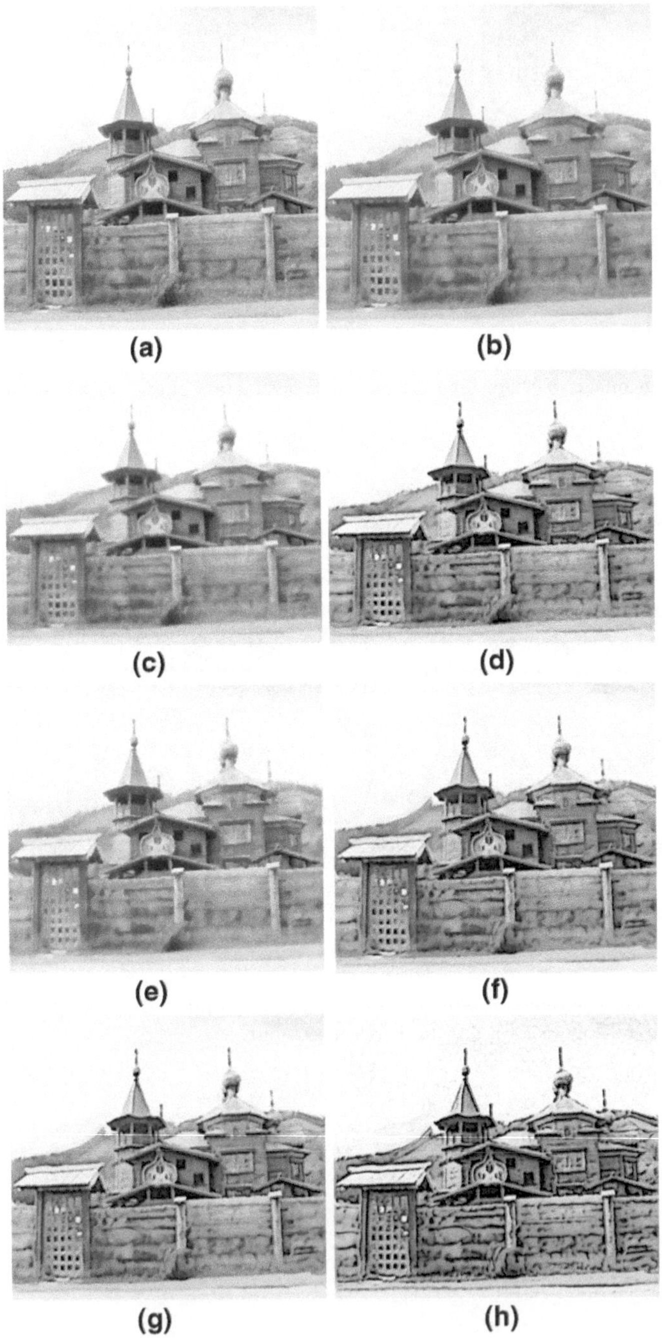


**Fig. 4** **a** Church image. **b** Anisotropic filter output. **c** DoG filter output. **d** Shock filter (5 iterations) output. **e** Shock filter (10 iterations) output. **f**. Mean curvature flow (MCF) output. **g** 24 bit Dithering output. **h** 8 bit Dithering output

## 5 Conclusion

this paper, we have presented an integrated filter-based approach to produce abstracted and stylized images. The proposed method incorporates the features of various filters to produce the better result. The method does not require any individual brush strokes to produce abstraction and stylized effect. The proposed method is found to be effective for all types of blurred images, high contrast images, and complex background images. The experimental results obtained for various types of images are highly encouraging and are comparable with the other abstraction techniques. The approach is found to be computationally efficient in producing effective cartoon like images being simple in terms of its implementation.

## References

1. Hertzmann A (2001) Paint by relaxation. In: Proceedings of computer graphics international, pp 47–54
2. Decarlo D, Finkelstein A, Rusinkiewicz S, Santella A (2003) Suggestive contours for conveying shape. In: Proceedings of ACM Siggraph'03, pp 848–855
3. Nagao M, Matsuyama T (1979) Edge preserving smoothing. Comput Graph Image Process 9:394–407
4. Haeberli P et al (1990) Paint by number: abstraction image representation. Comput graph 24:207–214
5. Hays J, Essa IA (2004) Image and video based painterly animation. In: Proceedings of the ACM NPAR 2004, pp 113–120
6. Decarlo D, Santella A (2002) Stylization and abstraction of photographs. In: Proceedings of ACM SIGGRAPH '02, pp 769–776
7. Comaniciu D, Meer P (2002) Mean shift: a robust approach toward feature space analysis. IEEE Trans Pattern Anal Mach Intell 24:603–619
8. Collomosse JP, Rowntree D, Hall PM (2005) Stroke surfaces: temporally coherent non-photorealistic animations from video. IEEE Trans. Visual Comput Graphics 11(5):540–549
9. Wang J, Xu Y, Shum HY, Cohen MF (2004) Video tooning. ACM Trans Graphics 23(3):574–583
10. Wen F, Luan Q, Liang L, Xu YQ, Shum H-Y (2006) Color sketch generation. In: Proceedings of non-photorealistic animation and rendering (NPAR'06), pp 47–54
11. Lecot G, lévy B (2006) Ardeco: automatic region detection and conversion. In: Euro graphics symposium on rendering 2006
12. Winnemoller H, Olsen S, Gooch B (2006) Real-time video abstraction. In: Proceedings of the ACM Siggraph'06, pp 1221–1226
13. Orzan A. Bousseau A, Barla P, Thollot J (2007) structure-preserving manipulation of photographs. In: Proceedings of the non-photorealistic animation and rendering (npar'07), pp 103–110
14. Gooch B, Coombe G, Shirley P (2002) Artistic vision: painterly rendering using computer vision techniques. In: Proceedings of the non-photorealistic animation and rendering (NPAR'07), pp 83–90
15. Kang H, Lee S (2008) Shape-simplifying image abstraction. IEEE Trans Comput Graphics 28
16. Kang H et al (2009) Image and video extraction by anisotropic kuwahara filtering. pp 866–872

17. Kang H, Lee S, Chui CK (2009) Flow based image abstraction. In: Proceedings of the non-Photorealistic animation and rendering, vol 15(1), January/February 2009
18. Weickert J (1998) Anisotropic diffusion in image processing. Teubner-verlag, stuttgart
19. Greenberg S, Kogan D (2006) Improved structure-adaptive anisotropic filter. Pattern Recogn Lett 27(1):59–65
20. Perona P, Malik J (1990) Scale-space and edge detection using anisotropic diffusion. IEEE Trans Pattern Anal Mach Intell 12(7):629–639
21. Ian T, Younga Lucas J (1995) Recursive implementation of the Gaussian filter, Signal Processing 44(2):139–151
22. Gomez G (2000) Local smoothness in terms of variance: adoptive Gaussian Filter. In: Proceedings of BMVC, vol 2, pp 815–824
23. Weickert J (1999) Coherence-enhancing diffusion filtering. Intern J Comput Vis 31(2-3):111–127
24. Osher S, Rudin L (1990) Feature-oriented image enhancement using shock filters. SIAM J Numer Anal 27(4):919–940
25. Grayson M (1986) The heat equation shrinks embed plane curves to round points. Differ Geom 26:285–314
26. Omohundro SM (1947) Floyd-steinberg dithering. International computer science institute, California, p 94704

# Progressive Filtering Using Multiresolution Histograms for Query by Humming System

**Trisiladevi C. Nagavi and Nagappa U. Bhajantri**

**Abstract** The rising availability of digital music stipulates effective categorization and retrieval methods. Real world scenarios are characterized by mammoth music collections through pertinent and non-pertinent songs with reference to the user input. The primary goal of the research work is to counter balance the perilous impact of non-relevant songs through Progressive Filtering (PF) for Query by Humming (QBH) system. PF is a technique of problem solving through reduced space. This paper presents the concept of PF and its efficient design based on Multi-Resolution Histograms (MRH) to accomplish searching in manifolds. Initially the entire music database is searched to obtain high recall rate and narrowed search space. Later steps accomplish slow search in the reduced periphery and achieve additional accuracy. Experimentation on large music database using recursive programming substantiates the potential of the method. The outcome of proposed strategy glimpses that MRH effectively locate the patterns. Distances of MRH at lower level are the lower bounds of the distances at higher level, which guarantees evasion of false dismissals during PF. In due course, proposed method helps to strike a balance between efficiency and effectiveness. The system is scalable for large music retrieval systems and also data driven for performance optimization as an added advantage.

**Keywords** Progressive filtering · Multiresolution histograms · Multiresolution analysis · Query by humming

T. C. Nagavi (✉)
Department of Computer Science and Engineering, S. J. College of Engineering, Mysore, Karnataka, India
e-mail: tnagavi@yahoo.com

N. U. Bhajantri
Department of Computer Science and Engineering, Government Engineering College, Chamarajanagar, Karnataka, India
e-mail: bhajan3nu@gmail.com

P. P. Swamy and D. S. Guru (eds.), *Multimedia Processing, Communication and Computing Applications*, Lecture Notes in Electrical Engineering 213,
DOI: 10.1007/978-81-322-1143-3_21, 

# 1 Introduction

Content based online music enabling systems are being developed and revamped in order to keep up with expectations of search and browse functionality. These approaches as a group describe the Music Information Retrieval (MIR) systems and have been the area under exhaustive research. The rationale of MIR research is to develop new theory and techniques for processing and searching music databases by its content. The QBH is a special branch of MIR and also a popular content based music retrieval method where the user enters a search query by humming.

Most of the research works on QBH [1–5] are based on the music processing and focused on components like melody extraction, representation, similarity measurement, size of databases, query and search algorithms. The strong literature supports the symbolic representation for melody in the form of zero-cross detection, energy, Modified Discrete Cosine Transform (MDCT) [6], pitch contour [5], rhythm [7] and quantized pitch change descriptor [3]. Also there is a remarkable amount of research work [8–10] in the broader areas of similarity measurement with reference to music patterns.

Most of the approaches proposed in the literature are not suited for real-world applications of music retrieval from a large music database. Perhaps, is due to either undue complexity in computation which leads to longer response time or performance degradation; subsequently leading to erroneous retrieval results. Striking a balance between computation and performance is the ultimate goal for such retrieval systems. As a result there are a few speeding up [11–13] mechanisms proposed for QBH.

Quite extensive literature [1–10] is available on QBH system, but there is no significant amount of literature [14–18] towards designing filtering procedures. Jang and Lee [15] have projected a mathematical analysis for a two-stage Query by Singing\Humming (QBSH) system, which is the first application of PF to QBSH. In another work authors [19] proposed the concept of iterative deepening Dynamic Time Warping (DTW), which is a special form of PF for speeding up DTW. Improvement in the form of multi phase PF for QBSH without much design analysis is presented in [12, 16]. Research work [19], proposes a simplified version of PF with a constant computation time with respect to survival rates for each comparison stage. However, most of the proposed methods still portray the deficit in meticulous investigation, efficiency and effectiveness.

Therefore, in this paper we have proposed to apply PF using MRH approach for QBH system to accomplish the improved retrieval accuracy. Real-world applications of music retrieval symbolize huge amount of non relevant songs with reference to user queries causing input imbalance problem. We expect that these two techniques are most applicable to mitigate the effect of input imbalance. The exhaustive experimentation substantiates the potential of proposed method to construct an effective music retrieval system based on humming input. In this paper, as explained above, we have motivated to use PF as a filtering procedure.

So, the next section gives a brief view of PF used for search space reduction. In Sect. 3 we have made diligent discussion on MRH framework for pattern matching in music retrieval systems. While Sect. 4 elaborates the details on similarity measure stratagem for QBH. In Sect. 5, experimental results are presented and discussed. The last section enumerates the conclusion.

## 2 Progressive Filtering

The inspiration behind PF is to apply a series of comparisons, in which each comparison will select a smaller set that is likely to contain the target of the input query. The process is repeated until final output contains list of songs with appropriate length, say 10 or 20. PF on QBH is performed by applying multiple stages of comparisons between a query and the songs in the database, using an increasingly more complicated recognition mechanism to the decreasing candidate pool. So that the correct song will remain in the final candidate pool with a maximum probability. Intuitively, the initial few stages are quick and impure such that the most unlikely songs in the database are eliminated. On the other hand, the last few stages are more sophisticated and time consuming such that the most likely songs are identified [15].

After each stage of PF, the number of surviving candidates in the candidate pool of the database becomes smaller, and the recognition technique turns into refined and effectual. The final output is the surviving candidate songs at the last stage. The multistage representation of PF is shown in Fig. 1, where there are $m$ stages, corresponding to different comparison methods with varying complexity.

For stage $i$, the input is the query and $n_{i-1}$ surviving songs from the previous stage. The output of stage $i$ is a reduced set of candidate songs of size $n_i = n_{i-1}s_i$ for the succeeding stage $i + 1$. In other words, each stage performs a filtering process that reduces the number of the candidate songs by a factor of the survival rate $s_i$. Each stage is characterized by its capability to select the most likely song candidates as the input to the succeeding stage. For a given stage, this capability can be represented by its recognition rate, which is defined as the probability that the target song of a given query is retained in the output song list of this stage. Intuitively, the recognition rate is a function of the survival rate.

## 3 Multi-Resolution Histograms

### 3.1 *Essence*

Over the past few years Multi-Resolution Analysis (MRA) is receiving major attention by researchers in the domain of computer graphics, geometric modeling, signal analysis and visualization. It is a most important approach for proficiently

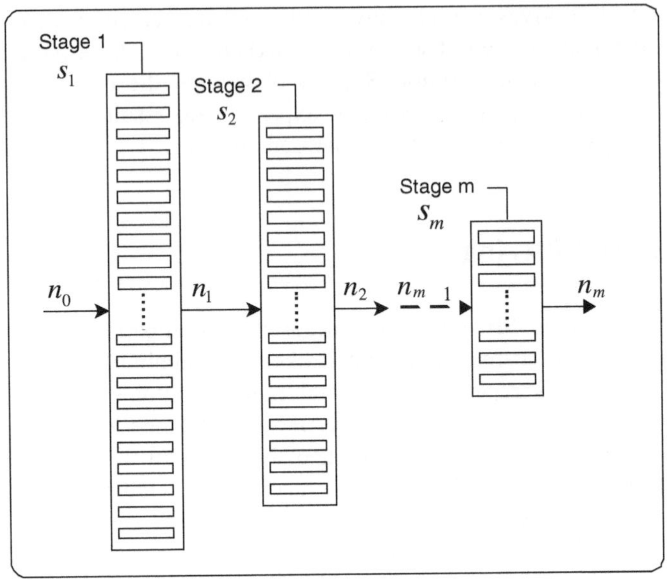


**Fig. 1** Multistage representation of progressive filtering

representing signals at many levels of detail with numerous advantages like compression, different layers of details display and progressive transmission [20]. The term multi-resolution is used in diverse perspective such as multi-resolution based wavelets, subdivisions, hierarchies and multi-grids.

Histograms provide a very effective means of data reduction and depict many attributes of the data like location, spread, and symmetry. It is also possible to decompose music signal and build histograms on the underlying cumulative data distributions. Histograms give better approximation for cumulative data distributions with less space usage. However, histograms provide a comprehensive analysis of the data distribution by excluding sequence details of values. MRH depiction is proposed for enhanced discrimination of music data based on their position fine points to assist effectual QBH system. The music signal is recursively decomposed and cumulative histograms are built. Together all these cumulative histograms of a music signal are remarked as MRH. The selection of number of levels $l$ is directly proportional to precision. Early phase cumulative histograms exhibit lesser amount of music information than later phases. These early phase MRH are used to provide quick approximate answers to music retrieval queries in the beginning. Later phase of searching with next level MRH gives us better estimates.

In this paper, a MRH based representation is proposed to approximate music signal that is invariant to shifting and scaling. The MRH representation detects existence of a pattern along with shape matching. In the early phases of searching

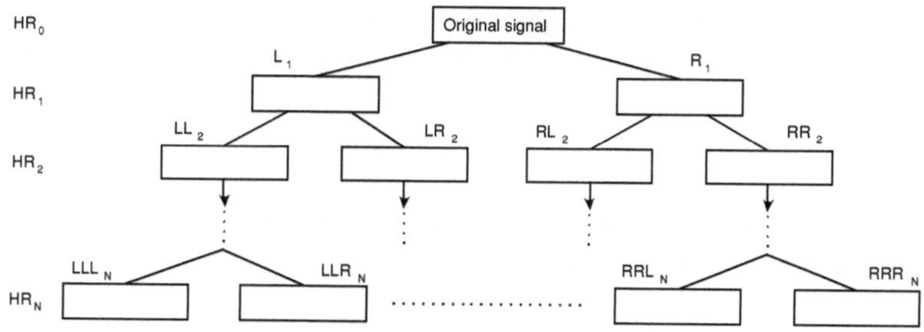


**Fig. 2** Multi-resolution histogram representation

music signals with specific specified pattern are retrieved, then search continues for shape matching yielding result set of music signals that are of interest to the user. The hierarchical MRH framework is shown in Fig. 2. The symbol $HR_i$ indicates the histogram representation at level $i$.

## 3.2 Connotation of Mathematics

Histogram function $h_i$ counts the number of samples that fall into each of the disjoint sets known as bins. Thus, if $n$ is the total number of samples and $t$ is the total number of bins, the histogram function $h_i$ is defined as following:

$$n = \sum_{i=1}^{t} h_i \tag{1}$$

A cumulative histogram function counts the cumulative number of samples in all of the bins up to the specified bin. In particular, the cumulative histogram function $hc_i$ of a histogram function $h_j$ is specified as:

$$hc_i = \sum_{j=1}^{i} h_j \tag{2}$$

Cumulative frequency distributions authorize users to approximate frequencies over numerous bins. There is no standard value for number of bins, and different number of bins exhibit different features of the samples. Based on the data distribution and the objective of the analysis, different bin widths are chosen. The numbers of bins $t$ are calculated from a recommended bin width $w$ as:

$$t = \left\lceil \frac{max(S) - min(S)}{w} \right\rceil \tag{3}$$

where $S$ = *samples to be histogrammed.* Also the equal sized bin widths are found by dividing the range with the number of bins $t$.

Primary objective of our research is to develop search criteria using similarity based queries over one dimensional music signal. Such music signal $S$ is defined as a sequence of values:

$$S = [s_1, s_2, \ldots s_N] \tag{4}$$

where $N$, the number of samples in $S$ and $s_i$ is a vector of values that was sampled at timestamp $t_i$. Given a music signal database

$$D = \{S_1, S_2, \ldots S_M\} \tag{5}$$

and a query $Q$, the aim is to find all the music signals in $D$ that contain the specified query $Q$ as well as histogram shape similar to that of $Q$. MRH are constructed by dividing the range [$min_D$, $max_D$] of music database $D$ into $t$ non-overlapping equal size sub-regions, identified as histogram bins. Histogram $H_s$ is computed by counting the number of data values $h_i$ ($1 \le i \le t$) that are located in each histogram bin $i$.

$$H_s = [h_1, h_2, \ldots h_t] \tag{6}$$

A cumulative MRH is a mapping that counts the cumulative number of observations in all of the bins up to the specified bin. That is, the cumulative histogram $HC_s$ of a histogram $H_s$ is defined as:

$$HC_s = \sum_{i=1}^{t} h_i \tag{7}$$

MRH at higher levels have enhanced discrimination power; however, the computation of MRH Distance (MRHD) at higher scales is more expensive than those at lower levels. So the number of levels trade-off should be established to balance complexity and precision.

### 3.3 Proposed Strategy

MRH construction system for database D is depicted in Fig. 2 and steps are shown in the following algorithm 1.

**Algorithm 1**: Procedure to Construct Multi-Resolution Histograms for Music Database

**Input**: a music database $D$, number of levels l and the number of histogram bins $t$
**Output**: a histogram data set $H_D$

level $l = 0$
repeat
for each $S_i$ of database $D$ do
  divide the $S_i$ into $2^l$ non overlapping equal segments $S_{il,l}$ and $S_{ir,l}$
  locate $max_D$ and $min_D$ values of the $D$
  divide the range $[min_D, max_D]$ into $t$ non-overlapping equal size bins $h_{il,l}$ and $h_{ir,l}$
  for each $S_{il,l}$ and $S_{ir,l}$ of $D$ do
   for each data point $s_{il,l}$ and $s_{ir,l}$ of $S_{il,l}$ and $S_{ir,l}$ respectively do
    for each bin $h_{il,l}$ and $h_{ir,l}$ do
     if $h_{il,lowerlimit} \leq s_{il,l} \leq h_{il,uppperlimit}$ then
     $h_{il,l} = h_{il,l} + 1;$
     end if
     if $h_{ir,lowerlimit} \leq s_{ir,l} \leq h_{ir,uppperlimit}$ then
     $h_{ir,l} = h_{ir,l} + 1;$
     end if
    end for
   end for
end for
insert generated $H_{S_{il,l}}$ and $H_{S_{ir,l}}$ to the result data set $H_D$
end for
$l = l + 1$ //increase level by 1//
until ($l = user\ specified\ levels$)
return the result data set $H_D$

# 4 Similarity Measure Stratagem for Query by Humming

## *4.1 Multi-Resolution Histograms Distance Measure*

In order to recognize the query pattern in the music database, we have attempted to develop a similarity function which separately considers signal frequency as well as positional information. Given a song $S$ of music database $D$ and humming query $Q$, feature vectors $H_{S_f}$ extracted from song MRH are matched with query MRH $H_{Q_f}$ by means of the MRHD measure:

$$MRHD(H_{S_f}, H_{Q_f}) = \sum_{i=1}^{t} min(H_{S_i}, H_{Q_i}) \times \frac{(\sqrt{2} - d(H_{S_i}, H_{Q_i}))}{\sqrt{2}} \quad (8)$$

where

$$d(H_{S_i}, H_{Q_i}) = \sqrt{\sum_{i=0}^{t} \left(h_{s_i} - h_{q_i}\right)^2} \tag{9}$$

is a Euclidean Distance function.

### 4.2 Database Pruning Using Threshold

MRHD for whole music database is calculated using Eqs. (8) and (9). The average of the MRHD considered as the upper limit and 0 as the lower limit of threshold as shown in Eqs. (10) and (11):

$$T_{upperlimit} = \frac{1}{M}\left(\sum_{i=1}^{M} MRHD(i)\right) \tag{10}$$

and

$$T_{lowerlimit} = 0 \tag{11}$$

where $M$ = *no of songs in the database*. Unlikely songs are quickly eliminated by comparing MRHD values of database songs with threshold range. The song whose threshold is not in the range will be eliminated from the pruned database. In other words, if the following condition is not satisfied such song may be purged:

$$T_{lowerlimit} \leq MRHD_S \leq T_{upperlimit} \tag{12}$$

This procedure is carried out at different histogram resolution level to form PF. The database pruning rate analysis is depicted in Fig. 4.

## 5 Results and Discussions

The relative performance of the proposed QBH method demonstrates several interesting trends and this section is dedicated to evaluate the proposed approach. Substantiation of feasibility of the proposed criteria is done through experimentation. In the sequel, three series of experiments were conducted with corresponding target and query corpus by varying the number of histogram bins from 100 to 1,000 and histogram resolution level from 1 to 5. Finally, comprehensive discussions of performances are portrayed in terms of error rate, database pruning, Mean Reciprocal Rank (MRR), Mean of Accuracy (MoA) and Top X Hit Rate.

## 5.1 Target Corpus

We are proposing a novel QBH system exclusively for Indian music songs, so the corpus chosen for this study consists of 1,000 Indian Kannada devotional monophonic MP3 songs. This collection is prepared from 39 subjects including songs from 22 males and 17 female singers. The corresponding training set includes a subset of 100, 200, 500 and 1,000 songs for different experiments. MP3 songs contain convoluted melody information and even noise. Thus preprocessing is applied on the MP3 songs database to extract information needed by the system. In music, human vocal part always plays an important role in representing melody rather than its background music therefore it is desired to segregate both [21].

## 5.2 Query Corpus

For system evaluation, we employ a monophonic query corpus containing total 200 sample queries from ten participants. Each participant was asked to hum beginning of the target song two or three times each. The participants were selected from variety of musical backgrounds like with and without considerable musical training. Also they were instructed to hum each query as naturally as possible using the lyrics of the target corpus.

## 5.3 Error Rate Analysis

Using the query and target corpus described above, the error rate is computed for the QBH system implementations presented in Sects. 2–4. Figure 3 displays the

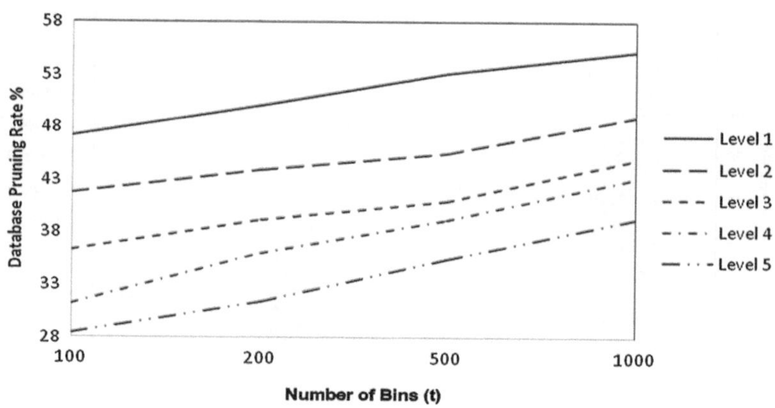


**Fig. 3** Error rate analysis

error rate for five histogram resolution levels. The target database number of histogram bins is represented along the horizontal axis and the error rate along the vertical axis. As expected, direct comparison of error rates with increasing histogram bin numbers, yields the better performance, this improvement diminishes as the number of bins decrease.

Through prominent observation it was found that fine grain level music signal approximation is possible with higher number of histogram bins, which yields better performance. However, error rate increases with the decrease in the number of histogram bins.

### *5.4 Database Pruning Rate Analysis*

Figure 4 displays the pruning rate analysis for QBH system across different sized target databases with five histogram resolution levels. The target database's number of bins are represented along the horizontal axis and the pruning rate along the vertical axis. In this figure, the pruning rate for histogram resolution level 1, 2, 3, 4 and 5 are shown with a line, dashed line, small dashed line, dash-dot line and dash-dot-dot line respectively.

Indeed, for increasing number of histogram bins and histogram resolution levels pruning rate is approximately 55 % as shown in Fig. 4. The first histogram resolution level representation yields the most robust performance of pruning around 55 %. For the target database with increasing number of histogram bins the best pruning rate is in the range 55.21–39.35 % across different histogram resolution levels. That is, the histogram representation with higher number of histogram bins yields good pruning rate, however, it is computationally domineering.

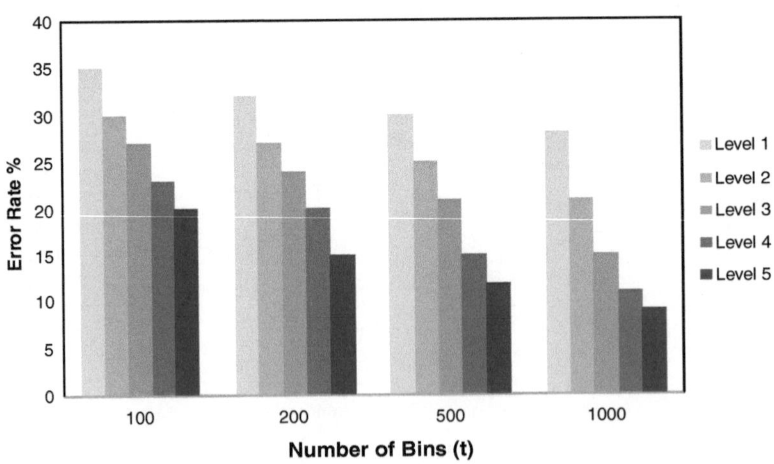


**Fig. 4** Database pruning rate analysis

## 5.5 Performance Analysis

Many different measures for evaluating the performance of QBH systems have been proposed [11, 15, 18]. The measures require a collection of training and testing samples for each test scenario and parameter combinations. The Mean Reciprocal Rank (*MRR*) is defined as:

$$MRR = \frac{1}{n}\sum_{i=1}^{n}\frac{1}{rank(t_i)} \tag{13}$$

MRR is a metric for estimating any system that generates list of potential responses to a query. Reciprocal rank of a query outcome is the multiplicative inverse of the rank of the first accurate response. That is, the MRR is estimated as the average of the reciprocal ranks of outcomes for a sample of queries. The reciprocal value of the MRR refers to the harmonic mean of the ranks. In other words frequency of the system estimating one of the first ranks is calculated through MRR [21]. We obtained MRR in the range 16.41–21.34 % for different histogram resolution levels. The proposed strategy reveals that the MRR increases with increase in histogram resolution level as portrayed in Fig. 5. In other words, frequency of occupying top five ranks increases as histogram resolution level increases.

Similarly for each test scenario and parameter combination the Mean of Accuracy (*MoA*) is defined as:

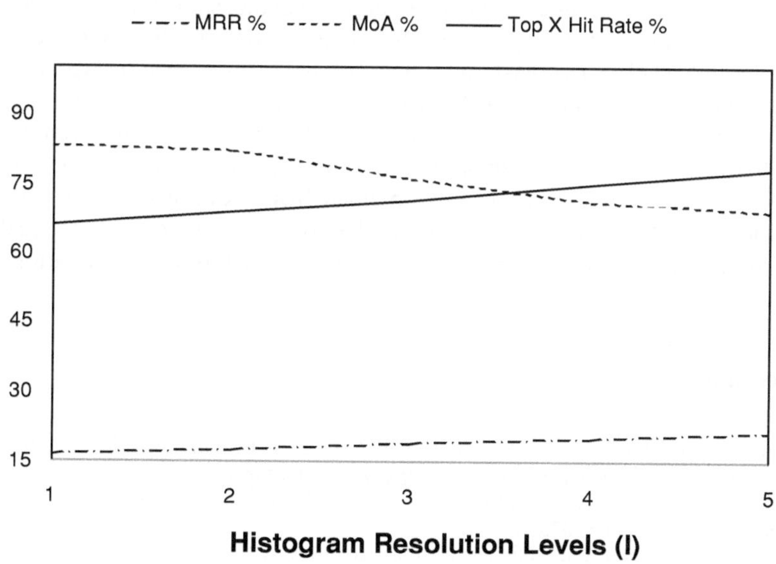


**Fig. 5** Performance analysis

$$MoA = \frac{1}{n}\sum_{i=1}^{n}\frac{n - rank(t_i)}{n - 1} \tag{14}$$

It demonstrates the average rank at which the target was found for each query. We obtained MoA in the range 68.84–83.21 % with histogram resolution levels one to five. From Fig. 5, it is found that the MoA decreases with increase in histogram resolution level. This indicates average rank of the retrieved song decreases with higher histogram resolution levels.

The Top X Hit Rate is defined as percentage of successful queries and it can be shown mathematically as:

$$Top(X) = \#\{rank(i) : rank(i) \leq X\}/N \tag{15}$$

where X symbolize top most songs and N indicates total number of songs. The impact of Top X Hit Rate for different histogram resolution level is portrayed in Fig. 5. The top X Hit Rate varied from 65.78 to 78.90 % for different histogram resolution levels. From the Fig. 5, X value 10 was found to be the best, at which system obtained retrieval accuracy in the range 65.78–78.90 % with increasing histogram resolution level.

Comparing Figs. 3, 4 and 5, the MRH based representations empirically yield relatively better performance in terms of MRR, MoA and Top X Hit Rate.

## 6 Conclusion

In this work, we have attempted to exploit advantages of MRA technique to progressively reduce search space for QBH applications. In these kinds of applications, initial result set consists of songs that have some specific patterns; subsequent steps perform relatively slow search in the small space to retrieve all songs whose histogram shape matches with query. MRH analysis is employed as database filtering procedure to support iterative search in the database to produce effective music retrievals. The results obtained from exhaustive experimentation are encouraging. Exhaustive exploration of the possibility of combining equal area bin histogram and MRA is to be considered as part of further investigation.

## References

1. Ghias A, Logan J, Chamberlin D, Smith BC (1995) Query by humming-musical information retrieval in an audio database. In: Proceeding ACM multimedia, pp 231–236
2. Tripathy AK, Chhatre N, Surendranath N, Kalsi M (2009) Query by humming system. Int J Recent Trends Eng 2(5):373–379
3. Fu L, Xue XY (2004) A new efficient approach to query by humming. International computer music conference, ICMC, Miami

4. Raju MA, Sundaram B, Rao P (2003) Tansen: a query-by-humming based music retrieval system. In: Proceedings of the national conference on communications (NCC)
5. Jang JSR, Gao MY (2000) A query-by-singing system based on dynamic programming. In: Proceedings of international workshop on intelligent system resolutions (8th bellman continuum), Hsinchu, pp 85–89
6. Liu CC, Tsai PJ (2001) Content-based retrieval of mp3 music objects. ACM, pp 506–511
7. Jeon W, Ma C (2011) Efficient search of music pitch contours using wavelet transforms and segmented dynamic time warping. In: Proceedings of IEEE international conference on acoustics, speech and signal processing (ICASSP), pp 2304–2307
8. Francu C, Nevill-Manning CG (2000) Distance metrics and indexing strategies for a digital library of popular music. In: Proceedings of IEEE international conference on multimedia and expo
9. Jang JSR, Lee HR (2001) Hierarchical filtering method for content based music retrieval via acoustic input. In: Proceedings of the 9th ACM multimedia conference, Ottawa, pp 401–410
10. Lu L, You H, Zhang HJ (2001) A new approach to query by humming in music retrieval. In: Proceedings of IEEE international conference on multimedia and expo (ICME), pp 595–598
11. Adams N, Marquez D, Wake_eld G (2005) Iterative deepening for melody alignment and retrieval. In: Proceedings of international symphony. Music information. Retrieval (ISMIR), pp 199–206
12. Wu X, Li M (2006) A top-down approach to melody match in pitch contour for query by humming. In: Proceedings of 5th international symposium on Chinese spoken language processing, Singapore
13. Zhu Y, Shasha D (2003) Warping indexes with envelope transforms for query by humming. In: Proceedings of SIGMOD, San Diego
14. Wang Z, Zhang B (2005) Quotient space model of hierarchical query-by-humming system. In Proceedings of IEEE Int Conference on Granular Computing 2:671–674
15. Jang JSR, Lee HR (2008) A general framework of progressive filtering and its application to query by singing/humming. IEEE Trans Audio Speech Lang Process, 16(2)
16. Adams NH, Bartsch MA, Shifrin JB, Wake_eld GH (2004) Time series alignment for music information retrieval. In: Proceedings of 5th ISMIR, pp 303–311
17. Addis A, Armano G, Vargiu v (2010) Using the progressive filtering approach to deal with input imbalance in large-scale taxonomies. In Proceedings of in large-scale hierarchical classification workshop of ECIR
18. Jang JSR, Lee HR (2006) An initial study on progressive filtering based on dynamic programming for query by singing/humming. In: Proceedings of 7th IEEE pacific-rim conference multimedia, Zhejiang, pp 971–978
19. Chu S, Keogh E, Hart D, Pazzani M (2002) Iterative deepening dynamic time warping for time series. In: Proceedings of 2nd SIAM international conference on data mining, CD-ROM
20. Bonneau G-P, Elber G, Hahmann S, Sauvage B (2008) Multiresolution analysis. In: De Floriani L, Spagnuolo M (eds) Shape analysis and structuring, mathematics+visualization, chapter 3. Springer, New York, pp 83–114
21. Nagavi TC, Bhajantri NU (2012) Perceptive analysis of query by singing system through query excerption. In: Proceedings of the 2nd international CCSEIT-2012, Avinashilingam University, Coimbatore

# Color and Gradient Features for Text Segmentation from Video Frames

**P. Shivakumara, D. S. Guru and H. T. Basavaraju**

**Abstract** Text segmentation in a video is drawing attention of researchers in the field of image processing, pattern recognition and document image analysis because it helps in annotating and labeling video events accurately. We propose a novel idea of generating an enhanced frame from the R, G, and B channels of an input frame by grouping high and low values using Min–Max clustering criteria. We also perform sliding window on enhanced frame to group high and low values from the neighboring pixel values to further enhance the frame. Subsequently, we use k-means with k = 2 clustering algorithm to separate text and non-text regions. The fully connected components will be identified in the skeleton of the frame obtained by k-means clustering. Concept of connected component analysis based on gradient feature has been adapted for the purpose of symmetry verification. The components which satisfy symmetric verification are selected to be the representatives of text regions and they are permitted to grow to cover their respective region fully containing text. The method is tested on variety of video frames to evaluate the performance of the method in terms of recall, precision and f-measure. The results show that method is promising and encouraging.

**Keywords** Min–Max clustering · Sliding window · K-means · Connected component analysis · Symmetry verification · Text detection

P. Shivakumara (✉)
Multimedia Unit, Faculty of Computer Science and Information Technology, University of Malaya, Kuala Lumpur, Malaysia
e-mail: hudempsk@yahoo.com

D. S. Guru · H. T. Basavaraju
Department of Studies in Computer Science, University of Mysore, Mysore, Karnataka, India
e-mail: dsg@compsci.uni_mysore.ac.in

H. T. Basavaraju
e-mail: basavaraju.com@gmail.com

P. P. Swamy and D. S. Guru (eds.), *Multimedia Processing, Communication and Computing Applications*, Lecture Notes in Electrical Engineering 213,
DOI: 10.1007/978-81-322-1143-3_22, 

# 1 Introduction

The retrieval of text information from images and videos is a very hot research area and it has gained increasing attention in the recent years. Text in images and video sequences can provide very useful semantic information which would be a good key to describe the image content and help a machine to understand it. This would help in bridging the gap between the low level and high level features [1–3]. Therefore, text segmentation and localization in video plays a vital role in several applications such as indexing video retrieval, video event identification, video understanding and video tracking etc. However, text segmentation is still a challenging problem because of the low resolution, complex background, unconstrained colors, sizes, and alignments of the characters [4, 5].

The methods [5–7] in document analysis cannot be used directly for video text segmentation as these methods require complete shape of the character components and high resolution images with plain background. These constraints may not be true for video as video usually have low resolution and complex background images. Therefore, the document analysis methods are not suitable. It is also observed from the literature on text detection from natural scene images that the methods [8–10] one or other ways extract features based on shape of the characters. Though these methods work for complex background images but they require high resolution images. Therefore, the scene text detection from natural scene methods cannot be applied on directly on video images.

In general, the existing methods on video text detection can be classified as connected component based [11, 12], texture based [13–15] and edge and gradient based [16–19] methods. The connected component based methods are same as documents analysis methods. Therefore, they work only for high resolution and simple background images. Texture based methods work well for complex background images but they require expensive classifier to classify text and non-text components. In addition, these methods are sensitive to font, font size and multi-script. To overcome the problems of the above categories, the edge and gradient based methods are proposed. These methods are fast compared to the above two categories. Recently, Eigen based method [20] is proposed to detect text in video where the method explores Eigen value analysis to classify text and non-text pixels for the both low and high resolution images. Since the method involves Eigen value analysis and gradient information, it is said to be expensive and give low precision for complex background images. In the same way, the method [21] based on run-lengths between inter and intra text component is proposed for video text detection. This method said to be simple and effective for both low resolution and high resolution video images. However, the methods produce more false positives due to text like edges in the background and hence the methods report high false positive rate and low precision. It is noted from the literature review that none of the existing methods give perfect solution to video text segmentation. In addition, none of the methods explore the combination of color values for text enhancement and symmetry based on stroke width for finding text representatives.

Hence, in this paper, we propose a novel method to explore the color values and symmetry concepts to accurate video text segmentation. The main contribution of the paper is that sharpening text information and widening gap between text and non-text using color values mapping and proposing new symmetry to identify the text representatives, which eliminates almost all background information since the symmetry is derived based on characteristics of text components. With the help of Sobel edge of the enhanced text image, we propose region growing to segment the complete using text representatives. In this way, the proposed method is different from the literature and is effective for video text segmentation.

## 2 Proposed Methodology

We observe that the color values of text pixel in R, G and B sub-bands usually are high values compared to its background values because the fact that text pixel have high contrast value compared to non-text pixel. To extract such observation, we propose Max–Min clustering on three values in R, G and B for each pixel to identify high contrast value to replace the pixel value in the image. As a result, we get high values for text pixels and low values for non-text pixels which is called enhanced image. Further to increase the gap between text and non-text pixels, we again propose Max–Min cluster with same criteria to identify the high contrast values from the neighbors. This results in sharpen image where text pixel are brighter than the pixel in the enhanced image. Since sharpened image increases the gap between the text and non-text pixels, we apply k-means with k = 2 to obtain text cluster. To analyze the components in the text cluster, the method obtains skeleton image and the skeletons are checked whether they are fully connected components or not. We believe that at least one of the text components in skeleton image satisfies the fully connected component condition. This operation eliminates most of the background information as they do not satisfy the fully connected component criteria. We call them as text candidates. Due to complex background, there are chances that non-text components are considered as text candidates. Therefore, we propose novel symmetry criterion which is computed based stroke width of the text components. The stroke width is estimated by analyzing the gradient direction of each pixel of text components. The output of this operation is called as text representative image. For each text representative, the method applies region growing to grow the contour of the text representative along text direction in Sobel edge image of the enhanced image to segment the complete text. The flow diagram can be seen in Fig. 1 where it shows all the steps of the proposed method.

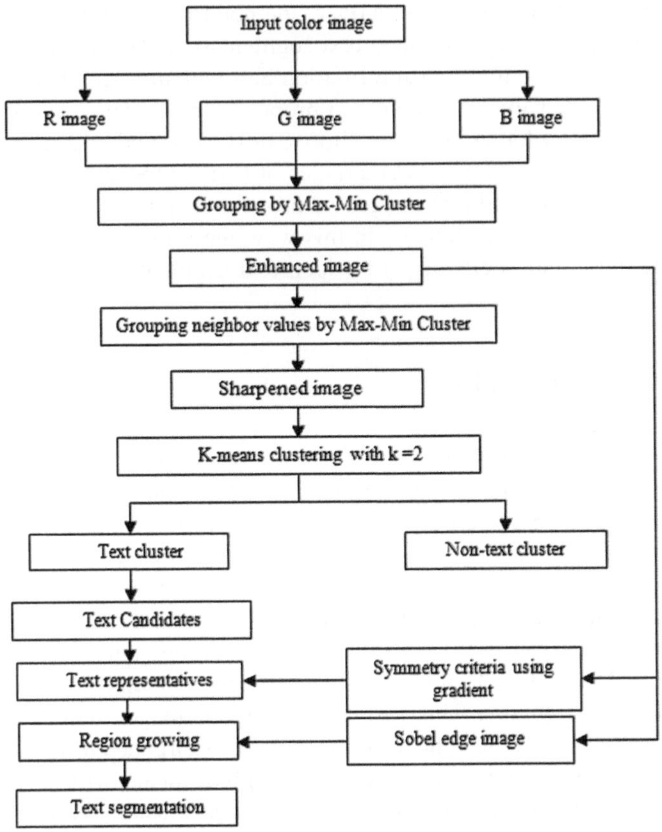


**Fig. 1** Flow diagram of the proposed method

## *2.1 Grouping Color Values for Text Enhancement*

For the color input image shown in Fig. 2a, the method obtains three color sub-band images that are R-image, G-image and B-image shown respectively in Fig. 2b–d. In order to identify the high intensity value in three sub-bands, we propose Max–Min clustering criteria which select Maximum (Max) and Minimum (Min) value from R, G, B sub-bands for each pixel. Then the third value is compared with Max and Min values to find its closest value. If the third value is close to Max value then it forms a Max cluster with Max value. The method selects maximum value in the Max cluster to replace actual pixel value. Similarly, the actual pixel will be replaced by the minimum value if the third value is close to Min value and it forms a Min cluster. In this way, the Max–Min cluster does grouping to identify the high intensity values for each pixel in the input image which results in enhanced image as shown in Fig. 2e where one can see the text pixels are brightened compare to the pixels in three sub-bands and input image.

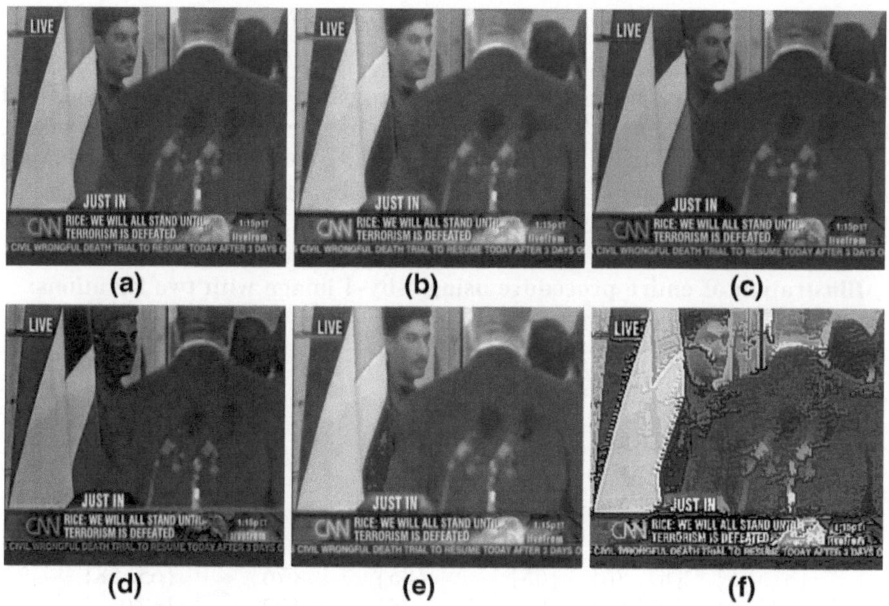

**Fig. 2** Steps to widen the gap between text and non-text pixels. **a** Input image. **b** R sub-band. **c** G sub-band. **d** B sub-band. **e** Enhanced image. **f** Sharpened image

The basis for this is that usually text pixel has high intensity value in any one of three sub-bands compared to its background.

**Illustration of entire procedure using 4-by-4 image:**

Input:

| | | | | | | | | | | | | | | |
|---|---|---|---|---|---|---|---|---|---|---|---|---|---|---|
| **R:** | 73 | 64 | 65 | 60 | **G:** | 79 | 70 | 69 | 66 | **B:** | 85 | 76 | 83 | 81 |
| | 75 | 74 | 78 | 86 | | 81 | 79 | 86 | 93 | | 89 | 82 | 87 | 97 |
| | 58 | 48 | 47 | 45 | | 65 | 55 | 57 | 55 | | 75 | 61 | 66 | 64 |
| | 43 | 71 | 48 | 40 | | 49 | 78 | 56 | 50 | | 63 | 86 | 67 | 63 |

| Maximum values: | | | | Minimum values: | | | | Third values: | | | |
|---|---|---|---|---|---|---|---|---|---|---|---|
| 85 | 76 | 83 | 81 | 73 | 64 | 65 | 60 | 79 | 70 | 69 | 66 |
| 89 | 82 | 87 | 97 | 75 | 74 | 78 | 86 | 81 | 79 | 86 | 93 |
| 75 | 61 | 66 | 64 | 58 | 48 | 47 | 45 | 65 | 55 | 57 | 55 |
| 63 | 86 | 67 | 63 | 43 | 71 | 48 | 40 | 49 | 78 | 56 | 50 |

Clusters formed upon comparisons:

| Max cluster: | | | | Min cluster: | | | |
|---|---|---|---|---|---|---|---|
| {85, 79} | {76, 70} | {83} | {81} | {73} | {64} | {69, 65} | {66, 60} |
| {89} | {82, 79} | {87, 86} | {97, 93} | {81, 75} | {74} | {78} | {86} |
| {75} | {61, 55} | { 66, 57} | {64, 55} | {65, 58} | {48} | {47} | {45} |
| {63} | {86} | {67} | {63} | {49, 43} | {78, 71} | {48, 56} | { 50, 40} |

After max-cluster and min-cluster method

| | | | |
|---|---|---|---|
| 85 | 76 | 65 | 60 |
| 75 | 82 | 87 | 97 |
| 58 | 61 | 66 | 64 |
| 43 | 71 | 48 | 40 |

## 2.2 Grouping Neighbor Values for Sharpening Text

It is true that text pixel must have high value compared to its neighbors because of high contrast of text pixels. To increase the gap between text and non-text pixels, we propose sliding window operation where we use the above process to sharpen the text pixel and to suppress the non-text pixel based on neighbor information. As a result, we get sharpened image as shown in Fig. 2f where the text pixel are still brighter than the pixel in the enhanced image shown in Fig. 2e.

**Illustration of entire procedure using 4-by-4 image with two iterations:**

Input: Modified gray image:

| | | | |
|---|---|---|---|
| 85 | 76 | 65 | 60 |
| 75 | 82 | 87 | 97 |
| 58 | 61 | 66 | 64 |
| 43 | 71 | 48 | 40 |

Max = 87 Min = 58

| Max cluster: | | | Min cluster: | | |
|---|---|---|---|---|---|
| {87, 85} | {87, 76} | {65} | {85} | {76} | {65, 58} |
| {87, 75} | {87, 82} | {87, 87} | {75} | {82} | {87} |
| {58} | {61} | {66} | {58, 58} | {61, 58} | {66, 58} |

Updated matrix after first iteration and input for next iteration:

| | | | |
|---|---|---|---|
| 87 | 87 | 58 | 60 |
| 87 | 87 | 87 | 97 |
| 58 | 58 | 58 | 64 |
| 43 | 71 | 48 | 40 |

Max = 97 Min = 58

| Max cluster: | | | Min cluster: | | |
|---|---|---|---|---|---|
| {97, 87} | {58} | {60} | {87} | {58, 58} | {60, 58} |
| {97, 87} | {97, 87} | {97, 97} | {87} | {87} | {97} |
| {58} | {58} | {64} | {58, 58} | {58, 58} | {64, 58} |

Updated matrix after iteration2 and input for next iteration:

| | | | |
|---|---|---|---|
| 87 | 97 | 58 | 58 |
| 87 | 97 | 97 | 97 |
| 58 | 58 | 58 | 58 |
| 43 | 71 | 48 | 40 |

## 2.3 Text Candidates

The above two methods presented in Sects. 2.2 and 2.3 help in widening gap between text and non-text pixels. This clue inspired us to use k-means clustering algorithm with k = 2 on sharpened image shown in Fig. 2f to separate the text

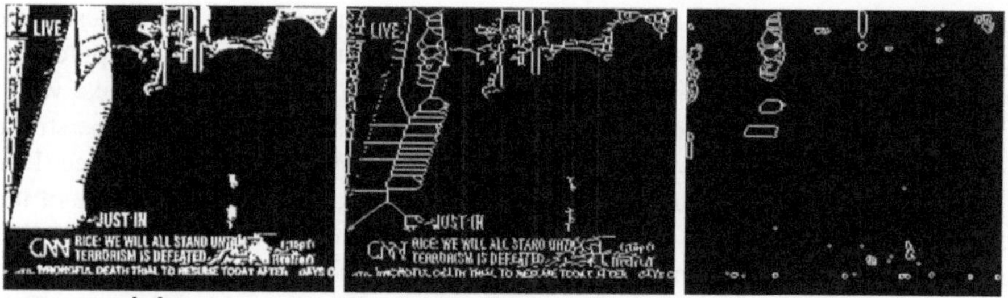

(a) (b) (c)

**Fig. 3** Steps to obtain text candidates. **a** K-means output. **b** Skeleton image. **c** Fully connected components

cluster. Since k-means clustering is unsupervised method, we consider cluster that gives high mean value compared to other mean of cluster as a text cluster. It is valid because high values will be classified into one cluster and that must be text cluster. The output of k-means can be seen in Fig. 3a where almost all text pixel are classified as text including few high contrast non-text pixel.

As it is noted that text cluster given by k-means provides binary information of both text and non-text components as shown in Fig. 3a. To reduce the pixel width to single pixel and preserve the shape of the text component, we get skeleton of the components using skeleton method as shown in Fig. 3b where due to low resolution and complex background, skeleton may not preserve the shape of the components but it gives significant information to study the characteristics of text and non-text. For the purpose of removing non-text components from the skeleton image, we test whether text components satisfies fully connected component condition or not because it is fact that text components must be connected without any disjoint compared to non-text components where most of them are disjoint. The output of performing fully connected component testing is shown in Fig. 3c where one can see most of the non-text components are removed and we call this output as text candidate image. The fully connected component is defined as starting and ending of the contour of the component should meet at one point. However, still we can see some non-text components in the results shown in Fig. 3c. We propose novel feature for text component verification in the following section.

## 2.4 Symmetry Criteria for Text Representative

We are inspired by the work presented in [10] where the stroke width concept is used for text detection and text component separation successfully. It works based on the fact that the stroke width is constant throughout the character while for non-text the stroke width distance is arbitrary. For each pixel in the fully connected component, the method computes stroke width distance that is traversing in perpendicular direction to the gradient direction of the pixel till it reaches white pixel which we call reached pixels. In this way, the stroke width is computed for each pixel in the fully connected component. Then to find dominant stroke width distance, we plot as histogram for the stroke width distances to choose the stroke width distance which gives highest peak as a dominant stroke width distance. Due to low resolution and complex background, it is hard to get complete shape of the characters. Therefore, one cannot expect constant stroke width for the character and hence we choose dominant stroke width distance for verification. For each reached pixel of the component, we obtain dominant stroke width distance. Then the method compares those two dominant stroke width distances to test the symmetry criteria. If both the distances are same then it is considered as the text component as it satisfies the symmetry criteria. This is true because according to stroke width concept presented in [10], the stroke width distance of the starting pixel and the reached pixel should be same. Note that the gradient image is obtained by performing the vertical and horizontal mask operation on the enhanced image as shown in Fig. 4a and b and the combined gradient image is shown in Fig. 4c to estimate stroke width distance for the pixels in the fully connected components. The result of symmetry verification is shown in Fig. 4d where almost all non-text components are removed and these components are called as text representatives.

## 2.5 Region Growing for Text Segmentation

The text representatives obtained by the previous section are considered as seed points to segment full text in the image. Therefore, we propose region growing method using Sobel edge map of the enhanced image as shown in Fig. 4e for the purpose of text segmentation. We believe that the symmetry verification gives at least one seed point for one text line. The contour of the seed point grows pixel by pixel till it reaches white pixel of the neighbor component along text direction in Sobel edge map of the enhanced image. This process continues till end of the text line. The end of the text line is determined based on experimental study on space between the words and characters. The region growing works based on the assumption that the space between text lines is greater than the space between the words and characters. The main advantage of the region growing is that it segments text line of any orientation because it works based on nearest neighbor concept. For example, nearest neighbor for first character in the text line would be second character and for second character, third character is the nearest neighbor. Therefore, generally the region growing segment text individual text lines but sometimes due to noisy pixel between the text lines and disconnections, the method combines two to three adjacent text lines as shown in Fig. 4f where the method segment four text line as one segment. This is the main drawback of this method. As a result, there is a scope for improving the region growing method so that it segments text lines properly.

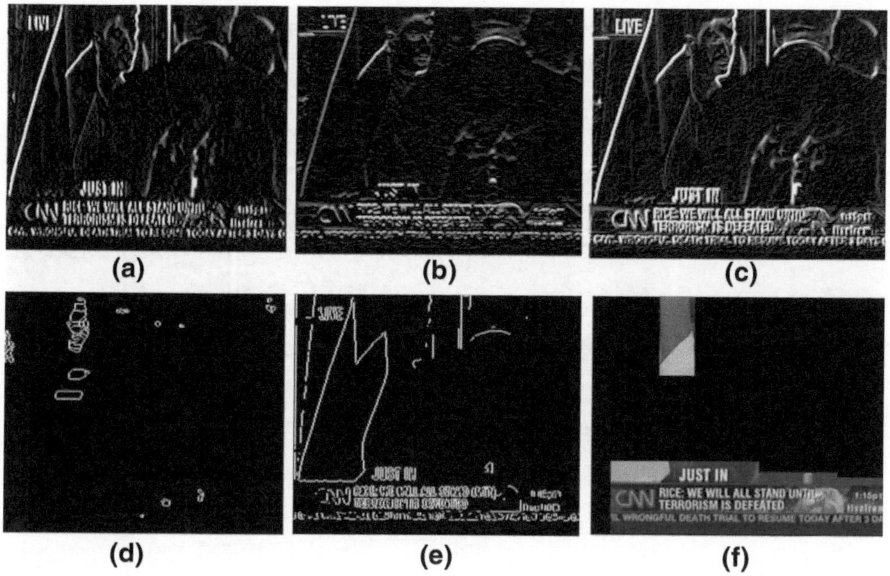


**Fig. 4** Steps for text segmentation. **a** Vertical gradient image. **b** Horizontal gradient image. **c** Combined gradient image. **d** Symmetry verification. **e** Sobel edge of Fig. 2e. **f** Text segmentation

## 3 Experimental Results

The proposed method is tested on variety of video frames to evaluate the performance of the method in terms of recall, precision and f-measure. Here we consider dataset provided by National University of Singapore (NUS) which contains video frames, extracted from news programmers, sports videos and movie clips. In this dataset, there are both graphic text and scene text of different languages, e.g. English, Chinese and Korean [20]. We consider 50 video text images to determine the recall, precision and f-measure in this work. Since we use small dataset to test the effectiveness of the proposed method, we test our method on large data by comparing with the existing methods in future.

To judge the correctness of the text blocks detected, we manually count the Actual Text Blocks (ATB) in the frames in the dataset and are considered as a ground truth. Since the main objective of the method is to segment text line in video frame, we consider each segmentation result given by the method as text block if it contains full text information else non-text block. Note that it is not in the line of text line detection in video. Based on this view, we define the following.

Truly Detected text Block (TDB): a detected block that contains text fully. Falsely Detected text Block (FDB): a detected block that does not contain text. The recall, precision and f-measures are defined as the following.

Recall (R) = TDB/ATB, Precision (P) = TDB/(TDB + FDB) and f-measure = 2RP/(R + P)

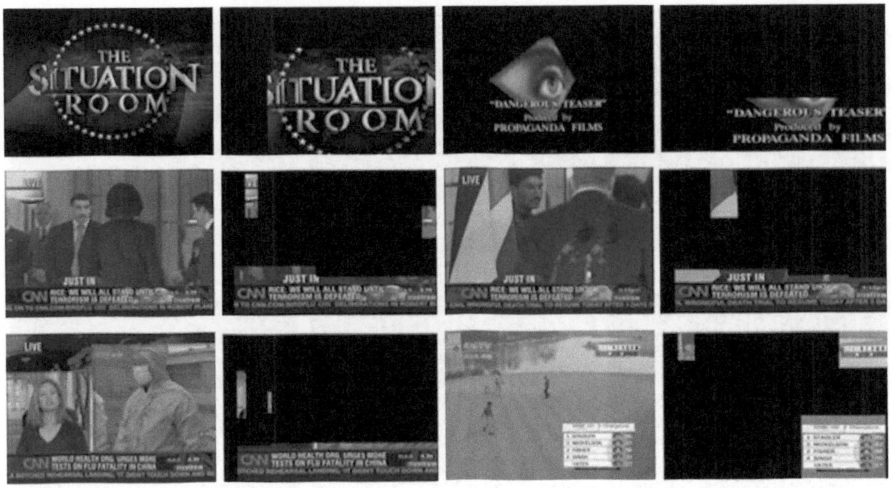

**Fig. 5** Sample results of the proposed method

**Table 1** Quantative measures of the proposed method

| Recall | Precision | F-measure |
|---|---|---|
| 0.85 | 0.84 | 0.82 |

Sample results of the proposed method is shown in Fig. 5 where first column and third column shows input images, and the second and the fourth column shows respective output images given the proposed method. It is observed from the results shown in Fig. 5 that the proposed method works well for low resolution, complex background, different fonts and font sized images. For the image in third row and first column, and third row and third column in Fig. 5, the proposed method segments text properly though the images contain complex background and small fonts. This is the advantage of the proposed method compared to the existing methods [20, 21] where generally the methods fail to segment low contrast, low resolution and small font text.

The quantative results of the proposed method is reported in Table 1 where it is noticed that the proposed method gives promising and encouraging results as f-measure is more than 80 % accuracy.

# 4 Conclusion and Future Work

In this work, we have proposed new enhancement method by exploring color value in different sub-bands. We propose Max–Min clustering method to obtain Max and Min clusters to identify the high color value of the pixel. We also use neighbor information to enhance the enhanced image further to increase the gap between text and non-text pixel based on sliding window operation. We propose new criteria that checks whether the components obtained by the k-means clustering and skeleton are fully connected component are not. This step helps in eliminating most of the non-text components. Further, we introduce new symmetry verification based on stroke width distance of the text components to eliminate the non-text components and to obtain seed points for each text line. The region growing method is proposed to segment text lines by referring Sobel edge map of the enhanced image. The experimental result shows the proposed method is good for text segmentation from video frames. It is observed from the experimental results that the method gives low accuracy because the region growing merges two lines as one line while growing. We are planning to modify the region growing to segment text lines of any orientation in future and to improve the accuracy. The robustness of the method for handling complex backgrounds, different arbitrary orientation will be further investigation as well.

## References


1. Sharma N, Pal U, Blumenstein M (2012) Recent advances in video based document processing: a review. In: Proceedings of DAS, pp 63–68
2. Zang J, Kasturi J (2008) Extraction of text objects in video documents: recent progress. In: Proceedings of DAS, pp 5–17
3. Jung K, Kim K, Jain K (2004) Text information extraction in images and video: a survey. Pattern Recogn 37:977–997
4. Doermann D, Liang J, Li J (2003) Progress in camera-based document image analysis. In: Proceedings of ICDAR, pp 606–616
5. Jung K (2001) Neural network-based text location in color images. Pattern Recogn Lett 22:1503–1515
6. Ye Q, Huang Q, Gao W, Zhao D (2005) Fast and robust text detection in images and videos frames. Image Vis Comput 23:565–576
7. Chen D, Odobez JM, Bourlard H (2004) Text detection and recognition in images and video frames. Pattern Recog 37:595–608
8. Neumann L, Matas J (2012) Real-time scene text localization and recognition. In: Proceedings of CVPR, pp 3538–3545
9. Yao C, Bai X, Liu W, Ma Y, Tu Z (2012) Detecting texts of arbitrary orientations in natural images. In: Proceedings of CVPR, pp 1083–1090
10. Epshtein B, Ofek E, Wexler Y (2010) Detecting text in natural scenes with stroke width transform. In: Proceedings of CVPR, pp 2963–2970
11. Jain AK, Yu B (1998) Automatic text location in images and video frames. Pattern Recogn 31:2055–2076 (1998)
12. Mariano VY, Kasturi R (2000) Locating uniform-colored text in video frames. In: Proceedings of ICPR, pp 539–542
13. Wu V, Manmatha V, Riseman EM (1999) Text finder: an automatic system to detect and recognize text in images. IEEE Trans PAMI 21:1224–1229
14. Kim KL, Jung K, Kim JH (2003) Texture-based approach for text detection in images using support vector machines and continuous adaptive mean shift algorithm. IEEE Trans PAMI 25:1631–1639

15. Shivakumara P, Phan TQ, Tan CL (2011) A laplacian approach to multi-oriented text detection in video. IEEE Trans PAMI 33:412–419
16. Shivakumara P, Sreedhar RP, Phan TQ, Lu S, Tan CL (2012) Multi-oriented video scene text detection through bayesian classification and boundary growing. IEEE Trans CSVT 22:1227–1235
17. Zhou J, Xu L, Xiao B, Dai R (2007) A robust system for text extraction in video. In: Proceedings of ICMV, pp 119–124
18. Lu C, Wang C, Dai R (2005) Text detection in images based on unsupervised classification of edge-based features. In: Proceedings of ICDAR pp 610–614
19. Wong EK, Chen M (2003) A new robust algorithm for video text extraction. Pattern Recogn 36:1397–1406
20. Guru DS, Manjunath S, Shivakumara P, Tan CL (2010) An eigen value based approach for text detection in video. In: Proceedings of DAS, pp 501–506
21. Basavanna M, Shivakumara P, Srivatsa SK, Hemantha Kumar G (2011) A new run-length based method for scene text detection. In: Proceedings of IICAI, pp 1730–1736

# A Review on Crop and Weed Segmentation Based on Digital Images

**D. Ashok Kumar and P. Prema**

**Abstract** Apparently weed is a major menace in crop production as it competes with crops for nutrients, moisture, space and light which resulting in poor growth and development of the crop and finally yield. Yield loss accounts for even more than 70 % when crops grown under unweeded condition with severe weed infestation. There are several weed control measures being practiced in crop production, they are physical, mechanical, biological and chemical methods. Weed Management plays vital role in agriculture and horticulture production and economic benefits derived by agricultural industry. Weed is controlled mainly by application of herbicides. Weeds are not uniformly distributed in the crop and uncropped fields and mostly they are found in patches. With the help of Color and growth parameters, the weeds and crops may not be distinguished in the fields for the reasons of imbalance in availability of nutrients, water and other environmental resources. Weed control need to be done at the early stage of the crop growth. The management of weeds with in the field is imperative. Weed management practices using chemical tools propose to apply herbicide in the dosage strictly necessary based on weed infestation and location or position. Currently research is carried out relating to identification of weed spices and the location of the weed occurrence with the aims to allow accurate weeding and apply herbicides based on the weed density. Machine vision system, remote sensing and aerial imaging techniques are used for control weeds. Sensor attached electromagnetic system, imaging spectra radiometer and spectrometer can also be used to identify weeds for effective weed control. Almost all the existing weed detection methods process the captured image by segmentation of vegetation against background (soil), detection of weed vegetation pixels. Further, classification of feature extraction of

D. Ashok Kumar
Government Arts College, Tiruchirappalli 620022 Tamil Nadu, India
e-mail: akudaiyar@yahoo.com

P. Prema (✉)
Agricultural College and Research Institute, Madurai 625104 Tamil Nadu, India
e-mail: pp76@tnau.ac.in

P. P. Swamy and D. S. Guru (eds.), *Multimedia Processing, Communication and Computing Applications*, Lecture Notes in Electrical Engineering 213,
DOI: 10.1007/978-81-322-1143-3_23, © Springer India 2013

weeds is done by color, shape and texture. The various methods studied and concepts used for crop and weed discrimination by the various researchers are discussed in this paper.

**Keywords** Digital image segmentation · Crop and weed · Weed detection · Feature extraction

## 1 Introduction

Image segmentation in general is defined as a process of partitioning an image into homogenous groups such that each region is homogenous but the union of no two adjacent regions is homogenous. Efficient image segmentation is one of the most critical tasks in automatic image processing [1]. Image segmentation has been interpreted differently for different applications. In agriculture, computer machine system were used for automatic identification of crop and weed, disease and pest. The motivation for this work is to discover the effective and efficient weed discrimination method for site specific herbicide application. This paper organized as image segmentation techniques and limits its analysis to crop and weed image segmentation.

In agriculture, researchers and farmers have recognized that weed compete with crop for water, sunlight, nutrients and space. Controlling of weed is a critical farm operation and can significantly affect the crop yield. Herbicides applications have vital importance in weed control and high crop yield. Weeds are not uniformly distributed in the field, they clumped together in patches [2] herbicides are applied with the same dose to the whole field, representing significant portion of the variable cost of agricultural production. In these days, there is a clear tendency of reducing the use of chemical agents in agricultural cultivations. The main goals of computer vision techniques developed towards this objective try to obtain products of a better quality and the saving of costs related to the crop field treatments. This tendency has been established over the recent years among various countries, creating a growing interest. The development of computer vision capabilities allows a reliable and fast identification and classification of weeds [3]. Automatic systems nowadays provide techniques to easily process the crop fields to obtain the necessary data to classify and to distinguish the crop from the weeds [4]. The developments of these systems are mainly based on the computation of geometrical characteristics of the weeds (shape factor, aspect ratio, length/ratio, etc.) [5]. Machine vision system to analyse the weed image based on the digital image captured by the imaging device and process the extracted information for decision-making to identify weed and crop and then apply the herbicide in the selected position [6]. Thus the image recognition task is important for crop and weed segmentation. Research work in this area is difficult to classify and compare due to the variations among different crop and weed species and to the different approaches taken to collect field data.

Almost all existing weed detection methods process the image based on the following steps [7] such as segmentation of vegetation against the background (soil and/or harvest residues), Detection of the vegetation pixels that represent weeds and feature extraction and classification. Efficient and automatic segmentation of vegetation from images of the ground is an important step for many applications such as weed detection for site specific treatment. With the numerous recent developments of new segmentation methodologies, the requirement of their categorizations based on successful applications have become essential for real time precision herbicide applicator. This paper contains categories the color space, image segmentation method and review of crop and weed segmentation.

## 2 Color Space

A color space is defined as a means by which the specification, creation and visualization of colors is performed. Color is perceived by humans as a combination of tristimulus R(red), G(green) and B(blue) which are usually called three primary color. From R,G,B representation, we can derive other kinds of color representation (spaces) by using either linear or nonlinear transformation. Several color spaces, such as RGB, Normalized RGB, HSI, CIE XYZ, CIE YUV, CIE L*a*b*, YCbCr, HSV are utilized for color image segmentation, but none of them can dominate the others for all kinds of color image. Selecting the best color space still is one of the difficulties in color image processing [8]. HSL (Hue Saturation and Lightness) represents a wealth of similar colour spaces, alternative names include HIS (intensity), HSV (value), HCI (chroma/colourfulness), HVC, HSD (hue saturation and darkness) etc. The separation of the luminance component from chrominance (colour) information is stated to have advantages in applications such as image processing. The YCbCr colour spaces separate RGB into luminance and chrominance information and are useful in compression applications (both digital and analogue). The CIE space of visible color is expressed in several common forms: CIE xyz, CIE L*a*b*, and CIE Lu'v'. CIE xyz is based on a direct graph of the signals from each of the three types of color sensors in the human eye. These are also referred to as the X, Y and Z tristimulus functions. CIE Lu'v' was created to correct for the CIE xyz distortion by distributing colors roughly proportional to their perceived color difference. A region that is twice as large in u'v' will therefore also appear to have twice the color diversity making it far more useful for visualizing and comparing different color spaces. CIE L*a*b* remaps the visible colors so that they extend equally on two axes conveniently filling a square. Each axis in the L*a*b* color space also represents an easily recognizable property of color, such as the red-green and blue-yellow shifts. These traits make L*a*b* as useful color space for editing digital images. However, almost all existing weed detection methods process the image in two steps: (1) segmentation of vegetation against the background (soil and/or harvest residues) and (2) detection of the vegetation pixels that represent weeds. The procedures for the segmentation of vegetation usually assume that all pixels belonging to vegetation can be easily extracted by some combination of the color planes on the RGB model [9]. HSI color model resolve the problem of under segmentation [10]. Other approaches propose the use of the HIS color model combined with classification methods such as Bayes networks and clustering [11]. To discriminate vegetation pixels, a linear combination of the RGB planes with coefficients (r = −0.884, g = 1.262, b = −0.311) and mean pixel intensity thresholding approaches were used. Extract the greenness by combining Normalized RGB indices computation methods such as ExG, CIVE, ExGR and VEG based on the uniformity of the corresponding histograms [12]. Offset excess green (OEG) combined with Non Green Subtraction (NGS) algorithm address the over segmentation problem and accurately segment vegetation under different illumination condition [13].

## 3 Image Segmentation

Computer vision is a rapidly expanding area that is dependent on the capability to automatically segment, classify and interpret images. Segmentation is central to the successful extraction of image features and their subsequent classification. Image segmentation techniques can be grouped into six categories [14] amplitude thresholding, component labeling, boundary based segmentation, region based segmentation, template matching and texture segmentation. During segmentation, an image is preprocessed, which can involve restoration, enhancement, or simply representation of the data [8]. Certain features are extracted to segment the image into its key components. The segmented image is routed to a classifier or an image-understanding system. The image classification process maps different regions or segments. Each object is identified by a label. The image understanding system then determines the relationships between different objects in a scene to provide a complete scene description. This session discusses the categorization of image segmentation algorithms.

Amplitude thresholding, or window slicing, is useful whenever an object is sufficiently characterized by the amplitude features. Component labelling is a simple and effective method of segmenting binary images by examining the connectivity of pixels with their neighbours and then labelling the connected sets. Boundary extraction techniques segment objects on the basis of their profiles. Therefore, such techniques as contour following, connectivity, edge linking, graph searching, curve fitting, Hough transform, and others are applicable to image segmentation. Region-based segmentation techniques are primarily used to identify various regions with similar features in one image. Region-based approaches [15] are generally less sensitive to noise than the boundary based methods. Many region based segmentation techniques are available, including region-growing and merging, relaxation labelling, symmetric nearest neighbor, hierarchical segmentation, and shadow boundary segmentation, several well-known image processing techniques are offered in the context of region-based segmentation, such as clustering, pattern recognition, edge-detection, noise reduction, and three-dimensional object recognition. Clustering refers to a class of algorithms used extensively for image segmentation. Clustering assembles unlabelled data by sets. Data point values represent characteristic features of interest such as grayscale, color brightness, contrast etc. During the cluster operation, the clusters are assigned labels that arc mapped back into the image, so that the original pixel values are replaced. The basic clustering operation examines each pixel individually and assigns it to the cluster that best represents the value of its characteristic vector. This assignment is done according to the selected measure of similarity between the data point and the criterion function that measures clustering quality. The process is repeated until some condition is satisfied by the current grouping of data points. Texture segmentation becomes important when objects in a scene have a textured background [16]. Since texture often contains a high density of edges. Clustering and region-based approaches applied to textured features can be used to segment textured regions. In general, texture segmentation and classification is a complicated problem. Use of a priori knowledge about the existence and kinds of textures that may be present in a scene can be beneficial when applied to practical problems.

## 4 Categorisation Based on Homogeneity Measure

Next stage of categorization corresponds to the homogeneity measures used for image segmentation. The primary homogeneity measure is spectral/tonal feature. Secondary homogeneity measures are spatial, texture, shape and size. Tertiary homogeneity measures are contextual, temporal and prior knowledge [17]. The most primitive measures of homogeneity are spectral and textural features. Texture features points to spatial pattern represented by spectral values [18]. A textured image may have various texture patterns. However, quantitatively characterizing texture is not simple [17]. Due to this fact texture segmentation has been studied widely in combination with other features like shape, spectral and contextual and various models till today. Texture segmentation is mostly used after segmentation technique. This is mainly because of the presence of highly textured regions in high resolution satellite imagery. Currently, the research has shifted from texture to multiresolution model. The importance of shape and size measure could be understood when the natural object are to be identified. The state of art use of shape and size refers to multi-scale/multiresolution approach to image segmentation. Shape and size measures are especially helpful when delineating complex objects in high resolution satellite imagery. Prior knowledge refers to photo interpreter knowledge regarding the regions/objects of the image [17]. It may be the knowledge of classes of the image region or about some specific area, building or trends etc. Incorporating prior knowledge in image analysis is one steps towards developing artificial intelligence in the machine [19]. Prior knowledge is specifically useful when for segmentation of complex landscape object indistinguishable using texture and context.

## 5 Literature Review for Crop and Weed Segmentation

In computer vision, segmentation is a process by which an image is partitioned into multiple regions (pixel clusters). The aim of segmentation is to obtain a new image in which it is easy to detect regions of interest, localize objects, or determine characteristic features such as edges. The image obtained by the segmentation process is a collection of disjoint regions covering the entire image whereby all the pixels of a particular region share some characteristic such as color, intensity, or texture. Lists of models generally used for image segmentation are Object Background/Threshold Model, Neural Model, Fuzzy Model, Multi-resolution and Wavelet model.

### *5.1 Thresholding and Neural Network Based Approaches*

Artificial Neural Networks (ANN) are widely applied for pattern recognition. Their extended parallel processing capability and nonlinear characteristics are used for classification and clustering. In crop and weed discrimination, the acquired input image are preprocessed using filters. The processed color images were converted into grey level images and later binary images were created for easier identification of weed [5]. Image segmentation was initially focused to detect weed seedlings based on geometrical measurements such as shape factor, aspect ratio, and length/area [5, 20]. In traditional plant taxonomies plants are identified based on shape features, colour, texture, etc. Although there are methods to identify individual shapes, the major challenge is to separate one green leaf from another green leaf. It becomes more difficult when two different shaped leaves overlap.In crop and weed detection, problem also arises when individual leaves have similar boundary characteristics and it becomes difficult to define and perform subsequent shape analysis. Various studies conducted on image based identification of weed and various classification features are listed in Table 1. Colour images were successfully used to detect weeds, seeds and other types of pests [21].

To identify and detect weeds and crop plants under uncontrolled outdoor illuminations condition normalized excessive green conversion, statistical threshold value estimation, adaptive image segmentation, median filter was applied to segmented images to eliminate random noise. Morphological features of plants and Artificial Neural Network (ANN) techniques are used for better classification [20]. RGB planes with coefficients (r = −0.884, g = 1.262, b = −0.311) and mean pixel intensity thresholding and Robust Crop Row Detection system [22] successfully detects an average of 95 % of weeds and 80 % of crops under different

**Table 1** Studies on vegetation detection using imaging techniques

| Crop/weed | Features | | | Statistical methods | Reference |
|---|---|---|---|---|---|
| | Shape | Color | Texture | | |
| Blueberry and weed | – | Excess green | – | Statistical and spatial frequency Hough transform | Fangming et.al. [38] |
| Crop and weed | Height | Excess green | Wavelet statistical features, second-order statistical features | Multi resolution combined statistical and spatial frequency | Sabeenian and Palanisamy [34] |
| Barley and wild oat | | RGB Planes with coefficient(r=-0.884, g=1.262,b=-0.311) | – | Mean percentage of crop an weed pixels | Xavier et al. [22] |
| Canola and narrow leaf weed | Area, Perimeter, eccentricity, circularity | Excess green | Radial spectral energy | Fourier and Bayesian classifier | Mathanker et al. [37] |
| Cabbage, carrot and weed | Area, eccentricity, convexity, roundness | Hue, saturation, intensity | – | Fuzzy logic | Hemming and Rath [11] |

illumination, soil humidity and weed/crop growth conditions. For the effective classification of crops and weeds in digital images [5] Otsu thresholding, morphological methods and feature space as colour features, size dependent object descriptors, size independent shape features and moment invariants are used in support vector machine (SVM).

Weed species retard the growth of the crop and reduce farm yields. To control the growth of weed species, a large number of herbicides are used in agriculture fields. Discrimination between corn seedlings and weeds is an important and necessary step to implement spatially variable herbicides application [23]. Otsu's threshold was applied to segment weeds images based on the modified excess green feature, it could distinguish the plant objects from the background effectively. The probabilistic neural network classifier was created for recognition of corn seedlings and weeds according to the shape features. Comparing the probabilistic neural network (PNN) method with the back propagation neural network, the BP method is better than the PNN seeing from the experimental results.

Weeds are general green color, a highly irregular leaf shape and varying surface texture, and an open plant structure which contributes to its being a challenging task to identify weeds in the field [24]. The Anisotropic Diffusion Based Weed Classifier is based on anisotropic diffusion also called Perona-Malik diffusion. This classifier classifies the images in four categories i.e. Broad Leaf, Narrow Leaf, Low Weed and Mixed. The Anisotropic diffusion enhance the image by considering the local structures in the images to filter noise, preserve edges and significantly increasing the signal-to noise ratio (SNR) with no major quantitative distortions of the signal. Since information about weed numbers in unit area and average weed size (age) could be used to make the decision to skip some low weed density control zones or to decide between multiple application rates for different weed infestation levels. This algorithm give is 95 % accuracy in classification of different leaf textures.

Several methods have been implemented for accurate weed detection: spectral reflectance of plants with artificial neural networks [25], or statistical analysis [6, 26] such as Principal Component Analysis. Gray Level Co-occurrence Matrix (GLCM) [27], statistical properties of the histogram, texture features and Support Vector Machine (SVM). Other researchers have investigated texture features [28] or biological morphology such as leaf shape recognition [27]. However, for use in real-time, there have been fast methods are implemented to identify crop rows in images [29]. Most are based on Hough transform [30, 31], Kalman filtering [32] and linear regression [21]. Moreover, Hough transform is usually implemented for automatic guidance in crop fields [33]. Consequently, there are now various vision systems available on autonomous weed control robots for mechanical weed removal.

## 5.2 Wavelet Based Approach

Wavelet transform is the best trade-off to represent both time and frequency content of a signal. There are a number of ways to separate the low (smooth variations in colour) and the high frequency components (the edges which give details). One way is decomposition of the image using the discrete dyadic wavelet transform. A multiresolution analysis (MRA) based on the well known Mallat algorithm gives image details at various scales. With MRA and separable wavelet basis functions, we can extract the details contained in various parts of the image from different levels of resolution. A new method based on Gabor wavelets (GW), Lie group structure of region covariance (LRC) representation and texture characteristics of the weed image at different directions and scales was applied for classification of broadleaf weed images on Riemannian manifolds [15]. Multiresolution Combined Statistical and Spatial Frequency (MRCSF) and texture features [34] are used to classify weed images as broad and narrow weed. Bossu et al. [35] tested and validating the accuracy of four wavelet algorithms (Daubechies 25, Symlet, Coiflet, Biorthogonal, Reverse Biorthogonal Meyer) for crop/weed discrimination in synthetic and real images. The accuracy of these algorithms for different Weed Infestation Rates (WIR) was compared to Gabor filtering, which is currently implemented for real-time site-specific sprayer from vision system [36]. The best results were with Daubechies and discrete approximation Meyer wavelets. They provided better results than Gabor filtering not only for crop/weed classification but also in processing time.

The crop and weed segmentation techniques are summarized in Table 2 based on the above literature. The literature has revealed that researchers have followed many different ways for weed detection and decision-making but the most common steps include image acquisition. After acquiring an image, it is processed for band separation, or an excess green image is generated and removing blobs, holes, shades, etc. from the image. The processed image is then converted into binary image using threshold. Various methods of such thresholding have been utilised by different researchers. Based upon threshold values regions of similar values is segmented and each segment is treated as a region of interest (ROI). Once the ROI is defined, various geometrical measurements are obtained such as height, length, minimum bounding rectangle, major and minor axis, perimeter, area, textural parameter such as energy, entropy, contrast, moments, etc. These features can be distinctive to a particular species of crop or weed and can be utilised for differentiating between two species of the crops or crop and weed. These features could also be utilised for further analysis.

**Table 2** Crop and weed segmentation techniques

| Crop | Color space | Segmentation techniques | Features | Classification method | Reference |
|---|---|---|---|---|---|
| Corn and weed images | Excessive green | Adaptive image segmentation | Morphological features | Artificial neural network | Hong et al. [20] |
| Maize | RGB planes with coefficients | Fast image processing, robust crop row detection | Mean | support vector machine (SVM), Bayesian classifier | Xavier et al. [22] |
| Chilli, pigweed | Excessive green | Global thresholding | Colour features, size dependent object descriptors, size independent shape features and moment invariants | Support vector machine | Ahmed [5] |
| Cereals | RGB | Thresholding and hough transform lines detection | Shape and size features | Support vector machine | Tellaechea [3] |
| Cotton | Excessive green | Thresholding | Size, shape features | Statistics of time-consuming | Yin Donu 2011 |
| Paddy sugarcane, sunflower, onion and tomato | Excessive green | Wavelet based MRCS and spatial frequency | Texture features | Statistics features | Sabeenian and Palanisamy [34] |
| Corn | Excessive green | Gray level co-occurrence matrix | Texture features | Support vector machine, back-propagation (BP) neural network | Wu [27] |
| Crop and weed | Excessive green | Wavelet daubechies 25, symlet, coiflet, biorthogonal reverse | Color | Confusion matrix | Bossu [35] |
| Corn | Excess green | Color segmentation and thresholding | Shape features | Probabilistic neural networks | Chen [23] |

## 6 Conclusion and Future Work

It could be concluded from the above study that various color space methods are used for foreground and background extraction. With the numerous amounts of crop and weed segmentation techniques presented above. The color space RGB planes with coefficients (r = −0.884, g = 1.262, b = −0.311), mean pixel intensity thresholding and Robust Crop Row Detection system successfully detects an average of 95 % of weeds and 80 % of crops under different illumination, soil humidity and weed/crop growth conditions. For real time precision herbicide application, Offset excess green and Non-Green Subtraction algorithm address the over segmentation problem. This method is detect weeds under different lighting conditions and it is suitable for using in real-time application. Other vegetative methods HSI, YCbCr methods, thresholding techniques, median filter, morphological operation are used for vegetative segmentation. Homogeneity measures are spectral, spatial, texture, shape, size, contextual, temporal and prior knowledge used for crop and weed feature extraction. The widely applied homogeneity measure is based on color and texture. Wavelet segmentation algorithm is currently used crop and weed discrimination. Wavelet and texture segmentation is more successful because it inherits spectral and spatial properties in itself. The selection of segmentation approach depends on what quality of segmentation is required. Further, it also depends on what scale of information is required. For qualitative and quantitative comparison confusion matrix, discriminate analysis, neural network, Bayesian classifier and support machine vector are used. For classification of crop and weed neural network approach gives good result. Having gone through the techniques above, still it need further improvement for effective method of weed segmentation, it could be achieved by involving hybrid techniques (a combination of two or more of the segmentation techniques) in future course of work.

## References


1. Haralick RM, Shapiro LG (1985) Image segmentation techniques. Comput Vis, Graph Image Process 20(2):100–132
2. Dieleman JA, Mortensen DA, Buhler DD, Cambardella CA, Moorman TB (2000) Identifying associations among site properties and weed species abundance. I. Multivariate analysis. Weed Sci 48(5):567–575
3. Tellaechea A, Pajaresa G, Xavier P, Burgos-Artizzu RA (2011) A computer vision approach for weeds identification through support vector machines. J Appl Soft Comput 11(1):908–915
4. Hague T, Tillet N, Wheeler H (2006) Automated crop and weed monitoring in widely spaced cereals. Precision Agric 1:95–113
5. Ahmed F, Hossain ASM, Bari A, Hossain E, Al-Mamun HA, Kwan P (2011) Performance analysis of support vector machine and bayesian classifier for crop and weed classification from digital images. World Appl Sci J 12(4):432–440
6. Vrindts E (2000) Automatic recognition of weeds with optical techniques as a basis for site specific spraying. Unpublished Ph.D. Thesis. Katholieke University Leuven. Belgium, p 146

7. Karan Singh KN, Agrawal G, Bora C (2011) Advanced techniques for weed and crop identification for site specific weed management. Bio Syst Eng 109:53–64
8. Cheng HD, Jiang XH, Sun Y, Wang JL (2000) Color image segmentation: advances and prospects
9. Woebbecker DM, Meyer GE, Von Bargen K, Mortensen DA (1995) Shape features for identifying young weeds using image analysis. Trans Am Soc Agric Eng 38(1):271–281
10. Zheng L, Zhang J, Wang Q (2009) Mean-shift-based color segmentation of images containing green vegetation. Comput Electron Agric 65:93–98
11. Hemming J, Rath T (2001) Computer vision-based weed identification under field conditions using controlled lighting. J Agric Eng Res 78(3):233–243
12. Guijarro M, Pajares G, Riomoros I, Herrera PJ, Burgos XP, Ribeiro A (2011) Automatic segmentation of relevant textures in agricultural images. Comput Electron Agric 75:75–83
13. Muangkasem A, Thainmitt S (2012) A real-time precision herbicide applicator over between-row of sugarcane field. International conference on emerging trends of computer and information technology
14. Shaw KB, Lohrenz MC (1992) A survey of digital image segmentation algorithms
15. Chen Y, Lin P, He Y, Xu Z (2011) Classification of broadleaf weed images using Gabor wavelets and Lie group structure of region covariance on Riemannian manifolds. BioSyst Eng 109:220–227
16. Meyer G, Metha T, Kocher M, Mortensen D, Samal A (1998) Textural imaging and discriminant analysis for distinguishing weeds for spot spraying. Trans ASAE 41(4):1189–1197
17. Richards JA, Jia X (2006) Remote sensing digital image analysis: an introduction. Springer, New York pp 67–68, 128–130, 342–352
18. Haralick RM, Shanmugam K, Dinstein I (1973) Textural features for image classification. IEEE Trans Syst, Man, Cybern 3(6):610–621
19. Srinivasan A, Richards JA (1993) Analysis of GIS spatial data using knowledge based methods. Int J Geogr Inf Syst 7(6):479–500
20. Jeon HY, Tian LF, Zhu H (2011) Robust crop and weed segmentation under uncontrolled outdoor illumination. Sensors 11:6270–6283
21. Søgaard HT, Olsen HJ (1999) Crop row detection for cereal grain. In: Stafford JV (ed) Proceedings of the second European conference on precision agriculture. Odense, Danemark, pp 181–190
22. Burgos-Artizzu XP, Ribeiro A, Guijarro M, Pajares G (2011) Real-time image processing for crop/weed discrimination in maize fields. Comput Electron Agric 75:337–346
23. Chen L, Zhang JG, Su HF, Guo W (2010) Weed identification method based on probabilistic neural network in the corn seedlings field. In: IEEE Proceedings of the ninth international conference on machine learning and cybernetics July 2010, vol 11–14. Qingdao, pp 1528–1531
24. Khan SA, AM Naeem (2010) Anisotropic diffusion based weed classifier. IEEE international conference on educational and network technology. pp 11–15
25. Moshou D, Vrindts E, De Ketelaere B, De Baerdemaeker J, Ramon H (2001) A neural network based plant classifier. Comput Electron Agric 31(1):5–16
26. Vrindts E, De Baerdemaeker J, Ramon H (2002) Weed detection using canopy reflection. Precis Agric 3(1):63–80
27. Wu L, Wen Y (2009) Weed/corn seedling recognition by support vector machine using texture features. Afr J Agric Res 4(9):840–846
28. Meyer GE, Hindman TW, Lakshmi K (1998) Machine vision detection parameters for plant species identification. SPIE, Bellingham
29. Fontaine V, Crowe TG (2006) Development of line-detection algorithms for local positioning in densely seeded crops. Can Biosyst Eng 48(7):19–29
30. Jones G, Gee C, Truchetet F (2009) Modelling agronomic images for weed detection and comparison of crop/weed discrimination algorithm performance. Precis Agric 10:1–15

31. Leemans V, Destain MF (2006) Application of the Hough transform for seed row location using machine vision. Biosyst Eng 94(3):325–336
32. Hague T, Tillet ND (2001) A bandpass filter-based approach to crop row location and tracking. Mechatronics pp 1–12
33. Marchant J (1996) Tracking of row structure in three crops using image analysis. Comput Electron Agric 15:161–179
34. Sabeenian RS, Palanisamy V (2010) Crop and weed discrimination in agricultural field using MRCSF. Int J Sig Imaging Syst Eng 3(1):61–67
35. Bossua J, Géea C, Jones G, Truchetet F (2009) Wavelet transform to discriminate between crop and weed in perspective agronomic images. Comput Electron Agricu 65:133–143
36. Søgaard HT, Olsen HJ (2003) Determination of crop rows by image analysis without segmentation. Comput Electron Agric 38:141–158
37. Mathanker SK, Weckler PR, Taylor RK (2008) Canola weed identification for machine vision based patch spraying. ASABE annual international meeting, RI, USA, Paper No.085134, June 29–July 2
38. Fangming Z, Qamar Z, Arnold DCP, Dainis NWS, Travis E (2009) Detecting weeds in wild blueberry field based on color images. ASABE Annual International Meeting, Reno, NV, USA. Paper No. 096146, June 21–June 24

# Classification and Decoding of Barcodes: An Image Processing Approach

**R. Dinesh, R. G. Kiran and M. Veena**

**Abstract** Barcodes are being widely used in many fields of applications of great commercial value, which provide a means of representing data in machine readable format. Various symbologies exist to map the data into barcodes. Image based barcode readers provide many advantages over laser scanners in terms of orientation independence, image archiving and high read rate performance even when barcodes are damaged, distorted, blurred, scratched, low-height and low-contrast. The availability of imaging technology provides a platform for decoding barcode rather than the use of the conventional laser scanner which is lack of mobility. In this paper, image based technique for classification of the given 1-D barcode into respective symbology and its decoding have been proposed. The proposed method first localizes the barcodes and subsequently, a classifier which classifies the given 1-D barcode into respective symbology is applied. Decoding is then performed based on the specification of the symbology. To establish the superiority of the proposed approach in classification and decoding of barcodes, we have conducted extensive experiments on various datasets, both standard as well the images captured from low resolution cameras (mobile camera). The results reveal the superiority of the proposed method in terms of better read rate for standard and even for blurred barcode images.

**Keywords** Barcode · Image acquisition · Localization · Tree classifier · Decoding standards

R. Dinesh (✉) · R. G. Kiran · M. Veena
HCL Technologies, Bangalore 560068, Karnataka, India
e-mail: dinesh.ramegowda@hcl.com

R. G. Kiran
e-mail: kiran.gachchi@hcl.com

M. Veena
e-mail: veena.m@hcl.com

P. P. Swamy and D. S. Guru (eds.), *Multimedia Processing, Communication and Computing Applications*, Lecture Notes in Electrical Engineering 213,
DOI: 10.1007/978-81-322-1143-3_24, © Springer India 2013

# 1 Introduction

A barcode is an optical machine-readable representation of data relating to the object to which it is attached. There are different types of barcodes, often referred to as barcode symbologies [1], but all have the common purpose of encoding a string of alphanumeric information as a set of bars and spaces of varying widths printed on a product. Barcodes can be one dimensional or two dimensional. 1-D barcodes are referred as linear barcodes as they are made up of a collection of bars and spaces frequently known as *elements* or *modules*. In these barcodes, the height of the barcode provides added redundancy to the system if part of the symbol becomes damaged or occluded. Processing the barcode to find the data encoded involve several stages.

- Barcode localization: Searching the image to find the barcode containing region
- Decoding (Reading): Extracting the information encoded in the barcode.

# 2 Related Work

Several methods have been proposed for localization and decoding the barcodes. Alexander [2] used DCT (Discrete Cosine Transform) properties to distinguish bar code from other texture. The precondition is that a bar code has to occupy at least 10 % of the whole image. Because the weighting matrix coefficients are not determined self-adaptively, the robustness of the result is not as good as desired. This approach also brings a serious disadvantage of whole image being processed more than once. Hence hinders its use in real time applications. Jain and Chen [3] suggested the barcode localization using multi-channel Gabor filter.

Muniz, Junco and Otero [4] investigated using the Hough transform for locating barcodes. It leads to a computationally expensive and time consuming approach since Hough transform takes more time to detect lines in the image. Sheng et al. [5] proposed a new method for barcode localization and recognition to overcome the disadvantage of using Hough transform. Initially, Sobel edge detector is applied to remove irrelevant image background and preserve barcode edges. Later, angle of fitting lines is computed by using the turning points. The area of lines with the same or similar angles is considered the area of bar code. If the angle is not 90 degrees with respect to the horizontal direction, rectify the bar code by bilinear interpolation. Juett and Xiaojun Qi [6] proposed non-traditional localization algorithm using a bottom hat filter. In this approach, directional opening was performed at different orientation followed by density region analysis. All these approaches mainly focus on localization of barcode rather than decoding and add more overhead in terms of computations. Hence they are time consuming. To overcome the limitation of existing approaches, we use a simple approach for localization based on edge detection and morphological closing. Chai and Hock

[7] proposed a vision based technique for locating and decoding EAN13 barcodes captured from digital cameras. A block based approach is used to localize barcodes. An image is divided in to 32*32 blocks and finds each block's angle. Blocks with the same angles are selected to form a bar code area. Decoding is performed by obtaining bar widths. Only EAN13 barcodes were considered. Liyanage [8] described an edge detection based method to classify and decode only a subset of symbologies such as EAN13 and Code39. Localization of barcode in an image is performed manually. The barcode classifier used in this method fails for other symbologies. Wachenfeld et al. [9] used an image analysis and pattern recognition methods which rely on knowledge about structure and appearance of 1D barcodes. Symbologies such as UPC-A/EAN-13/ISBN-13 were considered for decoding using this approach.

The existing methods quoted for decoding the barcodes can decode only few symbologies. In this paper we present classification and decoding of most of the existing 1-D barcode symbologies. We first perform the localization of the barcode in the input image i.e. to find the location of the four corners of the barcode. And classify the localized barcode into respective symbology. Finally, decoding is then performed based on the specification of the symbology.

# 3 Proposed Methodology

This section presents the proposed method for Barcode localization and decoding of 1-D barcodes. The architecture of the proposed method for barcode localization, classification and decoding is given in Fig. 1. In the initial stage given image is processed to locate the region containing the barcodes. Subsequently a classifier is applied to find the symbology of the localized barcode. Later, decoding is performed according to the specification of its respective symbology.

## *3.1 Barcode Structure*

A typical 1-D barcode image is shown in Fig. 2. It consists of following parts: a quiet zone, a start character, data characters, an optional check digit, a stop character, and another quiet zone.

Quiet zones are non-printed zones immediately before and after the barcode. It is recommended that the quiet zone be at least 10 times the narrow bar width. A quiet zone smaller than this, can make the barcode unreadable. Start/Stop character is a pattern of bars and spaces that provide the scanner with start and stop reading instructions. An optional check digit included within a barcode whose value is used to perform a mathematical check that ensures the accuracy of the read. It is placed immediately after the barcode data. Length of label includes left and right quiet zones. A read cannot be made if the quiet zones are not large enough. The

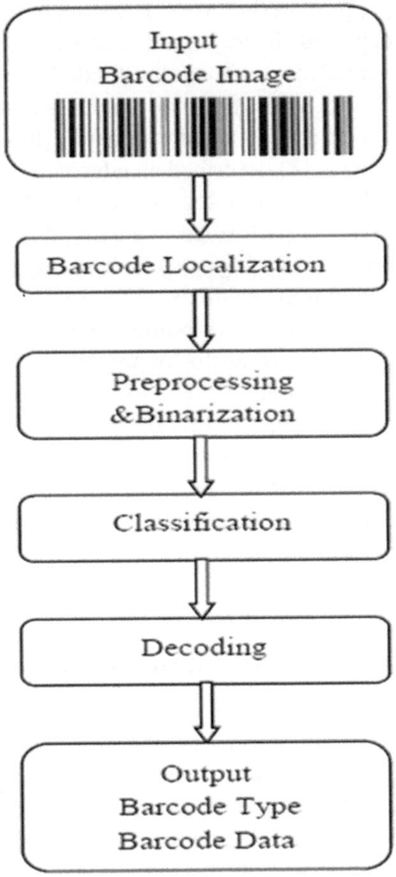


**Fig. 1** Stages of proposed methodology

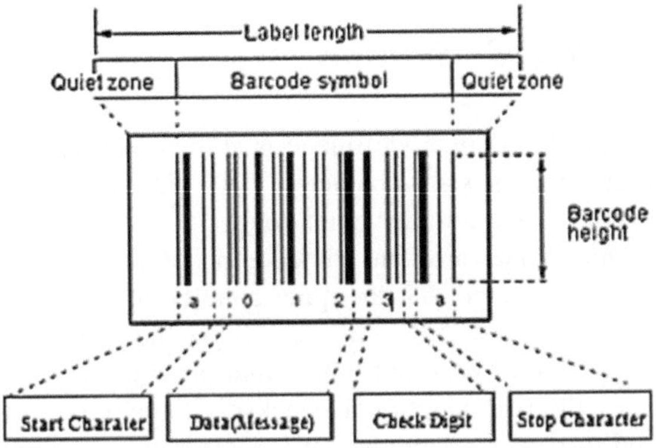


**Fig. 2** Typical 1-D barcode and its parts

height of the barcode provides added redundancy to the system if part of the symbol becomes damaged or occluded.

## 3.2 Barcode Localization

Barcode location within the given image is obtained by performing Barcode localization. Figure 3 shows an image containing the barcode. Initially edge detection is applied over the image to remove irrelevant background. First order gradient operator such as Sobel, shown in Fig. 4 is used to obtain the edges. Barcode area is retained since it is made up of black and white bars. The transitions from black bar to white bar or vice versa results in an edge in the image. The result of edge detection after applying the Sobel masks is shown in Fig. 5.

In the next stage, morphological closing operation is performed on the resultant image of edge detection. Closing expands the white regions in the image without altering the regions that are already white. A rectangular structuring element used for closing operation needs to be as wide as widest bar in the image. (In this work we have used the structuring element of size 15 × 10). The result of closing operation is shown in Fig. 6. This effectively highlights the barcode region. Other regions containing the dark on light patterns are also highlighted, which will be removed in the later stages.

The regions obtained by closing operation are then labeled using connected component labeling. Since closing operation retains other regions along with the barcode, such regions are removed in this stage. For each labeled region, the

**Fig. 3** Image containing the barcode

**Fig. 4** Sobel edge detection masks

| | | |
|---|---|---|
| -1 | -2 | -1 |
| 0 | 0 | 0 |
| 1 | 2 | 1 |

| | | |
|---|---|---|
| -1 | 0 | 1 |
| -2 | 0 | 2 |
| -1 | 0 | 1 |

**Fig. 5** Result of edge detection

**Fig. 6** Result of closing operation

boundary coordinates in *X* and *Y* directions are determined. From these coordinates width and height of each region is calculated. Since the barcode region appears rectangular or square in most of the images, other non-rectangular regions are removed by thresholding based on width and height. Rectangular region are thus retained. The resultant image obtained after this stage is shown in Fig. 7.

The resultant image may contain non barcode rectangular regions. To locate exact barcode region, the intensities are projected on to a horizontal profile. The barcode region will result in abrupt edges in the profile because of the black and white bar pattern as shown in Fig. 8. Thus the barcode region in the image is located.

### *3.3 Preprocessing and Binarization*

Preprocessing is an essential step in any image processing application and preprocessing can significantly increase the reliability of the vision system. Several filter operations which intensify or reduce certain image details enable an easier or

**Fig. 7** Result of thresholding regions based on width and height

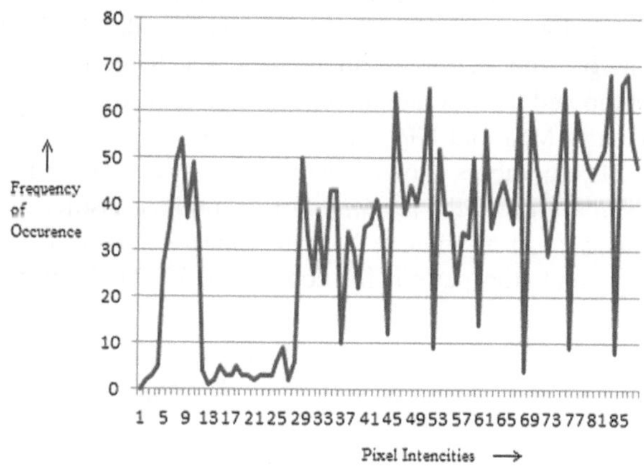


**Fig. 8** Horizontal profile of a barcode region

faster evaluation. Barcode preprocessing is the key step for the accurate recognition of the given barcode.

We use median filter and Gaussian filter for removal of noise in the barcode image. Median filter is a non-linear filter which replaces each entry by the median of neighboring entries. This filter performs well in denoising the salt and pepper noise, Speckle noise by preserving the sharp edges. Gaussian filter removes the impulse noise. The sigma values are varied to improve the denoising process. Later Binarization is performed automatically using Otsu's histogram shape-based image thresholding. This approach assumes that the image to be thresholded contains two classes of pixels or bi-modal histogram(e.g. foreground and background) then calculates the optimum threshold separating those two classes so that their combined spread is(intra-class variance) minimal. We assume the barcode image is corrected for skewness.

### 3.4 Classification

All barcodes are constructed from a series of bars and intervening spaces. The relative size of these bars and spaces and the number of them is decided by the specification of the symbology (or barcode type) which is being used. There are a number of 1-D barcode symbologies in common use. Such symbologies are shown in Fig. 9. Each symbology differs in the way data is encoded and often also in the type or amount of data encoded [1]. Generally, only one symbology is chosen for a particular application.

Given a 1-D barcode image, Tree classifier is applied to classify it into a specific symbology. Such classification is done based on its widths of bars and spaces appearing in start and/or stop patterns [1]. Following Fig. 9 shows the Tree classifier for 1-D barcode (numbers on arrow marks shows the start and/or stop pattern of the barcode). Each 1-D barcode has unique start and/or stop pattern. The initial and final bars and spaces contribute to form these patterns. From the widths of bars and spaces determined, we compare them to the unique patterns of each symbology as in [1]. Since code 128, code-39 and Codabar has the largest start and/or stop pattern among all other barcode symbologies, these patterns are compared first. Later code-93, Interleaved 2 of 5 and code 11 are compared. Finally MSI and UPC are compared.

The numbers shown beside the arrows represent the widths of bars and spaces as multiples of the *module*. For example, consider the codabar barcode symbology made up of alternative bars and spaces. Given the module width as 2 pixels, the start and stop pattern for codabar is 1122121 refers to 1*2 = 2 pixels wide bar,

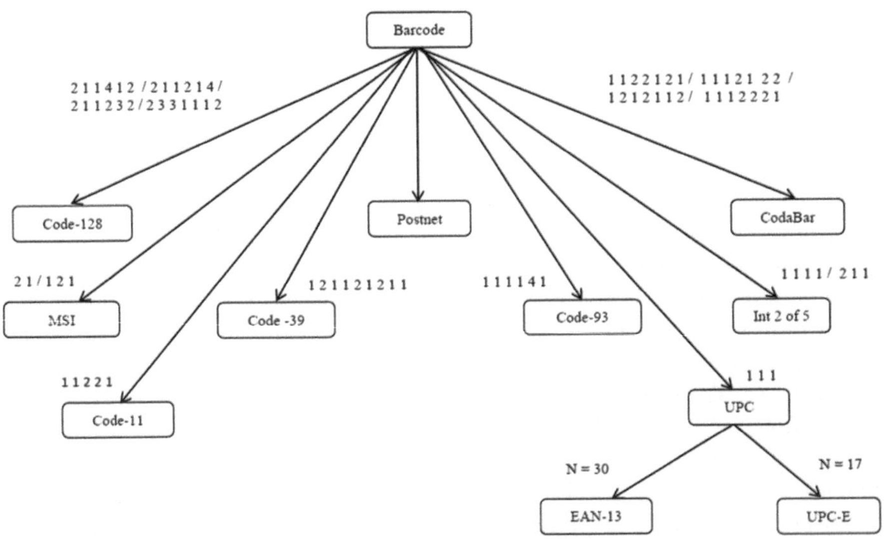


**Fig. 9** Tree-classifier for 1-D barcode based on start and/or stop pattern (*for Postnet barcode, height of the barcodes are considered*)

1*2 = 2 pixels wide space, 2*2 = 4 pixels wide bar, 2*2 = 4 pixels wide space, 1*2 = 2 pixels wide bar, 2*2 = 4 pixels wide space and finally 1*2 = 2 pixels wide bar. Similar calculations are followed for all types of symbologies.

For UPC family of barcodes, an additional measure in terms of number of bars is considered for classification. This is in regard of the start and/or stop pattern being same for EAN-13 and UPC-E symbologies. This is shown as 'N' in Fig. 9. For EAN-13 symbology N = 30 and for UPC-E symbology N = 17.

Postnet symbology differs from all other symbologies as it encodes the data by varying heights of bars rather than varying widths of bars and spaces. Hence given barcode can be classified as Postnet by determining the height of bars [1].

### *3.5 Decoding*

Decoding is the process of converting the bar and space patterns of the barcode into data characters. Our barcode decoding algorithm analyzes the barcode using multi scan line approach. Several scan lines contained in the detected barcode area are taken. This ensures that data loss in one scan line due to severe blurring, occlusion etc. can be recovered from decoding other scan lines. Each scan line results in decoded data characters. To find out the correct encrypted characters among these, at each position the character which is resulted maximum number of times by decoding several scan lines is taken as decode data.

Decoding of barcodes is performed according to the standard decoding algorithms [10] based on the type of the barcode detected. The decoding process requires the widths of the bars and spaces to be known. The thickness of bars and spaces in comparison with each other determines the digit they represent. The thinnest bar/space represents *module* or *element*. Estimate the widths of each bar and space of the barcode after start character till before the stop character. Get the data of the barcode from the respective barcode encoding table [1] using the obtained width of bars and spaces.

## 4 Experimental Results

We took images from various datasets, both standard as well the images captured from low resolution cameras (mobile camera). We conducted extensive experiments on clean and noisy barcode images. The noisy barcode images are generated by adding noise such as salt and pepper, Gaussian and Speckle noise.

The results of decoding few barcode images are presented below. Both the success and failure cases are listed.

| Barcode Image | Noise added | Result of decoding |
|---|---|---|
| A123$+56B | Salt and pepper noise<br>Noise density = 0.05 | Symbology: codabar<br>Data: A123$ + 56 |
| 123-4530 | Salt and pepper noise<br>Noise density = 0.05 | Symbology: code-11<br>Data: 123-4530 |
| *ABCD123%* | Salt and pepper noise<br>Noise density = 0.05 | Symbology: code-39<br>Data: ABCD123 % |
| 123456 | Salt and pepper noise<br>Noise density = 0.05 | Symbology: code-93<br>Data: 123456 |
| (12)34barcode | Salt and pepper noise<br>Noise density = 0.05 | Symbology: code-128<br>Data: 1234 barcode |
| 5 032423 434303 | Salt and pepper noise<br>Noise density = 0.05 | Symbology: EAN-13<br>Data: 5032423434303 |
| 1234456781122475847587<br>1234456781122475847587 | Salt and pepper noise<br>Noise density = 0.05 | Symbology: interleaved 2 of 5<br>Data: |
| 12345 | Salt and pepper noise<br>Noise density = 0.05 | Symbology: MSI<br>Data: 12345 |
|  | Salt and pepper noise<br>Noise density = 0.05 | Symbology: postnet<br>Data: 12345 |

(continued)

Fig. 2 Shows one example frame from each flower videos created

**Fig. 3** Set of flowers video frames with minimum *rectangular box* fixed by the proposed detection method on flower region

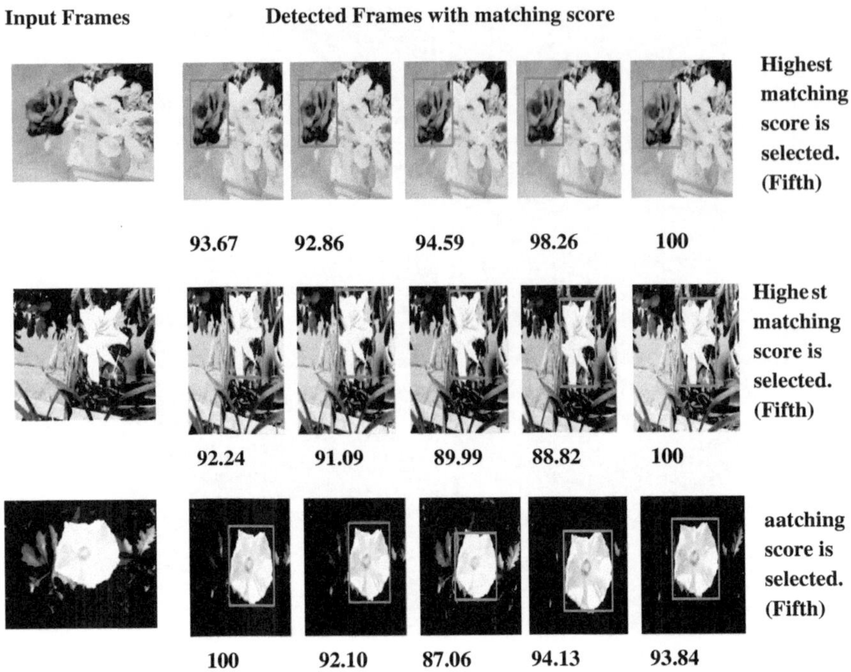


**Fig. 4** Frames detected along with their matching scores for different input frames

**Fig. 5** Misidentification of flowers. **a** Input frames. **b** Output of the proposed method with matching score

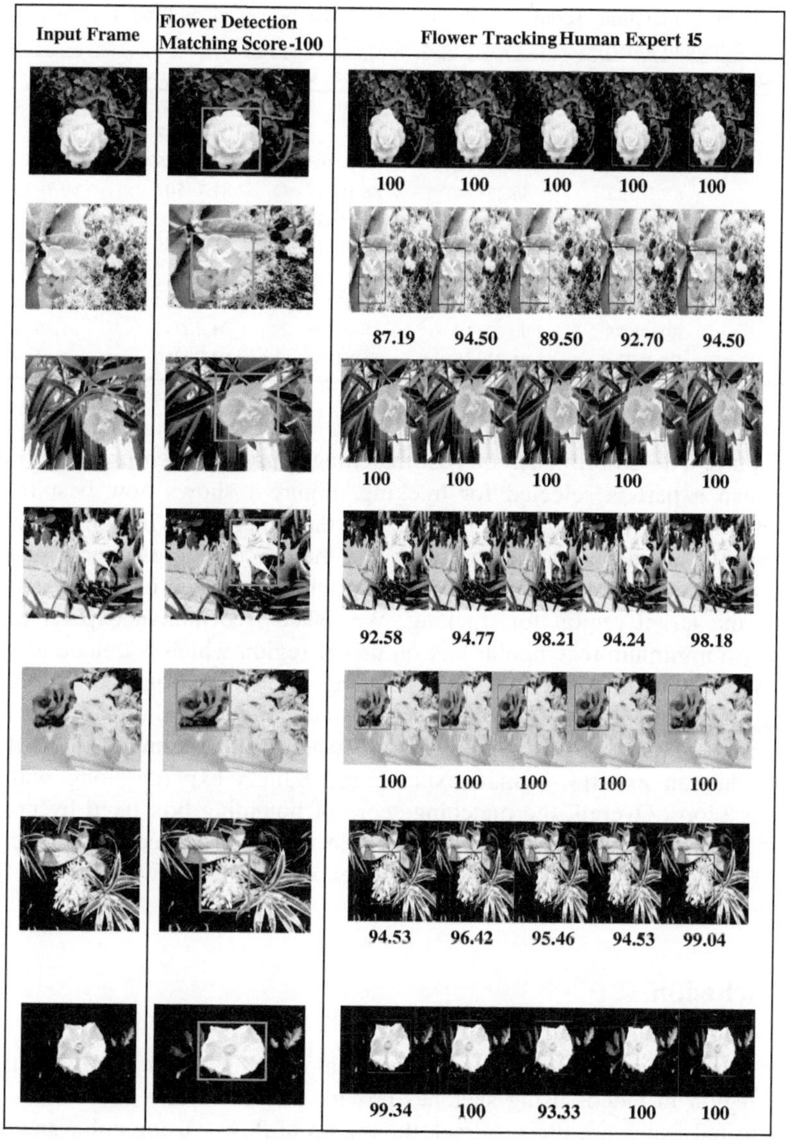


**Fig. 6** *Column 1* shows a set of frames. *Column 2* Shows the selected frame for tracking through maximum matching score. *Column 3* Tracking matching score between the coordinates obtained by the proposed method and corresponding human experts

rectangular box on flower region which is treated as ground truth for our experimentation. The sketch is given as input it is detected in the respective frames. Figure 3 shows some samples of flower images containing minimum bounding rectangular box fixed by the proposed method. We calculate percentage of matching by comparing the co-ordinates obtained by the expert and detection

**Table 1** Overall matching scores for the output of the proposed method w.r.t ground truth created by 5 human experts

| Flower videos no. | Human expert 1 | Human expert 2 | Human expert 3 | Human expert 4 | Human expert 5 |
|---|---|---|---|---|---|
| 1 | 95.63 | 96.04865 | 94.79122 | 96.73176 | 96.51785 |
| 2 | 100 | 100 | 100 | 99.75758 | 100 |
| 3 | 94.3199 | 95.45312 | 98.00295 | 96.60815 | 98.26666 |
| 4 | 85.58392 | 86.23154 | 88.19397 | 87.74507 | 89.79025 |
| 5 | 96.41151 | 100 | 93.56215 | 100 | 99.92304 |
| 6 | 90.83918 | 94.79543 | 89.83116 | 95.26265 | 94.8868 |
| 7 | 100 | 100 | 100 | 100 | 100 |
| 8 | 87.1321 | 85.45818 | 85.87476 | 91.15284 | 94.3773 |
| 9 | 89.89978 | 90.02119 | 88.89532 | 91.4034 | 91.69119 |
| 10 | 81.4677 | 86.31115 | 86.81036 | 84.20956 | 84.65697 |

co-ordinates. The co-ordinates of best matching score between proposed method and human expert is selected for tracking. Figure 4 shows how best frame is selected based on their matching score. On the other hand Fig. 5 shows some samples with misidentification of flower and the percentage of overlapping with human experts. In tracking phase the best matching score co-ordinates is used to identify the target region for tracking. We asked five human experts to draw manually a minimum rectangular box on flower region which is treated as ground truth. We calculate percentage of matching by comparing the co-ordinates obtained by the experts and tracking co-ordinates. Figure 6 shows some samples flower images bounding box fixed by the proposed method, human expert1, human expert2, human expert3, human expert4 and human expert5 along with their matching score. Overall, the matching score of bounding box fixed by proposed method and human experts is shown in Table 1. The Proposed method achieves 93.46 % matching score across five human experts for 10 flower videos.

## 5 Conclusion

In this work we have used the fast directional chamfer matching to identify the flower region in videos using sketches provided by user. The coordinates of the best detected frame are used to track the region of flower using color and texture information throughout the video. We have conducted experimentation on our own dataset. To corroborate the efficiency of the proposed method we have created ground truth where five human experts have identified the flower region by drawing rectangular bounding box manually. Later we matched the bounding box drawn by the proposed method with bounding box of human experts to study the error analysis. In future we intend to develop a multi tracking of flower naming system based on sketches of the flowers.

## References


1. Linneaus C (1759) Systemae naturae. Impensis Direct Laurentii Salvii
2. Das M, Manmatha R, Riseman EM (1999) Indexing flower patent images using domain knowledge. IEEE Intell Sys 14:24–33
3. Kass M, Witkin A, Terzopoulos D (1987) Snakes: active contour models. Int J Comp Vis 1(4):321–331
4. Chan TF, Vese LA (2001) Active contours without edges. IEEE Trans Image Process 10:266–277
5. Mortensen E, Barrett WA (1995) Intelligent scissors for image composition. In: Proceedings of ACM SIGGRAPH, pp 191–198
6. Saitoh T, Aoki K, Kaneko T (2004) Automatic recognition of blooming flowers. In: The Proceedings of 17th international conference on pattern recognition 1:27–30
7. Boykov Y, Jolly MP (2001) Interactive graph cuts for optimal boundary and region segmentation of objects in N-D images. In: Proceedings of international conference on computer vision (ICCV–01), vol 2, pp 105–112
8. Rother C, Kolmogorov V, Blake A (2004) Grabcut: interactive foreground extraction using iterated graph cuts. In: Proceedings of the ACM SIGGRAPH conference on computer graphics, vol 23(3), pp 309–314
9. Kumar MP, Torr PHS, Zisserman A (2005) OBJ CUT. In: Proceedings of the IEEE conference on computer vision and pattern recognition, San Diego, vol 1, pp 18–25
10. Nilsback ME, Zisserman A (2006) A visual vocabulary for flower classification. In: The proceedings of computer vision and pattern recognition, vol 2, pp 1447–1454
11. Nilsback ME, Zisserman A (2004) Delving into the whorl of flower segmentation. In: The proceedings of British machine vision conference, vol 1, pp 27–30
12. Calderero F, Marques F (2010) Region merging techniques using information theory statistical measures. IEEE Trans Image Process 19(6):1567–1586
13. Calderero F, Marques F (2008) General region merging approaches based on information theory statistical measures. In: The 15th IEEE international conference on image processing, pp 3016–3019
14. Ning J, Zhang L, Zhang D, Wu C (2010) Interactive image segmentation by maximal similarity based region merging. Pattern Recognit 43(2):445–456
15. Yoshioka Y, Iwata H, Ohsawa R, Ninomiya S (2004) Quantitative evaluation of flower color pattern by image analysis and principal component analysis of *Primula sieboldii* E. Morren Euphytica, pp 179–186
16. Nilsback ME, Zisserman A (2008) Automated flower classification over a large number of classes. In: The proceedings of sixth indian conference on computer vision, graphics and image processing, pp 722–729
17. Guru DS, Sharath YH, Manjunath S (2011) Textural features in flower classification. Math Comput Model 54(3–4):1030–1036
18. Guru DS, Sharath YH, Manjunath S (2010) Texture features and KNN in classification of flower images. IJCA, Special Issue on RTIPPR (1), pp 21–29
19. Guru DS, Sharath YH, Manjunath S (2011) Classification of flowers based on whorl region. In: 5th Indian international conference on artificial intelligence, pp 1070–1088
20. Varma M, Ray D (2007) Learning the discriminative power invariance trade-off. In: Proceedings of international conference on computer vision
21. Fukuda K, Takiguchi T, Ariki Y (2008) Multiple classifier based on fuzzy c-means for a flower image retrieval. In: Proceedings of international workshop on nonlinear circuits and signal processing, pp 76–79
22. Cho SY, Chi Z (2005) Generic evolution processing of data structures for image classification. IEEE Trans Knowl Data Eng 17(2):216–231

23. Goller C, Kuchler A (1996) Learning task-dependent distributed representations by back-propagation through structure. In: Proceedings of IEEE international conference on neural networks, pp 347–352
24. An-xiang H, Gang C, Jun-li L, Zhe-ru C, Dan Z (2004) A flower image retrieval method based on ROI feature. J Zhejiang Univ Sci, pp 764–722
25. Ming-Yu L, Oncel T, Ashok V, Rama C (2010) Fast directional chamfer matching. Comput Vision Pattern Recognit, pp 1696–1703
26. Comaniciu D, Ramesh V, Meer P (2003) Kernel-based object tracking. IEEE Trans Patt Anal Mach Intell 25(5):564–575
27. Ojala T, Pietikainen M, Maenpäa T (2002) Multiresolution gray-scale and rotation invariant texture classification with local binary patterns. IEEE Trans Patt Anal Mach Intell 24(7):971–987
28. Ojala T, Valkealahti K, Oja E, Pietikainen M (2001) Texture discrimination with multi-dimensional distributions of signed gray level differences. Pattern Recogn 34(3):727–739
29. Ning J, Zhang L, Zhang D, Wu C (2009) Robust object tracking using joint color-texture histogram. Int J Pattern Recognit Artif Intell 23(7):1245–1263

# Segmentation, Visualization and Quantification of Knee Joint Articular Cartilage Using MR Images

**M. S. M. Swamy and Mallikarjun S. Holi**

**Abstract** Knee is a complex and highly stressed joint of the human body. Articular cartilage is a smooth hyaline spongy material between the tibia and femur bones of knee joint. The change in cartilage morphology is an important biomarker to study the progression of osteoarthritis (OA). Magnetic resonance imaging (MRI) is the modality widely used to image the knee joint because of its non ionization effect and soft tissue contrast. In the present work a semiautomatic algorithm is developed for segmentation of articular cartilage from knee MR images. Segmented cartilage is visualized in 2D and 3D. Cartilage thickness is measured in different regions of femur and total volume of the cartilage is computed from the sequence of MR images. The cartilage thickness measurement and visualization is of diagnostic use for early detection and assessment of progression of the disease in case of OA affected patients.

**Keywords** Cartilage thickness · Image segmentation · Knee joint · MRI · Osteoarthritis

## 1 Introduction

The knee joint is the largest and most complex synovial joint of the human body. It is a major weight bearing joint of the body and is made up of three bones. The femur (thigh bone) is located on superior side, tibia (shin bone) on inferior and

M. S. M. Swamy (✉) · M. S. Holi
Department of Biomedical Engineering and Research Center, Bapuji Institute of Engineering and Technology, Davangere 577004 Karnataka, India
e-mail: ms_muttad@yahoo.co.in

M. S. Holi
e-mail: msholi@yahoo.com

P. P. Swamy and D. S. Guru (eds.), *Multimedia Processing, Communication and Computing Applications*, Lecture Notes in Electrical Engineering 213,
DOI: 10.1007/978-81-322-1143-3_26, 

patella (knee cap) on anterior part of the knee joint. Articular cartilage is a thin layer between the femur and tibia bones. It is a soft tissue at the end of bones that allows the joint to move easily. The knee joint contains a small amount of synovial fluid in a cavity that nourishes the cartilage and lubricates the joint. The menisci are C-shaped wedges of fibro cartilage located between the tibial plateau and femoral condyles which provides an additional stability to the knee joint. Osteoarthritis is a common disease of the knee joint affecting the elderly people. It occurs when cartilage becomes soft and gets eroded due to continuous wear and tear movements and with ageing. The OA affected knee joint often leads to inflammation, decrease in motion of joint due to stiffness, and formation of bone spurs (tiny growths of new bone). This decreases the ability of the cartilage to work as a shock absorber to reduce the impact of stress on the joints. The remaining cartilage wears down faster and eventually, the cartilage in some regions may disappear altogether, leaving the bones to rub against one another during motion and may further leads to formation of bone spurs. With OA, synovial fluid does not provide proper lubrication, which leads pain, inflammation and restriction of movements at the joints. In severe OA there will be complete breakdown of cartilage over a period of time, leading to severe pain and joint loss. There is no artificial material that can replace only the cartilage at the joint.

Osteoarthritis is most frequently occurring joint disease with prevalence of 22–39 % in India [1]. OA has affected nearly 27 million Americans according to the study in 2007 [2]. After the age 50, women are more often affected by OA than men [3]. Symptoms of OA typically begin after age 40 and progress slowly. Loss of joint function as a result of OA is a major cause of work disability and reduced quality of life. For every one pound of weight loss, there is a four pound reduction in the load exerted on the knee for each step taken during daily activities [4].

## 2 Background on the Imaging of Knee Joint

### *2.1 MR Imaging of Knee Joints*

MRI can visualize cartilage, bone and other surrounding soft tissues distinctly. High resolution gradient echo MRI sequences with fat suppression are the most useful techniques for quantization of cartilage dimensions. MRI is non-invasive and repetitive imaging study of an individual using MRI is possible without side effects. The assessment of cartilage dimensions is important for the study of the progression of cartilage damage due to OA. MR images are widely used for diagnosis of knee joint abnormalities. Figure 1 shows the MR images of normal and OA affected knee joints.

MRI can visualize knee joints in sagittal, coronal and axial planes. Sagittal images give better view of cartilage than the other two. Even though, MRI directly visualizes the cartilage, for quantitative and progressive assessments image

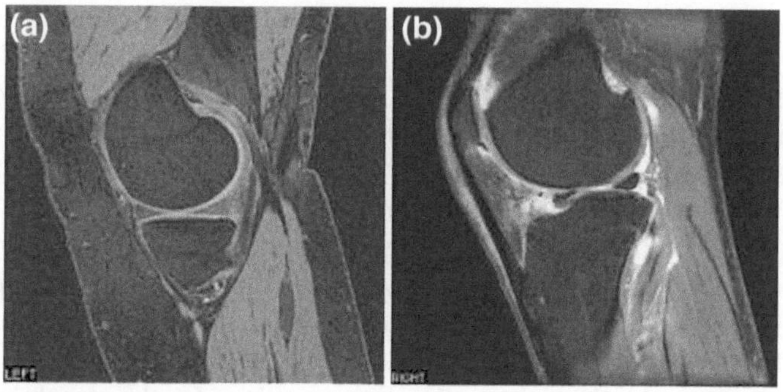


**Fig. 1** Knee joint MRI. **a** Normal. **b** OA affected

processing techniques are needed. Knee joint cartilage thickness measurement gives vital information in the diagnosis and treatment of OA. The three dimensional (3D) visualization gives complete information and insight on cartilage degradation and menisci tears. In specific cases, MR images with high signal to noise ratio with high resolution are used for in vitro studies. But in most of the in vivo, low contrast knee image slices are used. The 3D reconstruction algorithms are used to obtain the data missing in 2D images.

## *2.2 Knee Cartilage Segmentation Methods*

Knee joint image segmentation is a very challenging task because of the complexity of joint. Knee joint segmentation methods can be classified into three categories based on manual intervention required, namely manual, semiautomatic and fully automatic. The manual segmentation methods are observer dependent and time consuming for 3D reconstruction. Semi automatic methods are developed to increase the accuracy of processing. Fully automatic methods are also available with advanced processing techniques.

Cohen et al. [5] initially segmented cartilage manually by digitizing the consecutive points along the articular contour curves with a typical spacing of 0.5–1.0 mm. To reduce the time, a semi automatic segmentation method is adopted. Region of interest is obtained from each image as in the manual segmentation by fixing the points along the curves. An interpolated cubic B-spline curve is fitted for the points. The image gradient vector is evaluated using the Prewitt convolution kernel. The B-spline curve, which follows the contour of the desired surface, is then sampled at 0.5 mm intervals. Cashman et al. [6], developed algorithm, in addition to edge detection techniques, thresholding is performed to exclude all pixels below one half of the average intensity of the image. Boundary discontinuities are bridged using B-spline interpolation. Selecting a seed point inside the bone and using a recursive region growing procedure the interior of the bone is filled and labeled. In radial search method developed by Poh and Kitney

[7], an origin is fixed at center of femur and horizontal line is drawn. Each vector of radius R and angle $\theta$ with reference to horizontal line is defined. Pixel intensities and coordinates along the line of radius R are taken. A threshold method is used to detect the inner boundaries. This procedure is repeated for 0–180° to cover the entire cartilage. B-spline curve is used to get continuous contour. Segmented images are obtained by masking the original image with this contour. In the method developed by Gamio et al. [8] Bezier splines are used. The control points are placed inside the cartilage following its shape to create a Bezier spline. Rays perpendicular to the spline on the control points are traced to find the bone cartilage interface. The edges are found based on the first derivative of brightness using bicubic interpolation along the line profiles. In the graph cut method developed by Shim et al. [9], seeds are placed manually (curvilinear marks) over specific anatomic regions. The seeds are propagated to neighboring pixels and then segmented. The method is compared with manual method. A fully automatic method using voxel classification is developed by Folkesson et al. [10], is based on kNN classifier algorithm, which works with reduced processing time and low field MRI data.

In shape or model based methods, segmentation is based on cartilage shape or model developed. The development of the cartilage model is based on prior knowledge obtained for population under study. The active contours (snakes) are deformed to match the cartilage of the image. The deformed contour is used as cartilage map. Snakes or active contours are energy minimizing curves that deform under the influence of internal forces within curve itself and external force derived from the image to minimize the energy function [11]. In the 2D active contour algorithm developed by Kauffmann et al. [12], a local coordinate system (LCS) is developed for the femoral and tibial cartilage boundaries that provide a standardized representation of cartilage geometry, thickness and volume. The LCS can be registered in different data sets from the same patient so that results can be directly compared. Cartilage boundaries are segmented from 3D MRI and transformed into offset maps, defined by the LCS. Gaps in the offset map resulting from inter slice distance are filled using a bi-cubic interpolation scheme.

## 2.3 Disease Findings and Quantification using MRI

MR imaging findings in different stages of disease are correlated with clinical findings is well established. The cartilage lesions, bone marrow edema pattern and menisci lesions are well detected on MR images in patients with advanced OA [13]. The abnormalities in cartilage, menisci and subchondral cysts are found in MR imaging only [14]. MRI for diagnosis and assessment of cartilage defect repairs is studied by Stefan et al. [15]. MR sequences which offer high contrast between articular cartilage and adjacent structures like femur, tibia and menisci are used. MR imaging protocols like fat suppression, spoiled gradient echo sequence and the fast spin echo sequence are accurate and reliable for evaluating surface defects of articular cartilage. MR imaging findings are compared with radiographic severity measurements (scores) and pain in middle aged women [16]. It is found that significant association between pain, radiographic severity of OA of the knee and MR image findings. Patterns of femorotibial cartilage loss are studied in knees with neutral, varus and valgus alignments. Coronal MR Images of symptomatic OA patients is obtained and processed for computation of cartilage volume, surface area and thickness. Dependency of alignment to medial to lateral rate of cartilage is shown [17]. The articular cartilages were divided into 5 compartments including lateral and medial tibial, lateral and medial femoral and patellar compartments. The grades of articular cartilage were compared with cartilage volume measurements. Cartilage volume correlates well with MR grading of articular cartilage [18].

Knee image cartilage segmentation, thickness and volume quantification is a complex procedure. Even though, a number of algorithms are developed in this regard, there is a necessity of a simple method to segment cartilage of diseased knee joint and quantification of its thickness and volume.

## 3 Methodology

The knee joint images are obtained from National Institute of Health, OA Initiative (OAI) which includes normal and OA images. The MR images of this database include water excitation double echo steady-state (DESS) imaging protocol with sagittal slices at 1.5-T. The imaging parameters for the sequence are: TR/TE: 16.3/ 4.7 ms, matrix: 384 × 384, FOV: 140 mm, slice thickness: 0.7 mm, x/y resolution: 0.365/0.365 mm.

The obtained MR images are preprocessed for noise removal using median filter of (3 × 3). Median filter removes noise without affecting edges and boundary information in an image. Canny edge detection technique is used to obtain the location of femur cartilage boundary and cartilage synovial boundary. The edge information is used to manually mark the control points on knee cartilage boundary.

Image segmentation is defined as the partitioning of an image into nonoverlapping, constituent regions that are homogeneous with respect to some characteristic such as intensity or texture [19]. If the domain of the image is given by Ω then the segmentation problem is to determine the sets $S_k \in \Omega$ whose union is the entire domain Ω. Thus the sets that make up segmentation must satisfy

$$\Omega = \bigcup_{k=1}^{K} S_k \tag{1}$$

Where $S_k \cap S_j = \emptyset$ for $k \neq j$ and each $S_k$ is connected.

The Canny edge detection operator returns a value for the first derivative in the horizontal direction ($G_x$) and the vertical direction ($G_y$). From this the edge gradient and direction can be determined [20].

$$\nabla f(x, y) = \left(\frac{\delta f}{\delta x}, \frac{\delta f}{\delta y}\right) \tag{2}$$

The magnitude (edge strength) of the gradient is then approximated using the formula:

$$|G| = |G_x| + |G_y| \text{ and } \Theta = \tan^{-1}(G_y/G_x) \tag{3}$$

In the femur cartilage boundary (inner boundary) the contrast between the femur and cartilage is good and this is detected using canny edge detection. In the cartilage and synovial interface (outer boundary) the contrast is poor. The boundary obtained is incomplete after the edge detection. The control points are placed manually to mark the femur-cartilage and cartilage-synovial boundaries. The sample points are increased by interpolation. The marked points are interpolated using B-spine curve fitting technique. A B-spline of degree n is derived through n convolutions of the box filter, $B_0$. Thus, $B_1 = B_0 * B_0$ denotes a B-spline of degree 1, yielding the familiar triangle filter. The B-spline of degree 1 is equivalent to linear interpolation. The second degree B-spline $B_2$ is produced by convolving $B_0 * B_1$. The cubic B-spline $B_3$ is generated from convolving $B_0 * B_2$. That is $B_3 = B_0 * B_0 * B_0 * B_0$ [21]. The filter function of interpolation is given by Eq. (4). Figure 2 Shows images at different steps of processing.

$$h(x) = \frac{1}{6}\begin{cases} 3|x|^3 - 6|x|^2 + 4 & 0 \leq |x| < 1 \\ -|x|^3 + 6|x|^2 - 12|x| + 8 & 1 \leq |x| < 2 \\ 0 & 2 \leq x \end{cases} \tag{4}$$

From segmented cartilage lateral, medial and patellar regions are identified. The thickness of the cartilage is computed along the normal to inner boundary of the cartilage. The thickness in three different regions of cartilage is computed. The thickness of a cartilage layer is defined as the Euclidian distance between the bone-cartilage interface and the cartilage-synovial interface. Thickness measurements are made from the inner cartilage boundary in a direction normal to the inner boundary toward the outer cartilage boundary. The Euclidian distance $D_E$ between a point p belonging to the outer boundary of the cartilage and the nearest point r belonging to the inner boundary I is computed [22] using

$$D_E(p) = \min\left(\sqrt{(p_x - r_x)^2 + (p_y - r_y)^2 + (p_z - r_z)^2}\right) \quad \text{Where } r \in I \tag{5}$$

The method is further extended for 3D cartilage visualization and volume measurement. The 2D thickness maps of articular cartilage obtained after processing are stored. The pixels in between the MR slices are interpolated in 3D. The

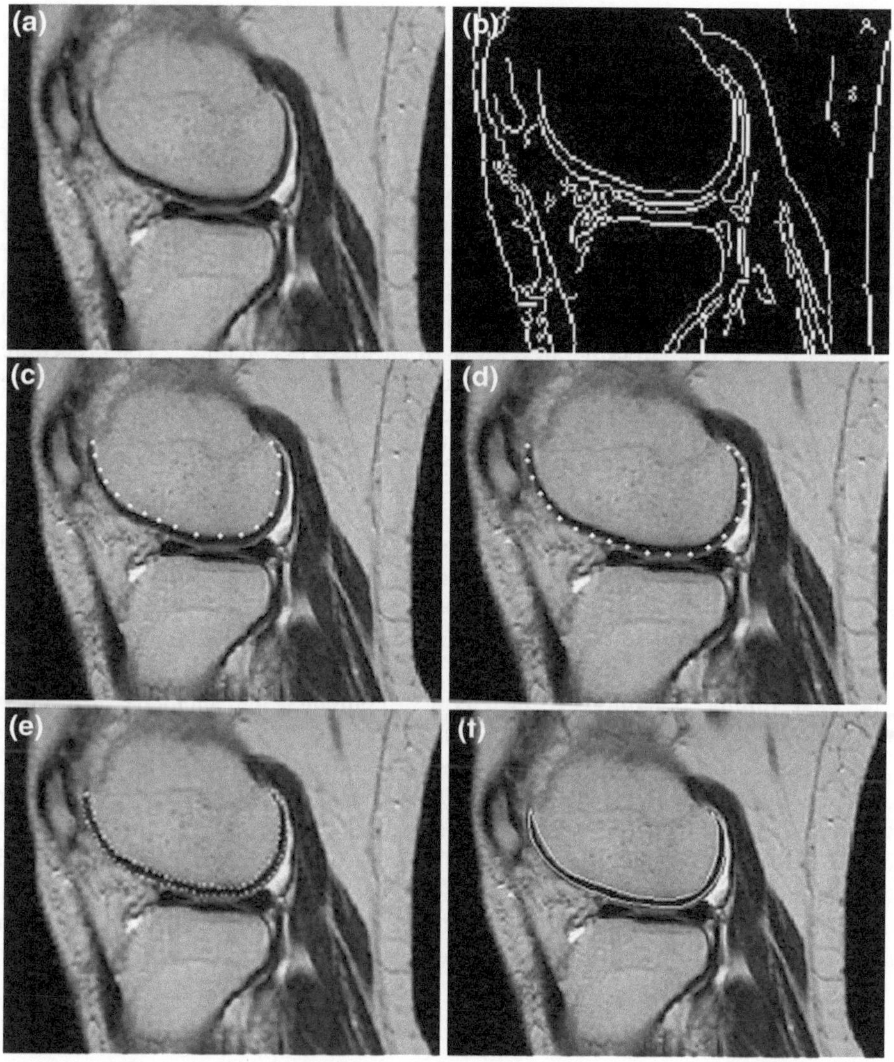


**Fig. 2** Processing steps of cartilage segmentation. **a** Input knee MRI. **b** Canny edge detected image. **c** Marking of inner boundary. **d** Marking of outer boundary. **e** Interpolated boundary. **f** Segmented cartilage

articular cartilage is reconstructed in 3D. The volume of the cartilage is computed by cumulatively adding the local volume computed in 2D. The total volume is computed as

$$V_{\Omega} = \sum V_{i,j} \qquad \forall V_{i,j} \in \Omega \tag{6}$$

Total volume $V_{\Omega}$ of articular cartilage is computed using local volumes $V_{(i,j)}$ obtained [12].

## 4 Results

Cartilage is segmented from knee joint MRI and visualized. Regions of segmented cartilage are identified as lateral, medial and patellar. Then thickness measurements at each region are averaged. The standard deviation (SD) and coefficient of variation (COV) is calculated for each compartment of the knee joint. The COV indicates variations in the thickness of cartilage in that region. The method of thickness computation is applied on normal and OA knee joint images at different level of disease. The thickness of cartilage is measured for segmented cartilage in lateral, medial and patellar regions of an image from MRI sequence. The mean thickness of cartilage in different regions of cartilage is computed. The procedure is repeated for all the images in MRI sequence. The average thickness of cartilage in particular region is computed for the set of images in MRI sequence. The SD is calculated for cartilage thickness of a region. The COV is computed for cartilage thickness of a region. This procedure is repeated for two normal and four OA affected knee joints of different level of severity. The measured cartilage thickness, SD and COV for different regions are tabulated in Table 1. The value of computed SD is less for thickness of cartilage in a region. The COV is increased in OA knee joint cases compared to normal knee joint cases.

The local volumes of cartilage in lateral, medial and patellar regions are computed. The mean volume in different regions with SD is tabulated in Table 2. The total volume of entire cartilage is also computed and tabulated. Cartilage volumes in normal and OA knee joints of different stages of severity are calculated.

The processed 2D thickness maps of cartilage of an MRI sequence are saved for volume visualizations. The voxels between the 2D images are obtained using 3D interpolation. The images are squeezed to construct a 3D image. The squeezed data of the image is volume rendered for 3D visualization. A rotatable 3D articular cartilage is visualized. The articular cartilage is visualized in 3D and rotated in different angular direction to view different compartments is shown in Fig. 3.

**Table 1** Mean cartilage thickness

| Clinical symptom | Thickness in mm | | | | | | | | |
|---|---|---|---|---|---|---|---|---|---|
| | Lateral | | | Medial | | | Patella | | |
| | Avg. | SD | COV | Avg. | SD | COV | Avg. | SD | COV |
| Normal 1 | 2.093 | 0.07 | 3.52 | 2.068 | 0.06 | 2.94 | 2.145 | 0.11 | 4.93 |
| Normal 2 | 2.125 | 0.08 | 3.99 | 2.052 | 0.05 | 2.63 | 2.115 | 0.10 | 4.83 |
| OA1 | 2.022 | 0.13 | 6.65 | 1.993 | 0.74 | 3.72 | 1.998 | 0.06 | 3.47 |
| OA2 | 1.998 | 0.13 | 6.55 | 1.975 | 0.12 | 6.17 | 1.962 | 0.13 | 6.75 |
| OA3 | 2.016 | 0.12 | 6.11 | 1.944 | 0.14 | 7.36 | 1.983 | 0.11 | 5.35 |
| OA4 | 1.935 | 0.18 | 9.59 | 1.933 | 0.17 | 8.85 | 1.885 | 0.21 | 11.02 |

**Table 2** Femur cartilage volume

| Clinical symptom | Lateral (mm$^3$) | | Medial (mm$^3$) | | Patellar (mm$^3$) | | Total volume mm$^3$ |
|---|---|---|---|---|---|---|---|
| | Mean | SD | Mean | SD | Mean | SD | |
| Normal 1 | 2,101 | 117.59 | 2,131 | 103.82 | 1,173 | 984.5 | 5,405 |
| Normal 2 | 2,151 | 119.59 | 2,181 | 108.42 | 1,154 | 914.5 | 5,486 |
| OA1 | 2,026 | 105.22 | 2,076 | 116.18 | 1,168 | 981.8 | 5,270 |
| OA2 | 2,011 | 106.81 | 2,021 | 129.51 | 1,127 | 1032.1 | 5,159 |
| OA3 | 1,991 | 142.75 | 1,966 | 174.94 | 1,105 | 1054.83 | 5,062 |
| OA4 | 1,902 | 221.81 | 1,911 | 218.03 | 985 | 1180.66 | 4,799 |

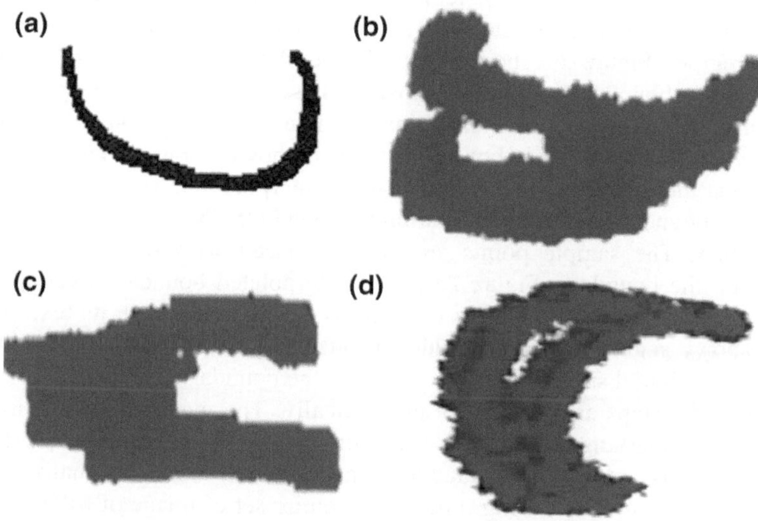


**Fig. 3** Cartilage visualization in 2D and 3D. **a** Cartilage in 2D view, **b** rotated 180° in Z, **c** rotated -90° in X, **d** rotated 180° in X and 45° in Z

# 5 Discussions

In this work knee joint MR images of normal and OA affected subjects are processed to segment the articular cartilage. The study involves total knee MR images of 58 subjects including 10 normal and 48 OA affected subjects with varying age group from 25 to 85 years. The segmentation method involves canny edge detection and B-spline interpolation. Cohen et al. [5] used these techniques and developed semi automatic segmentation method for detection of cartilage boundary. The objective of their work was to measure cartilage thickness in vivo and in vitro. The processing steps of their method differ from steps discussed in the present work. In their work, the initial set of points are marked on the cartilage boundary and region of interest (ROI) is extracted (cartilage portion). The interpolated cubic B-spline curve is fitted to these initial set of points. Then image gradient vector is evaluated to detect the boundary points along the normal to the curve. The limitation of the work is the gradient vector is evaluated to detect the boundary after manually marking the boundary and obtaining the ROI. In the present work, the processing steps of segmentation differ. After pre-processing of images, the edge detection is performed as a first step to detect the boundary. In the next step, using the edge information the manual marking of the boundary points is completed. After interpolation of points the complete cartilage boundary is obtained. The cartilage image is segmented using the mask of the cartilage boundary. The texture in the segmented cartilage is preserved for visualization. Cartilage is visualized in 2D and 3D. In the first step canny edge detection technique detects the inner boundary (femur cartilage boundary) well but with missing edges in the outer boundary (cartilage synovial interface) because of less contrast in that region. Figure 2b shows Canny edge detected image. The boundary (coordinates) obtained using gradient method is saved. In the next step, boundary points of previous step are used and points are manually marked on inner and outer boundary. Manual marking of boundary points is shown in Fig. 2c and d. Then the entire set is interpolated using cubic B-spline interpolation. The number of samples on the boundary before interpolation is around 60–75 and increased 150 after interpolation. The sample points are closely placed with separation less than 0.5 mm on the boundary. Figure 2e shows interpolated boundary overlapped on cartilage. The segmented cartilage is saved as a new image with its texture. Figure 3a shows segmented and visualized cartilage in 2D. In the segmentation procedure the initial step of marking points on the boundary of cartilage is manual and rest of the steps are performed automatically. The manual step requires less than 5 min for marking of boundary points on image of MRI sequence. Rest of the processing steps, interpolation, thickness measurement and 3D visualization are automated. The total processing time for the entire set of image of MRI sequence including manual and automatic steps takes less than 60 min. The thickness of cartilage is measured along the normal to the inner boundary curve till the outer boundary. The thickness measurements of more than 3.5 mm are discarded from the set. The thickness is computed in three regions of the cartilage and mean value of the cartilage thickness of the region is calculated. Table 1 shows cartilage thickness of few normal and OA cases. The local volume of cartilage is calculated region wise for all the images of a sequence. The procedure is repeated thrice on each image and average volume and SD is calculated. Table 2 shows volume of few normal and OA cases. For 3D visualization, the segmented cartilages of all the slices of a subject are saved to create a stack images. The stack of images is squeezed to make it as a 3D array of voxels representing the cartilage. The data is volume rendered using 3D interpolation and texture mapping technique. The cartilage is visualized in 3D and rotatable to visualize the different compartments (Fig. 3). Matlab 7.1 software is used for processing and 3D visualization.

The method detects thickness and volume decrease in OA affected patients. The measurements of thickness and volume are in agreeable range of other validated methods. The method shows good precision in measurements. The limitation of the work is the errors in the measurements are not quantified. However, the complexity of processing is less compared to advanced methods based on active contours or snakes. The normal and OA knee joints at different stages of disease are processed. The decrease in thickness and volume of cartilage is observed in OA knee joints. The COV is more for cartilage thickness measurements in OA affected knee joints and increases with the progression of the disease condition. The results are matching with the prior diagnosis of disease condition assessed by experts. The processing method is useful in diagnosis, treatment and in the assessment of progression of knee joint disease.

**Acknowledgments** Osteoarthritis Initiative (OAI), National Institute of Health, USA for providing knee MR Images.

## References

1. Mahajan A, Verma S, Tandon V (2005) Osteoarthritis. J Assoc Phys India 53:634–641
2. Reva CL, David TF, Charles GH, Lesley MA, Hyon C, Richard AD, Sherine G, Rosemarie H, Marc CH, Gene GH, Joanne MJ, Jeffrey NK, Hilal MK, Frederick W (2008) Estimates of the prevalence of arthritis and other rheumatic conditions in the United States. Arthritis Rheum 58(1):26–35
3. Lawrence RC, Helmick CG, Arnett FC, Deyo RA, Felson DT, Giannini EH, Heyse SP, Hirsch R, Hochberg MC, Hunder GG, Liang MH, Pillemer SR, Steen VD, Wolfe F (1998) Estimates of the prevalence of arthritis and selected musculoskeletal disorders in the United States. Arthritis Rheumatism 41(5):778–799
4. Stephen PM, David JG, Cralen D, Paul DV (2005) Weight loss reduces knee-joint loads in overweight and obese older adults with knee osteoarthritis. Arthritis Rheum 52(7):2026–2032
5. Zohara AC, Denise MM, Daniel KS, Perrine L, Fabian F, Edward JC, Gerard AA (1999) Knee Cartilage Topography. Thickness, and contact areas from MRI: in-vitro calibration and in-vivo measurements, osteoarthritis and cartilage 7:95–109
6. Cashman PMM, Kitney RI, Gariba MA, Carter ME (2002) Automated techniques for visualization and mapping of articular cartilage in MR images of the osteoarthritic knee: a base technique for the assessment of microdamage and submicro damage. IEEE Trans Nanobiosci 1(1):42–51
7. Poh CL, Richard IK (2005) Viewing interfaces for segmentation and measurement results. In: 27th annual conference IEEE engineering in medicine and biology, Shanghai, China, pp 5132–5135
8. Julio CG, Jan, SB, Keh-Yang L, Stefanie K, Sharmila M (2005) Combined image processing techniques for characterization of MRI cartilage of the knee. In: 27th annual conference IEEE engineering in medicine and biology, Shanghai, China, pp 3043–3046
9. Hackjoon S, Samuel C, Cheng T, Jin-Hong W, Kent KC, Kyongtae TB (2009) Knee cartilage: efficient and reproducible segmentation on high spatial resolution MR images with the semiautomated graph-cut algorithm method. Radiology 251(2):548–556
10. Jenny F, Erik BD, Ole FO, Paola CP, Claus C (2007) Segmenting articular cartilage automatically using a voxel classification approach. IEEE Trans Med Imaging 26(1):106–115
11. Thi-Thao T, Po-Lei L, Van-Truong P, Kuo-Kai S (2008) MRI image segmentation based on fast global minimization of snake model. In: 10th international conference on control, automation, robotics and vision, Hanoi, Vietnam, pp 1769–1772
12. Claude K, Pierre G, Benoît G, Alain G, Gilles B, Jean-Pierre R, Johanne MP, Jean PP, Jacques AG (2003) Computer-aided method for quantification of cartilage thickness and

volume changes using MRI: validation study using a synthetic model. IEEE Trans Biomed Eng 50(8):978–988

13. Thomas ML, Lynne SS, Srinka G, Michael R, Ying L, Nancy L, Sharmila M (2003) Osteoarthritis: MR imaging findings in different stages of disease and correlation with clinical findings. Radiology 226:373–381
14. Peter RK, Johan LB, Ruth YTC, Naghmeh R, Frits RR, Rob GN, Wayne OC, Marie PH, Le G, Margreet K (2006) Osteoarthritis of the knee: association between clinical features and MR imaging findings. Radiology 239(3):811–817
15. Stefan M, Tallal C, Mamisch GV, Christoph R, Siegfried T (2008) Magnetic resonance imaging for diagnosis and assessment of cartilage defect repairs, injury. Int J Care Injured 39(S1):S13–S25
16. Hayes CW, Jamadar DA, Welch GW, Jannausch ML, Lachance LL, Capul DC (2005) Osteoarthritis of the knee: comparison of MR imaging findings with radiographic severity measurements and pain in middle-aged women. Radiology 237:998–1007
17. Felix E, Wolfgang W, Martin H, Verena S, Verena L, September C, Meredith M, Pottumarthi P, Leena S (2008) Patterns of femorotibial cartilage loss in knees with neutral, varus, and valgus alignment. Arthritis Rheumatism (Arthritis Care Res) 59(11):1563–1570
18. Ozlem B, Tamer B, Alpay A, Zühal A, Saim Y (2004) Comparison of MRI graded cartilage and MRI based volume measurement in knee osteoarthritis. Swiss Med Wkly 134:283–288
19. Dzung LP, Chenyang X, Jerry LP (2000) Current methods in medical image segmentation. Annu Rev Biomed Eng 2:315–337
20. John C (1986) A computational approach to edge detection. IEEE Trans Pattern Anal Mach Intell 8(6):679–698
21. Póth M (2007) Comparison of convolutional based interpolation techniques in digital image processing. In: Proceedings of 5th international symposium on intelligent systems and informatics, 24–25 July, pp 87–90, Subotica, Serbia
22. Matej M, Anna V, Meister EG (2004) Interactive thickness visualization of articular cartilage. In: IEEE proceedings of visualization, pp 521–527

# Speaker Independent Isolated Kannada Word Recognizer

**G. Hemakumar and P. Punitha**

**Abstract** This paper addresses the problem of recognizing spoken Kannada words. The designed algorithm recognizes spoken Kannada words independent of speakers. The proposed method normalizes the original speech signal of every isolated word and extracts Linear-Predictive coding (LPC) coefficients, and converts them into Real Cepstrum Coefficient. These Real Cepstrum Coefficient values are subjected to dimensionality reduction through normal fit. These coefficients are used as the representatives of each spoken word. Euclidian distance measure is then used to compute the distance between the test samples to the model data in the database. The model datum in the database at a minimum distance is declared as the recognized word. For experimentation, we have used 294 unique Kannada words. Each of these words was recorded with 10 Speakers yielding 2,940 samples in total. Out of 10 speakers' data, 8 speakers' data i.e., 2,352 samples were used to compute the representative co-efficient for each word. Remaining 2 speakers' data along with re-recorded data of two speakers out of the 8 speakers is used for testing. Totally 2,352 signals are used for training and 1,176 signals are used for testing. The success rate of the proposed system- known speaker data is 98.29 % and unknown speaker data is 91.66 %.

**Keywords** Speech recognition · Speaker dependent and independent features · LPC co-efficient · Real Cepstrum · Normal fit

G. Hemakumar (✉)
Department of Computer Science, Government College for Women, Mandya, India
e-mail: hemakumar7@yahoo.com

P. Punitha
Department of MCA, PESIT, Banashankari 3rd Stage, 100 Feet Ring Road, Bangalore, India
e-mail: punithaswamy@pes.edu

P. P. Swamy and D. S. Guru (eds.), *Multimedia Processing, Communication and Computing Applications*, Lecture Notes in Electrical Engineering 213,
DOI: 10.1007/978-81-322-1143-3_27, 

## 1 Introduction

Automatic speech recognition is the process by which a computer maps an acoustic speech signal to text. The goal of speech recognition is to develop techniques and systems that enable computers to accept speech input and translate spoken words into text and commands. The problem of speech recognition has been actively studied since 1950s and it is natural to ask why one should continue studying speech recognition. Speech recognition is the primary way for human beings to communicate. Therefore it is only natural to use speech as the primary method to input information into computational device or object needing manual input. Speech recognition is the branch of human- centric computing to make technology as user friendly as possible and to integrate it completely into human life by adapting to humans' specifications. Currently, computers force humans to adapt to computers, which is contrary to the spirit of human- centric computing. Speech recognition has the basic quality to help humans easily communicate with computers and reap maximum benefit from them. The performance of speech recognition has improved dramatically due to recent advances in speech service and computer technology with continually improving algorithms and faster computing. However, the complexity of speech is not obvious to common man due to his innate sense of grammar which is not inherent to computer and other machines like robot. Hence speech recognition is still receiving more importance in research till today than in the past.

According to Census 2011, India has 122 major languages and 2,371 dialects. Linguistic Diversity is very rich and wide in India. Out of 122 languages 22 are constitutionally recognized languages. In these 22 languages Kannada language is also included and it is the administrative language of Karnataka State. Kannada language is one of the major Dravidian languages of India and it occupies 27th place in most spoken languages of the world. There is clear distinction between the spoken and written forms of language. Spoken Kannada varies from region to region. The written form is more or less constant throughout Karnataka. However, the Ethnologue reports 'about 20 dialects' of Kannada. Kannada language uses forty nine phonemic letters, divided into three groups: Swaras (thirteen vowels); Yogavaahakas (two); and Vyanjanas (thirty-four consonants). The character set is nearly identical with other Indian languages. The script itself, derived from Brahmi script, is fairly complicated like most other languages of India owing to the occurrence of various combinations of "half-letters" (glyphs), or symbols that are attached to various letters in a manner similar to diacritical marks in the Roman languages. Kannada script is an example for phonetic language, but for the sound of a "half n" (which becomes a half m). The number of written symbols, however, is far more than the forty-nine characters in the alphabet, because different characters can be combined to form ottaksharas (compound characters). Each written symbol in the Kannada script corresponds to one syllable, as opposed to one phoneme in languages like English. The script of Kannada is also used in other languages such as Tulu, Kodava, Takk and Konkani. Kannada writing is based on the concept of akshara or the 'graphic syllable', which has a vowel as the final constituent, i.e. V, VCV, CV, CCV, CCCV etc. (V indicate Vowels, C indicate Consonants). Word-initial V is written in its primary form; in the post consonantal position. The vowel is represented by a diacritic.

The isolated word recognizers usually require utterance of each word in isolation bounded by proper silence or pause on either sides of the sample window. Often, these systems have "Listen/Not-Listen" states, where they require the speaker to wait between utterances [1]. In any speech recognition system the recognizer has a training phase, where an individual speaker reads sections of text into the speech recognition system. The system then analyzes the person's specific voice and uses it to fine tune its learning and recognize that person's speech more accurately. There are some speech recognition systems that do not use training which are called 'Speaker- Independent' speech recognition systems. On the other hand, a system that uses many samples of a speaker to train the system to recognize his/her voice is called "Speaker- dependent" system. A speaker- dependent system is developed to operate for a single speaker or to recognize only the speech of users it is trained to understand. Speech recognition software that can recognize a variety of speeches recorded by different speakers, without training samples acquired by each individuals is known as speaker–independent speech recognition system [2, 3].

The remaining part of the paper is organized into five different sections; Sect. 2 deals with the related work on speech recognition of Indian languages, Sect. 3 deals with proposed method. This section is further sub-divided into 2 sub-sections which provide details of Feature extraction and Dimensionality reduction through normal fit which result in database creation consisting of representative values for each word. Section 4 deals with matching procedure. Section 5 deals with Experimentation. This section further sub-divided into 3 sub-sections which provide details of Speech Signal Data base creation, Evaluation procedure and Isolated Word Recognition. Section 6 deals with discussion and conclusion.

## 2 Related Works on Speech Recognition of Indian Language

In the area of Automatic Speech Recognitions, the work done with respect to Kannada language is rather negligible. However, a good deal of work has been done in speech recognition with respect to Hindi, Punjabi, Tamil, Telugu, Bengali and Marathi languages. Consortium Mode Project has been initiated for development of Automatic Speech Recognition systems for agricultural commodity prices in six Indian Languages: Hindi, Tamil, Telugu, Bengali, Assamese and Marathi languages. Phonetic Engine for Speech recognition system for Hindi and Telugu languages is being developed [4]. In paper [5] Isolated Word speech recognition system is built for most spoken 10 Indian languages, viz., Telugu,

Hindi, Urdu, Kannada, Marathi, Tamil, Malayalam, Bengali, Oriya and English using Hidden Markov Model tool kit (HTK) and it works as text dependent speaker recognition mode. Here they have chosen 10 speakers uttering unique words as passwords. In paper Kuldeep Kumar and Aggarwal [6] have experimented on Isolated Hindi words recognizer using acoustic word model and it is developed using HTK for small size of vocabulary and at the speech sampling rate at 16 kHz in room environment. In paper Anusuya and Katti [7] have designed Isolated Words Recognizer for Kannada Language speech, based on the Discrete Wavelet Transform (DWT) and Principal Component Analysis (PCA). First, the DWT of the speech is computed and then MFCC coefficients are calculated. For this, PCA procedure is applied for speech recognition. Here they create the database of Kannada isolated digits from 0 to 10. In paper Rajput et al. [8] present a Hindi Speech Recognition system which has been trained on 40 h of audio data and has a trigram language model that is trained with 3 million words. For a vocabulary size of 65,000 words, the system gives a word accuracy of 75–95 %. In paper Rao [9], a Voice Oriented Interactive Computing Environment (VOICE) has been implemented in the Hindi language. The system provides an interactive facility for visual and voice feedback. The 200 isolated word recognition systems are designed around a railway reservation inquiry task and uses acoustic–phonetic segments as the basic units of recognition.

## 3 Proposed Method

The proposed system works in offline mode, where the speech signal is prerecorded and stored for processing. Each analog signal is digitized and is standardized by normalizing the amplitude of the signals [10, 11]. The preprocessed signals are subjected to various phases such as Pre-emphasis, Frame Blocking, Windowing, Autocorrelation Analysis and LPC Analysis sequentially. The LPC Analysis values are converted into Real Cepstrum coefficients. The Real Cepstrum coefficients are subjected to dimensionality reduction by normal fit where we get mean hat and sigma hat values of that signal [11]. The central tendency of mean hat and sigma hat computed for a sample set is used as the representative of each word. During matching, Euclidean distance is used to compute the distance between the mean hat and sigma hat values of the query with that of the stored samples. The representative vector at a least distance will be declared as the recognized word. The architecture of the proposed system is shown in Fig. 1. The following subsections detail about each of these stages in the proposed Isolated Kannada Word Recognition System.

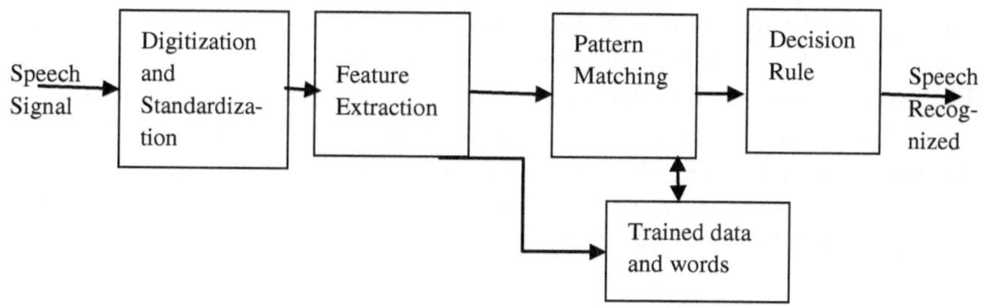


**Fig. 1** Architecture of the proposed isolated Kannada speech recognition system

## *3.1 Feature Extraction*

The feature extraction stage of the proposed system has six stages as mentioned above, Pre-emphasis, Frame Blocking, Windowing, Autocorrelation Analysis, LPC Analysis and LPC coefficients to Real Cepstrum. The sequence of feature extraction stage is as shown in Fig. 2.

**Pre-emphasize**: Here the speech signal is passed through low order digital system to spectrally flatten the signal and prevent the precision effects to affect the analysis. Thus, the signal is passed through the difference equation

$$\hat{s}(n) = s(n) - \tilde{a} \times s(n-1) \tag{1}$$

where the most common value of $\tilde{a}$ is 0.95 which has been used in our experiment [2]. Then standardization is done to entire set of values to have standard amplitude [11].

**Frame Blocking**: The pre-emphasized signal $\hat{s}(n)$ is divided into frames of N samples, with adjacent frames being separated by M samples. If we denote the lth frame of speech by $x_l(n)$ and there are L frames in the entire signal, then

$$x_l(n) = \hat{s}(M_l + n), \qquad n = 0, 1, \ldots, N-1, \quad l = 0, 1, \ldots, L-1. \tag{2}$$

Here each frame is checked by keeping some threshold value, if frame lies in this threshold value then it is allowed for next step otherwise frame is rejected

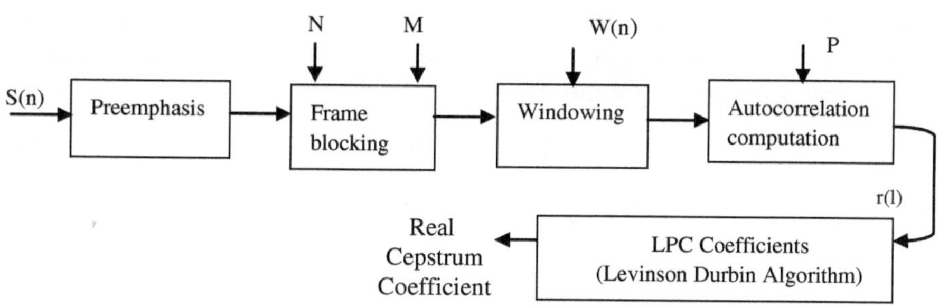


**Fig. 2** Different phases of feature extraction

considering there is no speech or it contains noise. In this paper we have trialed with N = 512, M = 128 and sampling rate at 8 kHz.

**Windowing**: Here each frame $x_l(n)$ samples is processed through a window to minimize the signal discontinuities at the beginning and end of each frame. The windowing is used to taper the signal to zero at the beginning and end of each frame. If the windows is defined as w(n), $0 \leq n \leq N - 1$ then the result of windowing each frame has following

$$\mathrm{x\_comp}_l(n) = x_l(n) \times w(n), 0 \leq n \leq N-1, \quad (3)$$

$$\text{Where} \quad w(n) = 0.54 - 0.46 \cos(2 \times \mathrm{pi} \times n / N - 1), \quad 0 \leq n \leq N-1 \quad (4)$$

is the Hamming window used in this work. In this paper we have used windowing size of N = 512, which is the same size of the frame length.

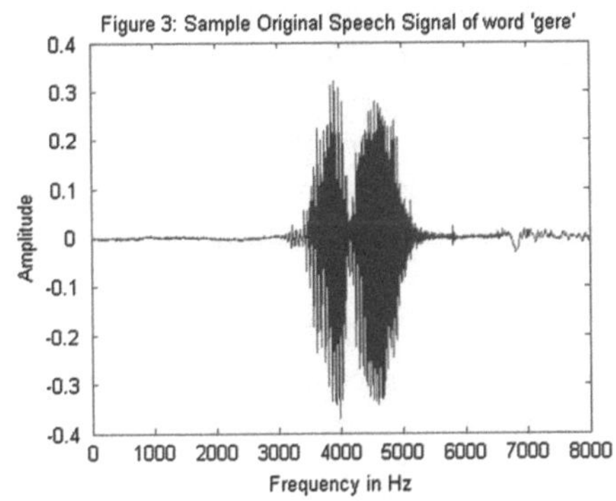


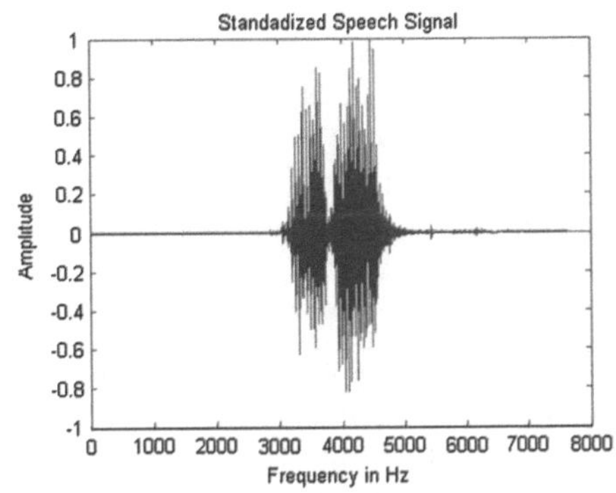


**Autocorrelation Analysis**: Here each windowed frame is autocorrelated to give:

$$r_l(m) = \Sigma_{n=0\ldots N-1-m} x_l(n)\, \mathrm{x\_comp}_l(n+m), \quad m = 0, 1, \ldots, p \quad (5)$$

Here p is the order of LPC analysis. In this case prediction order is trialed at $p = 24$. In this step *Signal is flattened due to reduction of noise, but the noise is not completely reduced at this stage.*

**LPC Analysis**: In this step, a formal method is used for converting autocorrelation coefficients to an LPC parameter set known as Durbin's method and is given by the following algorithm [2].

$$E(0) = r_l(0) \quad (6)$$

$$K_i = \frac{r_l(i) - \Sigma_{j=1..i-1}\, \alpha_j^{(i-1)}\, r_l(|i-j|)}{E^{(i-1)}}, 1 \leq i \leq p \quad (7)$$

$$\alpha_i^{(i)} = K_i, \quad 1 \leq i \leq p \quad (8)$$

$$\alpha_j^{(i)} = \alpha_j^{(i-1)} - K_i \alpha_{i-j}^{(i-1)}, \quad 2 \leq i \leq p \tag{9}$$

$$E^{(i)} = \left(1 - k_i^2\right) E^{(i-1)}, \quad 1 \leq i \leq p \tag{10}$$

$$\alpha_m = \alpha_m^{(i)}, \quad 1 \leq m \leq p \tag{11}$$

*At this stage feature set for speech recognition is computed, i.e. Nasality presence or absence, Frication, Voiced (periodic) and Unvoiced (aperiodic) classification.*

**LPC Coefficients to Real Cepstrum**: The *real cepstrum* of a signal *x*, sometimes called simply the cepstrum, is calculated by determining the natural logarithm of magnitude of the Fourier transform of *x*, then obtaining the inverse Fourier transform of the resulting sequence. The returned sequence is a real-valued vector of the same size as the input vector, here LPC order is P = 24, so output vector size is also 24.

### 3.2 Dimensionality Reduction Through Normal Fit and Database Creation

The real cepstrum values are p × L rows for each signal, where L is a valid frame and p is the LPC order. The sizes of L vary every time for same word spoken by same speaker or different speakers'. Whatever may be the length of L, for this dimensionality reduction is made through passing into normal fit where normal fit returns an estimate of the mean $\mu$, and an estimate of the standard deviation $\sigma$ of the normal distribution which gives the data with its roots in parent data (there is no need to worry about time wrapping, due to normal fit just reduce the L number of rows into 2 rows). With no censoring, the function normfit [11] computes sigma hat using the square root of the unbiased estimator of the variance. With censoring, sigma hat is the maximum likelihood estimate. So our data for one word will reduced form p × L rows to p × 2 rows (where LPC pth order is 24 is used in our experiment, the first row values are mean $\mu$ hat and second row values are standard deviation $\sigma$ hat) for each word. Here each word is uttered by 8 different speakers. Therefore the mean is calculated for 8 different sets of data to obtain the one set of representatives for one word. These results are considered as representative data sets and stored in database.

## 4 Matching Procedure

For testing the given speech signal, first digitalize and standardize the given speech signal and then Feature extraction is done using the above procedure. For this Feature extracted values dimensionality reduction is made through passing into

**Table 1** Speech database description

| | |
|---|---|
| Language | Kannada |
| Speech type | Read speech |
| Number of speakers | 10 speaker of different age categories |
| Recording conditions | Room environment |
| Number of signals used to training | 2,352 signals |
| Number of signals used to testing | 1,176 Signals |
| Total signals used in experiment | 3,528 Signals |

normal fit where normal fit returns an estimate of the mean $\mu$, and an estimate of the standard deviation $\sigma$ of the normal distribution which gives the data with its roots in parent data. These values are compared with each trained set of values using Euclidian distance method. Results are stored in array and then decision method is followed in three hits and word is declared as recognized by selecting the minimum scored value. In the first hit we directly check for minimum scored value and compare with word dictionary. If they are matched, then minimum scored value is declared as recognized word with the accuracy rate as 100 %. Otherwise, we go for second and third hit by grouping into 2 categories and by measuring distance keeping within the tolerance rate of 0.25 and 0.5 respectively; then they are compared with word dictionary. If there is matching, then the group which is within the tolerance limit of 0.25 and 0.5 is declared as a recognized word. Otherwise, it is declared as not found.

# 5 Experimentation

## *5.1 Speech Signal Database Creation*

In this work, pulse code modulation with a frequency of 8,000 Hz, 16-bit mono channel is used to develop the speech signal database. Each word signal contains a silence region before and after appropriate signal. The recordings are done in room environment using audio editing Gold wave software and microphone having frequency range of 20–20 kHz, sensitivity of 110 dB and Impedance of 32 Ω.

The details of the continuous speech segmentation techniques are beyond the scope of this paper. In this paper, for isolated word recognition, we have created our own speech database of unique 294 words of isolated Kannada from 10 speakers. Out of 10 speakers, 8 speakers' utterance are used for training and 2 speakers' utterances and 2 speakers' utterances among 8 speakers are rerecorded the words and used for testing. So altogether 4 speakers' utterances are used for testing; 2 speakers as known persons and 2 speakers as unknown persons. Details of database are shown in Table 1 and process of speech preparation for Feature extraction is shown in Fig. 2.

**Table 2** Speaker independent isolated Kannada word recognition accuracy rate and word error rate

| Speaker | Known speaker | | Unknown speaker | |
|---|---|---|---|---|
| | Accuracy rate (%) | Error rate (%) | Accuracy rate (%) | Error rate (%) |
| Male | 96.93 | 3.07 | 92.51 | 7.49 |
| Female | 99.66 | 0.34 | 90.81 | 9.19 |
| Average | 98.29 | 1.71 | 91.66 | 8.34 |

## 5.2 Evaluation Procedure

We calculated the accuracy rate by following method

$$\text{Total IWR Accuracy Rate} = (1 - (\text{N} - (\text{count} + \text{count1} + \text{count2}))/\text{N}) \times 100 \tag{12}$$

where N is Total Number of words in the dictionary; count is the number of words identified in the first hit, count 1 is the number of words found in the second hit and count 2 is the number of words found in the third hit.

Individual word accuracy rate will be 100 % if word is matched in the first hit. Otherwise, Individual word accuracy rate is

$$(1 - (\text{N} - (\text{N} - \text{C}))/\text{N}) \times 100 \tag{13}$$

Where N is total number of words in the dictionary, C is number of words which fall in the second or third tolerance rate.

## 5.3 Isolated Word Recognition

The success rate of recognition of total words form known speaker is 98.29 % and unknown speaker 91.66 %. In Table 2 shows the details about the results of word Recognition with respect to male and female of known speakers' and unknown speakers', this table clearly shows that for complete words recognition rate of known speakers is very high than compare to unknown speakers' and in the known speakers' the female voice recognition rate is high.

**Table 3** Speaker independent isolated Kannada individual word recognition accuracy rate in an average

| Speaker | Known speaker | | Unknown speaker | |
|---|---|---|---|---|
| | Accuracy rate (%) | Error rate (%) | Accuracy rate (%) | Error rate (%) |
| Male | 74.77 | 25.23 | 59.15 | 40.85 |
| Female | 68.74 | 31.26 | 55.97 | 44.03 |
| Total | 71.75 | 28.24 | 57.56 | 42.44 |

**Table 4** Speaker independent isolated Kannada word recognition accuracy rate shown by increasing vocabulary size

| Female | | | Male | | | Speaker | |
|---|---|---|---|---|---|---|---|
| Total words identified correctly | | In an average Individual word accuracy rate | Total words identified correctly | | In an average Individual word accuracy rate | | |
| 100 % | 50 words | 82.92 % | 100 % | 50 words | 81 % | 50 words | Known speaker word accuracy rate |
| 100 % | 100 words | 71.04 % | 99 % | 99 words | 73 % | 100 words | |
| 100 % | 150 words | 59.35 % | 99.33 % | 149 words | 71 % | 150 words | |
| 99.50 % | 199 words | 55.34 % | 99.50 % | 199 words | 68.80 % | 200 words | |
| 99.66 % | 293 words | 68.74 % | 99.66 % | 293 words | 74.77 % | 294 words | |
| 100 % | 50 words | 76.96 % | 92 % | 46 words | 68.28 % | 50 words | Unknown speaker word accuracy rate |
| 100 % | 100 words | 64.73 % | 96 % | 96 words | 61.76 % | 100 words | |
| 98 % | 147 words | 61 % | 97.33 % | 146 words | 59.03 % | 150 words | |
| 94 % | 188 words | 59.04 % | 96.50 % | 193 words | 55.32 % | 200 words | |
| 94.56 % | 279 words | 55.97 % | 95.92 % | 282 words | 59.15 % | 294 words | |

**Table 5** Showing the attempts taken to hit the target to recognize the word and its accuracy rate by known speaker and unknown speaker

| Attempts take to hit targeted words | Known speaker | | | | Unknown speaker | | | |
|---|---|---|---|---|---|---|---|---|
| | Male | | Female | | Male | | Female | |
| | Number of words correctly recognized | Individual word accuracy rate | Number of words correctly recognized | Individual word accuracy rate | Number of words correctly recognized | Individual word accuracy rate | Number of words correctly recognized | Individual word accuracy rate |
| First hit | 71.00 | 100 | 81.00 | 100 | 34 | 100 | 28.00 | 100 |
| Second hit | 44.00 | 98.37 | 152.00 | 94.13 | 63.00 | 95.62 | 38.00 | 96.81 |
| Third hit | 170.00 | 83.78 | 60.00 | 78.56 | 175.00 | 77.50 | 201.00 | 78.38 |
| Search failed words | 9.00 | 0.00 | 1.00 | 0.00 | 22.00 | 0.00 | 27.00 | 0.00 |
| Total | 294.00 | 70.54 | 294.00 | 68.17 | 294.00 | 68.28 | 294.00 | 68.80 |

Table 3 shows Individual word Recognition accuracy Rate in an average calculation for known and unknown speakers', it clearly defines that individual word recognition accuracy rate are high for male voices than the female voices in both known and unknown speakers'. Table 4 shows the comparison of Speaker Independent Isolated Kannada word Recognition accuracy Rate by increasing vocabulary size and comparison with male and female voice of known and unknown speakers. It gives the details of individual word accuracy rate and total number of words found correctly for known and unknown speakers' speech made by male and female. Table 5 shows the attempts taken to hit the target to recognize the word and number of words which have hit the target in each stage, its accuracy rate and missing rate by known speaker and unknown speaker.

## 6 Discussion and Conclusion

If we considered the Hidden Markov Model (HMM) system for Isolated word recognition, then probability computation step is generally performed using the Viterbi algorithm and requires on the order of $V * N^2 * T$ computations (Where $V =$ vocabulary, $N =$ state model and $T =$ observations for the unknown word). In our experiment $V = 294$ words, $T = 24$ and if $N = 5$ state model, then total 846,720 computations or if $N = 3$ then total 508,032 computations are required for recognition of one word. In our experiment we are only finding the Euclidian Distance between Trained Set of values (Normal fit values) and unknown signal values (Normal fit values) and then find the minimum scorer to declare the word recognized. So we required $10^3$ computation in the best case and around $10^4$ computations in the worst case for the recognition of one word. But HMM model works with high accuracy for all vocabularies; small, medium, large, out of Vocabulary (OOV) and Isolated and continuous speech recognition. So we need to experiment with large vocabulary Continuous speech recognition to accept this model as good.

In our experiment we found that Normal fit can be used to time wrapping and dimensionality reduction of feature values of speech signal and mean hat and sigma hat values can be used as representative set of data. While using normal fit no need to measuring the similarity between two sequences which may vary in time or speed. The dimensionality reduction will help in reducing the memory size and computation time. The standardization of speech signal will make it very easy to check for valid speech frame. It is also easy to fix tolerance rate and increase the amplitude of signal. The accuracy rate of individual word can increase by using more number of training sets and this can also lead to increase in the number of word recognition in the first hit. While the size of vocabulary increases, the accuracy rate of system decreases.

**Acknowledgments** The author would like to thank for all my friends who supported me in preparing the speech database and developing Kannada word list of covering all phonemes of the language, reviewers and Editorial staff for their efforts in preparation of this paper.

## References

1. Hemakumar G (2007) A study on hidden markov model for speech recognition. Submitted for the award of M. Phil in Computer Science, Bharathiar University, during Nov 2007
2. Rabiner L, Jung B-H (1993) Fundamentals of speech recognition. Pearson Education (Singapore) Private Limited, Indian Branch, 482 F.I.E Patpargans, Delhi 110092, India
3. http://en.wikipedia.org/wiki/Speech_recognition
4. Swarna Lata, Country Manager, W3C India and Head, TDIL Programme, Department of Information Technology, Government of India (2011) Challenges of multilingual web in India: technology development and standardization perspective. Reported-2011
5. Rajewara Rao R et al. JNT University, Hyderabad, India (2007) Text-dependent speaker recognition system for Indian languages. IJCSNS 7(11), Nov-2007
6. Kumar K, Aggarwal RK, Department of Computer Enginerring, National Institute of Technology, Kurukshetra (2011) Hindi speech recognition system using HTK. Int J Comput Bus Res ISSN (Online) 2(2-May issue):2229–6166
7. Anusuyal MA, Katti SK (2010) Mel frequency discrete wavelet coefficients for Kannada speech recognition using PCA. In: Proceedings of international conference on advances in computer science 2010
8. Rajput N, Verma A, Neti C, NCC, Mumbai (2002) A large vocabulary continuous speech recognition system for Hindi. 26–27 Jan 2002
9. Rao PVS (1993) VOICE: an integrated speech recognition synthesis system for the Hindi language. Speech Commun 13:197–205
10. Tan P-N, Steinbach M, Kumar V (2009) Introduction to data mining. Dorling Kindersley (India) Pvt. Ltd., Licensees of Pearson Education in South Asia, 4th Impression, 2009
11. Matlab R2009a help menu
12. Quatieri TF (2002) Discrete-time speech signal processing principles and practice. Pearson Education (Singapore) Private. Ltd, Indian Branch, 482 F. I. E Patparganj, Delhi 110092, India
13. Saeed K, Nammous MK (2007) A speech-and-speaker identification system: feature extraction, description, and classification of speech-signal image. IEEE Trans Ind Electron 54(2)
14. Umesh S (2010) Automatic speech recognition-research and standards. Department of Electrical Engineering, IIT, Madras, May 7th 2010
15. Three day's workshop on "Hands on experience in Sphinx and HTK for Speech Recognition" held on Feb-2011 at AU-KBC research center, MIT campus, Chennai